D0083904

Three Phases of Matter

SECOND EDITION

Alan J. Walton

Reader in Physics
The Open University

CLARENDON PRESS OXFORD

Oxford University Press, Walton Street, Oxford OX2 6DP

London New York Toronto
Delhi Bombay Calcutta Madras Karachi
Kuala Lumpur Singapore Hong Kong Tokyo
Nairobi Dar es Salaam Cape Town
Melbourne Auckland
and associate companies in
Beirut Berlin Ibadan Mexico City Nicosia

Oxford is a trade mark of Oxford University Press

Published in the United States
by Oxford University Press, New York

First edition published by McGraw-Hill, 1976
© First edition McGraw-Hill Book Company (UK)
© Second edition Alan J. Walton, 1983

First published 1983. Reprinted 1984

British Library Cataloguing in Publication Data
Walton, Alan J.
Three phases of matter.—2nd ed.
1. Matter—Properties
I. Title
530.4 QC173.3
ISBN 0-19-851957-5
ISBN 0-19-851953-2 pbk

Library of Congress Cataloging in Publication Data
Walton, Alan J.
Three phases of matter.
Includes index.
1. Matter—Properties. I. Title.
QC173.3.W34 1982 530.4 82-8303
ISBN 0-19-851957-5
ISBN 0-19-851953-2 (pbk.)

Printed in Northern Ireland at
The Universities Press (Belfast) Ltd.

To my parents

Preface to the second edition

THIS text has its origins in a first-year undergraduate course entitled 'The structure and properties of matter' which I gave while at the University of Sussex. Then my brief was to give students (who were potential chemists, physicists, biologists, and material scientists) a 'real feel' for what solids, liquids, and gases are like at the atomic level. My aim remains the same. Like the original lecture course, the text assumes that the reader has a working knowledge of elementary calculus and of classical mechanics. No prior knowledge of physical chemistry is assumed since it has been my experience that such concepts as the mole, the Avogadro constant, and relative atomic mass are often the victims of the assumption that someone else is 'doing' them.

The dominant theme of the text is the extent to which the properties of solids, liquids, and gases—both their structure and their transport properties—are determined by the competing effects of the atoms' kinetic and potential energies. Since a clear understanding of the conditions under which a substance exists as a solid, as a liquid, and as a gas is a necessary prerequisite, Chapter 1 provides this macroscopic background. Chapter 2 is also scene-setting. It shows how atomic masses and sizes can be measured and introduces the Mie and Lennard-Jones potentials. In discussing temperature in Chapter 3, I follow Jeans' example and look at how energy is transferred at the atomic level between a gas and a solid; this shows that empirical temperature is proportional to the mean atomic kinetic energy. Chapter 3 also examines Brownian motion and simple fluctuation phenomena. Chapter 4 (the approach presented here was developed jointly with Paul Clark and Andrew Millington) arrives at the Maxwell–Boltzmann distribution function via manual sorting of distinguishable particles among energy levels. This procedure introduces the density-of-states factor *ab initio* and opens the way for a more formal discussion. Before deriving the equation of state of a perfect gas (Chapter 5) I draw on atomic-beam studies to demonstrate that the usual assumption of zero dwell-time and of specular reflection are normally untrue. Neither such assumption is actually required in deriving the equation of state. The properties of real gases form the subject of Chapter 6, where van der Waals' and Dieterici's equations of state are deduced via arguments at the atomic level. In discussing transport processes in gases (Chapter 7), I pay particular attention to how energy and momentum are accommodated at the surface of a solid. This is anything but an academic nicety; the energy accommodation coefficient may be as low as 0·01.

A fairly traditional Chapter 8 shows how different crystal structures are produced by associating a basis with one out of a finite number of distinct lattices. This chapter also shows how the internal energy of a structure can be deduced and examines some of the factors which help determine the lattice in a particular compound. Chapter 9 looks at the mechanical properties of solids—the familiar elastic moduli—and at the role of structural imperfections in limiting the strengths of solids. In discussing the thermal properties of solids in Chapter 10, I have tiptoed into what is normally the preserve of advanced undergraduate texts. However, the arguments proceed gently and by induction: By examining how a series of coupled gliders behave on an air-track we can guess what should happen in a one-dimensional structure. We can further guess how a three-dimensional structure should behave (Debye's theory emerges in simplifying the very difficult problem of evaluating $C_{V,m}$ exactly). Having established the energy-equivalence of a vibrational mode and a harmonic oscillator (assumed to be one of Planck's), phonons appear naturally (Chapter 11). Once a phonon gas has been argued into existence the results of gas-kinetic theory can be taken over to derive the thermal conductivity of a non-metal. Since the Fermi–Dirac electron gas appeared during earlier discussions of the heat capacity of metals the way is now open to consider thermal and electrical conduction in metals. A nearly-free electron model is used to explain the existence of semiconductors and insulators (the Bragg reflection arguments echoing those introduced earlier during X-ray diffraction studies of crystals).

In discussing simple liquids (Chapters 12 and 13), I have adopted a cell model in which an atom behaves in gas-like fashion within its cell but which requires an activation energy to escape from its cell (an attribute reminiscent of solids). This allows the student to draw on what he has already learnt of the other two phases of matter. However, in arriving at the equation of state (via the Clausius virial theorem) and the internal energy function, I do point out how the results of molecular-dynamics simulations may be used to make accurate predictions of the behaviour of real liquids. My thanks go to Dr John Lewis for providing a series of stills from a molecular-dynamics simulation.

In the mid-1960s it was fashionable in certain institutions to denounce the type of macroscopic approach to solids, liquids, and gases typified by such texts as Newman and Searle's *The general properties of matter* and, instead, to go hell for leather at the 'real physics'. To say little, if anything, about how a quantity is measured surely misrepresents the nature of scientific enquiry: I therefore include 'corroborative details'. Where relevant, I also emphasize the importance of designing experiments to demonstrate *departures* from, rather than agreement with, the theoretical predictions.

Since learning is doing, a number of exercises have been built into the text. Some of these are retrospective, allowing the student to test what he has learnt; others are anticipatory to encourage the student to venture a little ahead on his own; yet others are there to provide the opportunity to make asides which might get out of perspective if made in the main text. Hopefully the reader will tackle the exercises as he meets them—at the very least he should study the answers and comments before proceeding with the text. I have tried to aim the, mainly retrospective, end-of-chapter problems at the average student.

For the most part, I have followed the recommendations of the Symbols Committee of The Royal Society as set out in *Quantities, units, and symbols* (1975), The Royal Society, London. Some few departures were thought desirable. The constant k seems very overworked; it is recommended for circular wavenumber, thermal conductivity, rate constant of a reaction, and the Boltzmann constant (all of which occur in this text). Confusion is very likely in certain areas, as, for example, in the gas kinetic relation for thermal conductivity (where I write $\kappa = \frac{1}{2} n \bar{u} \lambda k$).

In preparing this edition, I have rewritten the chapters on gas-kinetic theory to bring the level of treatment more into line with that adopted throughout the rest of the book. I have also made good certain omissions, notably the derivation of $\frac{1}{4} n \bar{u}$, Einstein's random-walk equation, density fluctuations, and adiabatic changes in gases. Real gases—now extended to include the virial equations and Dieterici's equation and to discuss techniques for cooling gases—have been hived off into a separate chapter. Previously the Maxwell–Boltzmann distribution came at the end of the gas-kinetic theory section. By placing it at the beginning of this section I have been able to draw on the distribution functions in certain gas-kinetic arguments. The treatment of solids has been extended to include more on the solid–gas interface; in particular, on surface diffusion, crystal growth, and on the surface energies of crystals. A simple discussion of semiconductors and insulators is also included. Dislocations are treated in somewhat greater depth than previously. The chapter describing the structure of liquids now introduces the main attributes of liquid crystals, while the final chapter has been extended to include the growth of liquid drops from a vapour. Overall, the treatment of the thermodynamics has been made much less cavalier. More end-of-chapter problems have been included. But perhaps the biggest change is a stylistic one: The arguments are now presented much more concisely and in less colloquial language. In fact, around eighty per cent of the English and thirty per cent of the physics has been rewritten.

My thanks go to many friends and colleagues for much useful discussion. Of these I name but four: Drs G. Brooker, P. Clark, M. O'Brien, and H. Rosenberg. Finally I am grateful to Mark Hodson and Ken Thomas for their help in preparing the index.

Oxford, 1982   A.J.W.

Contents

1. The p–V_m–T surface

To us lead is a solid, water is a liquid, and carbon dioxide is a gas. However, had we been born not on Earth but on another planet, our experiences as to what is a liquid, what is a solid, and what is a gas would have been quite different. An inhabitant of Venus would regard lead as a liquid. A Martian might, so to speak, have to shovel away the carbon dioxide.

Clearly, if we want to study the conditions under which a substance exists as a solid, a liquid, or a gas we must be less parochial in our choice of the experimental environment. So in this chapter we shall look at how the phase in which a substance is found is related to its temperature, pressure, and volume. We shall also look at the energy changes involved in altering the temperature, pressure, and volume of the substance. Finally, we define—and develop the interrelations between—the molar heat capacities at constant volume and at constant pressure; relations which will prove to be of great value in later chapters.

1.1. A closed system

We all know the salient properties of the solid, liquid, and gaseous *phases* of at least one substance—water. We also know that two phases may coexist in equilibrium—ice and water, for example. However, problems arise if we assert that our everyday experiences are experiments. Consider an ice cube resting in a tumbler of water and surrounded by water vapour. While it is indisputable that all three phases are present they are *not* in equilibrium: the ice will eventually all melt; the liquid will eventually all evaporate. An additional problem is that we have no real idea of the total volume occupied by the substance. Although it poses few problems to measure the volumes occupied by the liquid and solid, the volume occupied by the gas is very indeterminate. Furthermore, as the water evaporates we are no longer dealing with a pure substance—with a single *component*—but with a mixture of mainly gaseous water, nitrogen, oxygen, and carbon dioxide.

These three disadvantages—that of a non-equilibrium condition, of an indeterminate total volume, and of having unwanted components present—can be overcome by using the piston and cylinder arrangement shown in Fig. 1.1. Here a fixed amount of the substance under investigation is contained within the cylinder. We shall in fact assume that there is one mole of substance present. (This term will be defined formally in eqn (2.7) but for the moment you may take it that there is one mole—written

2 The p–V_m–T surface

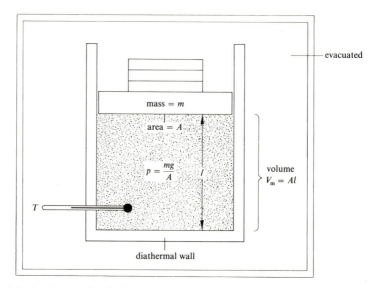

Fig. 1.1. A piston and cylinder arrangement for studying the p–V_m–T properties of a substance.

1 mol—of a substance present if it contains $6 \cdot 02 \times 10^{23}$ atoms when it is a monatomic substance like argon or $6 \cdot 02 \times 10^{23}$ molecules when it is a molecular substance like carbon dioxide.) The molar volume V_m occupied *in toto* by the substance (which may be present in one or more phases) is, of course, Al, where A is the cross-sectional area of the piston and l is the distance between the piston face and the end of the cylinder. We shall use the word *system* as shorthand for 'the substance under investigation contained within the cylinder'.

The pressure p of the system can be changed at will by altering the weights on the piston. If the combined mass of weights plus piston is m, then, provided there is no friction between the piston and the cylinder, $p = mg/A$, where g is the local acceleration due to gravity. Allowance must, of course, be made for the atmospheric pressure pushing on the cylinder unless it is placed in an evacuated vessel, as shown in Fig. 1.1.

The temperature T of the system (read with a thermometer—say, of the familiar mercury-in-glass type) may be set to any desired value by placing the *diathermal* wall (one that allows heat to pass) in contact with a suitable 'oven' or 'fridge'.

It is important always to remember that p, V_m, and T relate to the *system*. If two (or three) phases are present in equilibrium—say a liquid and a gas—the volume of each phase simply cannot equal Al! So we can only refer to the volume of the *system*. However, at equilibrium the temperature of each phase will be the same and equal to that of the system. Likewise the pressure of the liquid and gaseous phases will be the same and equal to that of the system. These strictures do not apply when a *single* phase is present; at equilibrium the pressure, temperature, *and* volume of the system are equal to the pressure, temperature, *and* volume of the substance.

1.2. The p–V$_m$–T surface

Collecting data

You might think it would be possible to adjust the system of Fig. 1.1 to conform to randomly chosen values of p, V_m, and T. Why not, for example, first set the piston position l (Fig. 1.1) so that $V_m = Al$ has the desired value? The temperature T could then be set at the value chosen for it by placing a fridge or oven, as appropriate, in contact with the diathermal wall. Finally, why not set p at its chosen value by placing the appropriate mass m on the piston?

If you try to carry out this experiment you will discover that (unless you have been particularly lucky in the values you chose for p, V_m, and T) it simply cannot be done! If you set V_m and T in the manner described you will find that in attempting to give p the desired value you change V_m. You might instead have settled the values for p and T hoping to then adjust V_m; what you discover now is that (provided there is only one phase present) the value of p changes as V_m is adjusted. (As we shall shortly see, when two or three phases are present there is some latitude in the choice of V_m when p and T are fixed.) Finally, you might decide first to give p and V_m their chosen values and then to fix T. Again, unless you have been lucky in your choice of T, you will not succeed! Changing T in the apparatus of Fig. 1.1 will change V_m. Thus, in general, we are only free to randomly choose the values of any two of p, V_m, and T.

One way to study the behaviour of the substance is indeed to select arbitrary values for any *two* of p, V_m, and T and to measure the value of the third once the system has come to equilibrium. A less haphazard way is to fix, say, T and then to study exhaustively how V_m depends on p. We would then choose a new value for T and repeat the study. Either way we acquire many triplets of (p, V_m, T)-values at which the system is in equilibrium. One such triplet might be, for example, $p = 5 \times 10^5$ Pa (where Pa, short for pascal, means newton metre^{-2}), $V_m = 7 \times 10^{-4}$ m^3 mol^{-1}, and $T = 200$ K (where K stands for kelvin). Although the experimental data can be left in tabular form it is often more instructive to plot it graphically using a set of p, V_m, and T axes. After plotting out all the experimental points we would draw in a surface to link them together as best we could.

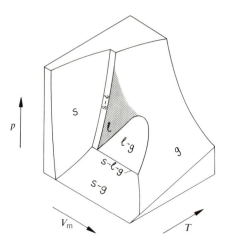

Fig. 1.2. The general form of the p–V_m–T surface of a simple substance which expands on melting.

What the surface tells us

The precise shape of the p–V_m–T surface depends on the particular substance under investigation but the general form of the surface is broadly similar for many different substances. For simple substances which expand on melting (e.g. carbon dioxide) the surface has the form shown in Fig. 1.2. Although drawn as a solid object in Fig. 1.2 (to give it perspective) the p–V_m–T surface is just that; it hangs in p–V_m–T space. It is worth emphasizing that the (p, V_m, T)-values which are measured when the system is truly in equilibrium *lie on the surface*. Should one attempt to give the system values of (p, V_m, T) other than values lying on the surface the system will *not* be in equilibrium; one or more of p, V_m, and T will change so as to bring the system to equilibrium. Once equilibrium is attained the values of (p, V_m, T) will lie on the surface.

If you were to carry out such an experimental study you would observe that for certain ranges of the independent variables (any two of p, V_m, and T), the substance exists wholly as a solid (s), as a liquid (l), or as a gas (g). The regions of the surface corresponding to these single phases are indicated in Fig. 1.2. In accord with our everyday experiences we see that the gaseous phase exists at high volumes and temperatures, the solid when the temperature and volume are low, and the liquid phase when the temperature and volume have intermediate values. Within other ranges of (p, V_m, T) two phases exist in equilibrium; the system consisting of solid plus liquid (s–l), solid plus gas (s–g), or liquid plus gas (l–g). At one particular pair of values of p and T, and only at these particular values,

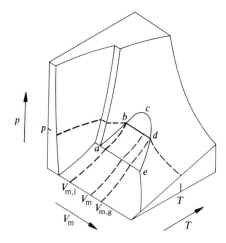

Fig. 1.3. Both the gas and liquid phases are present when the system volume has a value V_m between $V_{m,l}$ and $V_{m,g}$. Line ac is called the saturated liquid line. Line ce is called the saturated gas line (or saturated vapour line).

are all three phases (s–l–g) present in equilibrium. The line along which this occurs is called the *triple line* and such a line exists for all elements except helium.

It is worth noting that when two or three phases are present there is some latitude in the choice of V_m for a fixed (p, T). If, as shown in Fig. 1.3, the temperature is fixed at a value T and the pressure at p such that liquid and gas are present, the volume V_m of the system may have any value between $V_{m,l}$ and $V_{m,g}$. At a volume $V_{m,l}$ the substance is wholly in the liquid phase; at volume $V_{m,g}$ it is wholly in the gaseous phase. At some intermediate volume V_m—which, of course, measures the molar volume of the *system*—both liquid and gas are present. The line ac is called the *saturated liquid line*; line ce the *saturated gas line*.†

Because of this latitude in the choice of volume the value of V_m must always be specified when there are two or three phases present at a fixed (p, T). While this caution is unnecessary when there is only a single phase present, we can avoid possible pitfalls if we choose V_m and T, or V_m and p, as the independent variables.

Exercise 1.1

As was mentioned earlier, one is most likely to obtain the p–V_m–T surface of a substance by arranging to keep one of p, V_m, and T constant, varying another,

---

† Line ce is also called the *saturated vapour line*; the word *vapour* being used to denote a gas at temperatures lower than that at point c. In this text we shall often avoid using the word vapour, preferring instead to draw attention, where appropriate, to the temperature of the gas.

6 The p–V_m–T surface

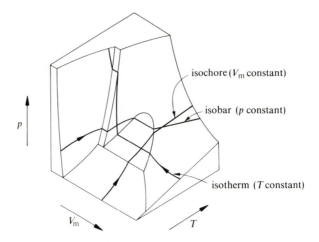

Fig. 1.4. Lines of constant pressure (isobars), constant volume (isochores), and constant temperature (isotherms) on the p–V_m–T surface.

and measuring the value of the third when equilibrium is attained. Fig. 1.4 shows the paths followed on the p–V_m–T surface in three such experiments. Describe in your own words what happens to the substance—in particular what phases are present—as each of these paths is followed. You may find it helpful to block off the solution to this, and all future, exercises when working through them. Should you decide to omit any exercise you are advised to read the question and the answer as if they are part of the text.

Solutions.

(a) *Temperature held constant.* At low pressures, where V_m is large, the substance is wholly gaseous. As p is increased liquid will start to condense on crossing the saturated gas line. Once this happens p remains constant while the volume of the liquid grows and that of the gas shrinks (as does the total volume of the system). On reaching the saturated liquid line the substance is wholly liquid. Large increases in p are required to reduce the volume of the liquid. Next, solid appears in equilibrium with the liquid and the pressure remains constant as the volume of the solid grows at the expense of the liquid. When the substance is wholly solid very large increases in pressure are required to compress it even slightly—solids are pretty incompressible.

(b) *Volume held constant.* Initially solid and gas are in equilibrium—no liquid is present. As T is increased there is one particular value at which solid, liquid, and gas coexist in equilibrium: on the triple line. As observed experimentally, T remains constant at the triple line value while the solid melts. When the solid is wholly melted T increases afresh, with only liquid and gas present until we reach the saturated gas line. Thereafter the substance is wholly gaseous.

(c) *Pressure held constant.* Initially only solid is present. Its volume expands slightly as T is increased. It then melts; the proportion of liquid to solid increasing as T remains constant. Next, the substance is wholly liquid. On crossing the saturated liquid line, gas appears which coexists in equilibrium with the liquid. The temperature remains constant while the liquid evaporates. Once the saturated gas line is crossed the substance is wholly gaseous and remains so with ever-increasing T.

    Comment. You might find it useful to memorize the meanings of *isotherm*, *isochore*, and *isobar* (Fig. 1.4). These terms will occur frequently in the text.

So far we have only considered the general form of the $p-V_m-T$ surface of a substance which expands on passing from the solid to the liquid phase. For a substance which contracts on melting the $p-V_m-T$ surface has the general form shown in Fig. 1.5. By studying the changes occurring along an isotherm we see that there is indeed a reduction in volume when the substance passes from the solid to the liquid phase. As always, the exact form of the $p-V_m-T$ surface depends on the substance; that for water is shown in Fig. 1.6, where the roman numerals I to VII refer to the seven forms of solid water which have been observed. As Fig. 1.6 makes plain, the volume decrease which occurs when 'ice' melts is only found with type-I ice. When type-VI ice, for example, melts there is an increase in volume. To give an idea of the scale of Fig. 1.6 it is convenient to introduce *standard temperature and pressure* (s.t.p.), also known as normal temperature and pressure (n.t.p.). Physicists usually define standard temperature as $273 \cdot 15$ K (0°C)—this is the value we shall adopt here—although engineers define it as either $15\frac{5}{9}$°C (60°F) or 20°C. Standard pressure is now defined by international agreement as $1 \cdot 01325 \times 10^5$ Pa. s.t.p. approximates to everyday atmospheric conditions at sea-level. Fig. 1.6 thus tells us that ice VII is only found at pressures of greater than 20 000 atmospheres.

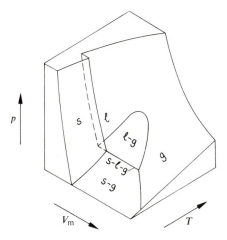

Fig. 1.5. The general form of the $p-V_m-T$ surface for a substance which contracts on melting.

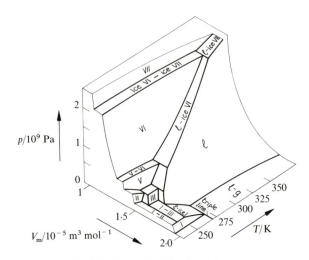

Fig. 1.6. The p–V_m–T surface of water.

The critical point

Experiments in which the volume of the system is kept constant are usually fairly difficult to perform if only because dangerously large pressures can all too easily build up as the temperature of the system is increased. However, if, as shown in Fig. 1.7, we start at point *1*, with gas and liquid present, and increase T the volume occupied by the liquid decreases, while that occupied by the gas grows. On reaching point *2* the

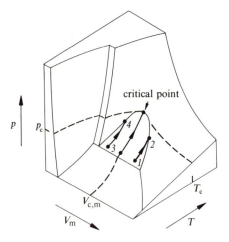

Fig. 1.7. When $V_m = V_{c,m}$ the isochore passes through the critical point (p_c, $V_{c,m}$, T_c).

substance is wholly gaseous. If we start instead at point *3* the volume occupied by the gas decreases with increasing T, while that occupied by the liquid grows, until at point *4* the substance is wholly liquid. What remains constant along both these paths is the total volume of the *system*. If we have chosen one particular value $V_{c,m}$ for the volume of the system (Fig. 1.7) then neither the liquid nor the gas grows in volume to fill the container. What we observe instead is that the usually clear boundary between the liquid and the gas above it becomes more and more indistinct until at $T = T_c$ we cannot tell one phase from the other. This is shown in Fig. 1.8 for carbon dioxide. In going from (a) to (e) we see the

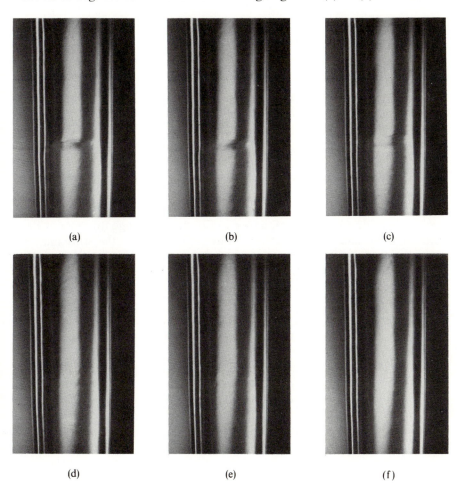

Fig. 1.8. A sealed tube ($V_m = V_{c,m}$) of carbon dioxide is heated through the critical point. The critical point (p_c, $V_{c,m}$, T_c) occurs around frame (d). (Reprinted with permission of the Open University.)

TABLE 1.1
Critical and triple point data

Substance	$p_c/10^5$ Pa	$V_{c,m}/10^{-6}$ m^3 mol^{-1}	T_c/K	$p_{tr}/10^5$ Pa	$V_{tr,m}^a/10^{-6}$ m^3 mol^{-1}	T_{tr}/K
Ne	27	42	44	0·43	16	24
Ar	49	72	151	0.68	28	84
Kr	55	92	209	0.73	35	116
N_2	34	90	126	0·12	17	63
CO_2	74	94	304	5·10	42	216
H_2O	221	59	647	0·006	18	273·16
O_2	50	73	155	9·0015	24	54
H_2	13	65	33	0·072	25	14

[a] $V_{tr,m}$ is the molar volume of the liquid phase which is in equilibrium with the solid and gaseous phases at the triple point.

boundary disappear without changing its position in the tube. We now have a problem on our hands. At $p = p_c$, $V_m = V_{c,m}$, $T = T_c$ has the liquid turned into a gas, or has the gas turned into a liquid? The answer is that it is purely a matter of convention whether we call this single phase a liquid or a gas. The point (p_c, $V_{c,m}$, T_c) is called the *critical point*; p_c is called the *critical pressure*; $V_{c,m}$ the *molar critical volume*; T_c the *critical temperature*. Some triple and critical point values for a variety of substances are given in Table 1.1.†

The problem of nomenclature exists elsewhere than at the critical point. Consider the two ways shown in Fig. 1.9 to change the state of the system from state *i* to state *f*. If route *1* is followed, the substance is initially present as liquid plus gas but as *T* is increased the liquid expands to fill the container. Thereafter only liquid is present. So we would confidently assert that the substance is a 'liquid' at point *f*. However following route *2* leads to a different conclusion. We see that the amount of liquid decreases with increasing temperature to vanish on crossing the saturated gas line. Over the rest of route *2* only one phase is present which we would assert is a 'gas'. Thus it becomes a quibble over words what we call the single phase present at point *f*. The convention we shall adopt is to use the word *liquid* to describe the single phase present when the state of the system is represented by points lying on the portion of the surface shown shaded in Fig. 1.2. This figure also records the fact that there is no critical point associated with the 'liquid' to solid transition. No

† To obtain the actual value of any quantity—say the critical pressure of N_2—you must equate the column heading, here $p_c/10^5$ Pa, to the appropriate numerical value for nitrogen, here 34. Cross-multiplying the equation so formed gives $p_c = 34 \times 10^5$ Pa. Exactly the same procedure is followed in reading the value of a physical quantity from a graph. For example, Fig. 1.6 tells us that at the point where the saturated liquid line meets the triple line $V_m/10^{-5}$ m^3 mol$^{-1} = 1·8$; that is $V_m = 1·8 \times 10^{-5}$ m^3 mol^{-1}.

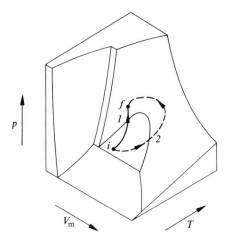

Fig. 1.9. Following route *1* from *i* to *f* leads to the conclusion that the substance is a liquid at *f*; following route *2* leads to the conclusion that it is a gas at *f*.

matter what route you follow from 'liquid' or 'gas' to solid you will observe a phase separation occurring somewhere along the route. You *cannot* pass unknowingly from the gas or liquid phase into the solid phase.

1.3. Metastable states

We have been at some pains to point out that the p–V_m–T surface represents the *equilibrium* states of the system and that values of p, V_m, and T not lying on the surface correspond to non-equilibrium conditions in the system.

Attainment of equilibrium is normally a very rapid process. However under certain conditions the approach to equilibrium may be very slow. A gas held at a temperature T below T_c as in Fig. 1.10 will normally start to condense to form a liquid when the applied pressure equals the value appropriate to the temperature T on the saturated gas line. However if the system is dust-free, it is often possible to increase p substantially beyond this value—pressures up to three or four times it have been realized—before condensation occurs and the system comes truly into equilibrium. A vapour existing in such a 'metastable state' is said to be *supercooled*. It is also often possible to reduce the pressure on a liquid held at a particular temperature to well below that at which evaporation should occur (Fig. 1.10). When this happens the liquid is said to be *superheated*.

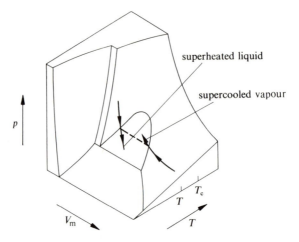

Fig. 1.10. Under certain conditions a substance may exist in metastable equilibrium as a supercooled vapour or as a superheated liquid.

1.4. Projections of the p–V_m–T surface

It is often useful to project the p–V_m–T surface on to the p–V_m, p–T, and T–V_m planes. In projecting the surface on to, say, the p–V_m plane we drop perpendiculars from each point on the surface on to this plane. All information about T is, of course, lost in carrying out this operation. Likewise we lose all information about V_m in projecting on to the p–T plane and about p in projecting on to the T–V_m plane. Fig. 1.11(a) shows the result of projecting the p–V_m–T surface on to the p–V_m plane. The plane is entirely peppered with points; the two independent variables p and V_m may be given any value we choose. So, as it stands, the p–V_m projection in Fig. 1.11(a) tells us *nothing*. However, by projecting the lines on the p–V_m–T surface which indicate phase changes (for example, the saturated vapour line) we obtain Fig. 1.11(b). It is even possible to restore some of the missing information as to the temperature of the system: An isotherm on the surface, as in Fig. 1.4, will project unaltered on to the p–V_m plane (Fig. 1.11(c)). Fig. 1.12 shows how isotherms on the p–V_m–T surface of argon appear in the p–V_m plane.

In an exactly similar way the surface can be projected on to the p–T plane as shown in Fig. 1.11(a). Notice how the surfaces on which two phases coexist in equilibrium project as lines and how the triple line projects as a point—the *triple point*. The line connecting the triple point to the critical point records how the boiling temperature (the temperature at which the liquid-to-gas phase transition occurs) varies with pressure. It is usually called the *vapour pressure* (or merely *vaporization*) *curve*. The line

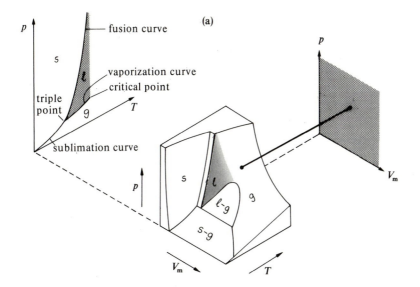

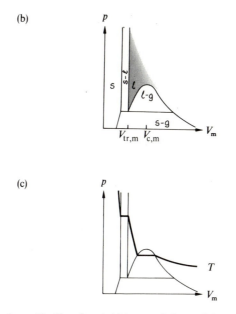

Fig. 1.11. (a) Projecting the p–V_m–T surface (which extends beyond that shown) on to the p–V_m and p–T planes. (b) Showing the regions of the p–V_m plane in which different phases are found. (c) An isotherm projects unaltered on to the p–V_m plane.

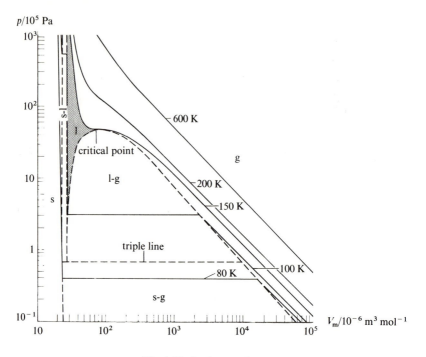

Fig. 1.12. Isotherms of argon.

recording the pressure dependence of the temperature at which solid and liquid coexist is called the *fusion curve*. When solid and gas coexist we refer to the *sublimation curve*. Finally, it is worth reminding ourselves that the entire p–T plane of Fig. 1.11(a) is peppered with points—the system may be given any values of p and T—and that the projection contains no information about the volumes of the system (although that could be made good, in part, by projecting isochores on to the p–T plane).

1.5. The internal energy

In all our discussions about changing the state of a substance from that represented by a point i on the p–V_m–T surface (Fig. 1.13) to that represented by a point f we omitted to say anything about the energy required to bring about this change. It turns out—perhaps unexpectedly—that the energy which is required is independent of the route which is followed. Exactly the same energy is required to take the substance from i to f along route *1* in Fig. 1.13, as along route *2* (to name but two possible routes).

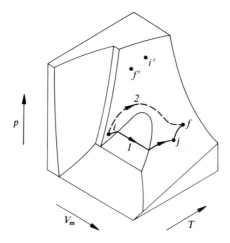

Fig. 1.13. Two different routes which take a system from a state *i* to a state *f*. State *f′* cannot be reached from state *i′* by any reversible adiabatic process.

Measuring the change in energy

Figure 1.14 shows the type of apparatus which would allow us to measure the change in energy of a system in following a route like route *1* in Fig. 1.13. It represents an idealized version of the sort of apparatus used by Joule in the early part of the last century. Here the substance—we shall suppose there to be 1 mol present—is contained within a piston and cylinder apparatus whose walls are *adiabatic* (heat insulating). Even

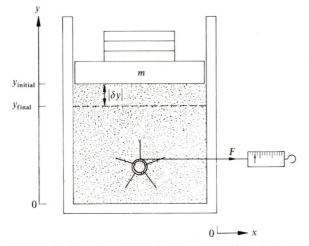

Fig. 1.14. Energy may be fed into the system (but not removed) via the stirrer. Energy may be fed in, and removed, via the piston. The system is contained within adiabatic walls and the piston is constructed of adiabatic material.

if we were to play a bunsen burner against either the piston or the cylinder, it would not change the state of the system. However, there are two ways in which energy—mechanical energy—can be fed into the system (also referred to as 'doing work on the system').

The first is via the paddle wheel (Fig. 1.14). This can be rotated by means of a piece of string which is connected via a frictionless, but leak-free, seal to a *Newton balance*; a spring balance calibrated in newtons. If it takes a constant force F to pull the string, then, on pulling it out through a distance Δx,

$$\text{energy change of the system} = F\,\Delta x. \tag{1.1}$$

In such a process the system always gains energy ($F\,\Delta x$ is positive). The opposite process in which the paddle wheel spontaneously rotates and pulls the string into the cylinder has never been observed.

Energy may also be fed into the system—and, furthermore, removed—via the piston. If it moves down, energy is fed in; if it moves up, energy is removed. In evaluating the work done via the piston we shall suppose that there is no friction between the piston and the cylinder (allowing us to equate the force on the piston to the force on the gas) and that the process is carried out slowly (only then will p have a common value throughout the gas; a value determined by V_m, T, and the form of the p–V_m–T surface). When both these conditions are satisfied we speak of a *reversible* process, here a *reversible adiabatic* process. If during such a process the piston moves from $y_{initial}$ to y_{final} (Fig. 1.14) and the combined mass of the piston plus weights is m, then

$$\text{energy change of the system} = -mg(y_{final} - y_{initial}). \tag{1.2}$$

The negative sign on the right-hand side of eqn (1.2) ensures that, when energy is added to the system by pushing down the piston, the energy change is positive, and that, when energy is removed from the system by allowing the piston to rise, the energy change is negative. Since $p = mg/A$, eqn (1.2) may also be written as

$$\text{energy change of the system} = -pA(y_{final} - y_{initial}) = -p\,\Delta V_m, \tag{1.3}$$

where $\Delta V_m = V_{final} - V_{initial}$ is the volume change of the system. By way of example, if the piston plus weight exert a pressure $p = 6 \times 10^5$ Pa and if $\Delta V_m = -3 \times 10^{-6}\ \mathrm{m^3\ mol^{-1}}$, the energy change in the system (which contains, remember, 1 mol of material) is $6 \times 10^5\ \mathrm{N\ m^{-2}} \times 3 \times 10^{-6}\ \mathrm{m^3\ mol^{-1}} = 1 \cdot 8\ \mathrm{J\ mol^{-1}}$. Eqns (1.2) and (1.3) assume that the entire apparatus is enclosed in an evacuated vessel (as in Fig. 1.1). If this is not the case, the work done by the atmosphere pushing on the piston must also be included (see problem 1.9).

It happens to be particularly easy to measure the energy change of the system in going from i to f along route 1 in Fig. 1.13. We first hold p constant, leaving the weights unchanged on the piston, and keep adding energy via the paddle wheel until we reach point j. Over this part of the route the energy change in the system is, from eqns (1.1) and (1.2), $F \Delta x - p \Delta V_m$, where Δx is the distance through which the string is pulled with force F, and ΔV_m is the change in the volume of the system in going from i to j. The remainder of route 1, as it is shown in Fig. 1.13, is followed by only feeding in energy via the piston (the paddle wheel is not rotated). Since p keeps changing from j to f we must evaluate $-\int p \, dV_m$ from $V_m = V_i$ to $V_m = V_f$. In practice we might do this by plotting a graph of p against V_m in going from V_i to V_f; the area under this graph giving the magnitude of the change in energy of the system.

It is somewhat more tedious to follow route 2 in Fig. 1.13. However, when followed experimentally the overall change in the system is found to be exactly the same as along route 1. Although we have been able to use the apparatus of Fig. 1.14 to go from a point i to a point f it is not possible to use this apparatus to go from *any* point i to *any* point f. Fig. 1.13 shows two points i' and f' located so that it is *impossible* to pass from i' to f'; no matter how we adjust the weights on the piston and/or rotate the paddle we will never reach f' (another such example is provided by points i and f in Fig. 5.20). This is really a statement of the *second law of thermodynamics* (Carathéodory's formulation). In words it says: In the neighbourhood of any equilibrium state of a system there are states which are inaccessible by a reversible adiabatic process. However, even if we cannot reach f' from i' we can use the apparatus of Fig. 1.14 to reach i' from f'; allowing us to find the difference in the energy of the system in states i' and f'.

The internal energy function

These experiments argue strongly that there is some 'function of state' U_m the difference between the value of which at f and i is the change in the energy of the system in going from i to f. In other words, the change in energy is $U_m(f) - U_m(i)$, where $U_m(f)$ is the value of the function at f and $U_m(i)$ is its value at i. The name *molar internal energy* is given to the function U_m. Knowing the value of $[U_m(f) - U_m(i)]$ from experiment the value of

$$U_m(f) = U_m(i) + [U_m(f) - U_m(i)] \tag{1.4}$$

follows immediately provided $U_m(i)$ is assigned some arbitrarily chosen value. We might, for example, decide to let $U_m(i) = 0$ at $T = 0$, $V_m = \infty$. (In practice we would have to start at a small, but finite, temperature and have a large, but finite, volume.) The measured value for $U_m(f)$ at a particular point (p, V_m, T) could now be 'written in' on the surface.

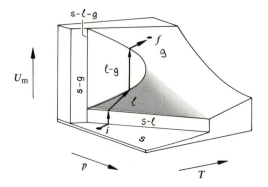

Fig. 1.15. The general form of the U_m–p–T surface of a substance which expands on melting.

There is, however, a much neater way to record the values of the internal energy. Once the values of the two variables V_m and T (or V_m and p) are specified for a given substance the p–V_m–T surface dictates the value of the third variable. The internal energy data may therefore be recorded as either a U_m–p–T surface, as a U_m–p–V_m surface, or as a U_m–V_m–T surface. Fig. 1.15 shows the general form of the U_m–p–T surface of a substance which expands on melting. You can see that this surface 'makes sense' by considering what happens as we follow an isobar. In terms of the apparatus of Fig. 1.14 we keep a constant weight on the piston and note how the temperature of the system changes as energy is fed in via the paddle wheel. Fig. 1.15 tells us that energy must be fed in as we raise the temperature of the solid, melt it (during which p and T of the system remain constant), raise the temperature of the liquid, evaporate it (again, p and T remain constant), and finally raise the temperature of the gas. This agrees with our everyday experiences.

1.6. The first law of thermodynamics

To summarize the arguments of §1.5, we have found that during a reversible adiabatic process in which a net amount of work W ($= F\,\Delta x - p\,\Delta V_m$) is done on the system, the internal energy of the system changes by

$$\Delta U = W. \tag{1.5}$$

Here the suffix m has been removed to emphasize the fact that the result holds true for any fixed amount of the substance.

Although we used a paddle wheel in the apparatus of Fig. 1.14, the same conclusion is reached when the wheel is replaced by a coil of resistance wire carrying a current I. If the potential difference across the wire is V, the electrical work done on the system in a time Δt is $VI\,\Delta t$, and this expression would, of course, feature in eqn (1.1) in place of $F\,\Delta x$. When V is measured in volts, I in amperes, and Δt in seconds, $VI\,\Delta t$ is again measured in joules.

If the adiabatic wall of Fig. 1.14 is replaced in whole or in part by a diathermal wall, an important change occurs: W is *not* the same in following different routes from i to f in Fig. 1.13. Under these conditions eqn (1.5) becomes

$$\Delta U \neq W. \tag{1.6}$$

But, since the left-hand side, which equals $U(f) - U(i)$, is determined solely by the initial and final states, it follows that the right-hand side of eqn (1.6) must contain another term Q with the property that

$$\boxed{\Delta U = Q + W} \; . \tag{1.7}$$

The quantity Q, so defined, is called *heat*. Like W it is labelled as positive when heat is transferred *to* the system. Eqn (1.7) is known as the *first law of the thermodynamics*; a law that is often given as the too-condensed statement that energy is conserved if heat is taken into account. This statement fails to do full justice to the fact that ΔU depends only on the initial and final states and is independent of the route connecting these states.

Differential form of the first law

Since all the arguments outlined in §1.5 will be equally true when i and f are close together in Fig. 1.13, we can rewrite eqn (1.7) as follows to describe infinitesimal changes occurring in a system containing 1 mol of the substance:

$$\boxed{dU_m = đQ - p\,dV_m} \tag{1.8}$$

As just written this assumes that no work is being done by the stirrer in Fig. 1.14. The symbol đ indicates that $đQ$ is not an exact differential; unlike dU_m the value of $đQ$ depends on how the system is changed from its initial to its final state (making it impossible to define Q as a function of state alone). Because U_m can be expressed as a function of any two of p, V_m, T, say $U_m = U_m(T, V_m)$, we can always write

$$dU_m = \left(\frac{\partial U_m}{\partial T}\right)_{V_m} dT + \left(\frac{\partial U_m}{\partial V_m}\right)_T dV_m, \tag{1.9}$$

where the partial differential notation, as in $(\partial U_m/\partial T)_{V_m}$, for example,

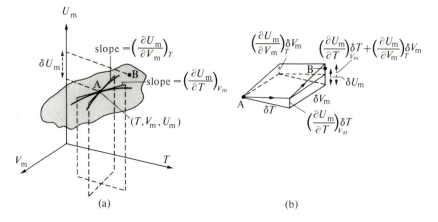

Fig. 1.16. (a) Showing $(\partial U_m/\partial V_m)_T$ and $(\partial U_m/\partial T)_{V_m}$ as the slopes of the curves of intersection of the surface $U_m(T, V_m)$ by the planes $T=$ constant and $V_m=$ constant, respectively. (b) Showing that, starting off from A, one (almost) reaches B by making consecutive moves, first parallel to the T-axis and then parallel to the V_m-axis.

means that we are to differentiate U_m with respect to T, keeping V_m constant. Fig. 1.16 should at least make eqn (1.9) plausible; a rigorous proof will be found in most calculus texts.

Substituting eqn (1.9) into eqn (1.8) gives

$$\dj Q = \left(\frac{\partial U_m}{\partial T}\right)_{V_m} dT + \left[p + \left(\frac{\partial U_m}{\partial V_m}\right)_T\right] dV_m. \tag{1.10}$$

Exercise 1.2

One mole of a certain substance is contained within a piston and cylinder containing a stirrer like that shown in Fig. 1.14. The walls of the cylinder are made of copper (that is, they are diathermal) but they can be lagged (making them adiabatic). By suitably transferring mechanical energy to and from the system and by adding or substracting heat it is possible to make the system follow the path *abcd* shown in Fig. 1.17.

(a) How much energy is transferred to the system in changing its state from *a* to *b*, assuming the walls are adiabatic and the stirrer is not rotated? (b) In going from *b* to *c* the walls are adiabatic and the piston is fixed. If the stirrer feeds in 8 J what is the total energy fed into the system in going from *a* to *c*? (c) In going from *c* to *d* no energy is added via the stirrer and the walls are adiabatic. Evaluate $U_m(d) - U_m(a)$. (d) The state of the system is restored to *a* by removing some of the lagging. How much heat will flow out through the diathermal wall? (e) If instead of 1 mol of the substance there is 5×10^{-2} mol how will this change your answers to (a) to (d)?

Method. In answering these questions we need only apply the first law of thermodynamics (eqn (1.7), with W defined by eqns (1.1) and (1.3).

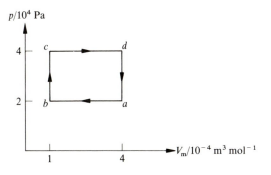

Fig. 1.17. Showing the route followed in the p-V_m plane by one mole of a certain substance.

Solutions. (a) In going from a to b the only energy entering the system does so via the piston. Since p is constant at 2×10^4 Pa, eqn (1.3) tells us that $(2 \times 10^4 \, \text{N m}^{-2}) \times (3 \times 10^{-4} \, \text{m}^3 \, \text{mol}^{-1}) = 6 \, \text{N m mol}^{-1} = 6 \, \text{J mol}^{-1}$ of energy *enters* the system. (b) In going from b to c the only energy *entering* the system is the 8 J fed into the 1 mol of substance via the stirrer. Hence the total energy fed in while going from a to c is $(6 + 8) \, \text{J mol}^{-1} = 14 \, \text{J mol}^{-1}$. (c) In going from c to d energy of amount $p \, \Delta V_m = (4 \times 10^4 \, \text{Pa}) \times (3 \times 10^{-4} \, \text{m}^3 \, \text{mol}^{-1}) = 12 \, \text{J mol}^{-1}$ *leaves* the system. Hence, from eqn (1.7), $U_m(d) - U_m(a) = Q + W = 0 + [8 + (6 - 12)] \, \text{J mol}^{-1} = 2 \, \text{J mol}^{-1}$. (d) In going from a to d around any closed path there can be no overall change in U_m. Hence 2 J of heat must leave the cylinder in this final stage. (e) With only 1/20 as much material present, the answers to parts (a) to (d) must be multiplied by 1/20.

Comments. A plot of p versus V (here V_m) is known as an *indicator diagram*. As we have just seen, it allows the mechanical work done on or by the system, via the piston, to be evaluated. Anyone aware of Joule's work would possibly wish to refer to the energy fed in via the stirrer in Fig. 1.14 as *heat*. Although permissible to do so, someone pulling the string from outside the cylinder would probably have insisted that he was 'doing work' on the system. However, had we chosen to write $Q = 8 \, \text{J mol}^{-1}$ in (b) none of the answers to parts (c) to (e) would have altered. Thus in part (c) we would have written $Q = 8 \, \text{J mol}^{-1}$ and $W = (6 - 12) \, \text{J mol}^{-1}$, leaving $U_m(d) - U_m(a)$ unchanged.

1.7. The molar heat capacities

An obvious experiment is to feed an infinitesimal amount of heat đQ into the system of Fig. 1.1 (in practice we would probably do this by placing the diathermal wall of the cylinder in contact with an electrical heater) and to measure the resulting temperature rise dT. We may perform the experiment in two different ways; either we may keep the piston fixed in the apparatus, thereby keeping the volume of the system fixed; or we may allow the piston to move, leaving the weights unaltered, thereby keeping the pressure fixed.

The *molar heat capacity at constant volume*, $C_{V,m}$, is defined as

$$C_{V,m} = \frac{\text{d}Q_V}{\text{d}T},$$ (1.11)

where the suffix V denotes that the volume is held constant. Substituting for $\text{d}Q$ from eqn (1.10) with $\text{d}V_m = 0$ gives

$$\boxed{C_{V,m} = \left(\frac{\partial U_m}{\partial T}\right)_{V_m}}$$ (1.12)

which can, of course, be substituted back into eqn (1.10) to give

$$\boxed{\text{d}Q = C_{V,m}\text{d}T + \left[p + \left(\frac{\partial U_m}{\partial V_m}\right)_T\right]\text{d}V_m}\,.$$ (1.13)

The *molar heat capacity at constant pressure*, $C_{p,m}$, is likewise defined as

$$C_{p,m} = \frac{\text{d}Q_p}{\text{d}T}$$

which, from eqn (1.10), can be written as

$$\boxed{C_{p,m} = \left(\frac{\partial U_m}{\partial T}\right)_{V_m} + \left[p + \left(\frac{\partial U_m}{\partial V_m}\right)_T\right]\left(\frac{\partial V_m}{\partial T}\right)_p}$$ (1.14)

or, substituting for $(\partial U_m/\partial T)_{V_m}$ from eqn (1.12), as

$$\boxed{C_{p,m} - C_{V,m} = \left[p + \left(\frac{\partial U_m}{\partial V_m}\right)_T\right]\left(\frac{\partial V_m}{\partial T}\right)_p}\,.$$ (1.15)

Eqns (1.12) and (1.14) show that the heat capacities at constant volume and at constant pressure can be predicted from a knowledge of $U_m(T, V_m)$ and the *equation of state* (the functional relation between p, V_m, and T) of the substance. In later chapters we shall see how these two functions can be deduced if we know the laws governing the behaviour of the atoms of the substance. Should there be any discrepancy between the predicted and the measured heat capacities—and such discrepancies will be found—we must suspect the correctness of our assumptions as to how the atoms behave rather than the correctness of the first law of thermodynamics.

PROBLEMS

1.1. Fig. 1.5 shows the general form of the p–V_{m}–T surface of a simple substance which expands on freezing. Describe in your own words what happens as (a) the pressure is decreased at a fixed temperature $T < T_{tr}$, (b) the temperature is increased at fixed pressure $p_{c} > p > p_{tr}$, and (c) the temperature is increased at a fixed volume intermediate in value between those at which the saturated liquid line and the saturated solid line (the line giving the state of a solid in equilibrium with a liquid) meet the triple line.

1.2. Make a rough sketch of the projection of the p–V_{m}–T surface of Fig. 1.2 on to the V_{m}–T plane. Record on this projection the phases which are present throughout. Also superimpose the results of experiments performed under conditions of constant pressure (that is, draw in isobars on the V_{m}–T plane).

1.3. You may well find that the mere act of sketching a p–V_{m}–T surface helps your understanding of what it represents. After tracing it a few times you will probably be able to draw it freehand (this makes quite a good party trick). Now draw in various routes over the surface and ask yourself how the phase of the substance changes as these various routes are followed.

1.4. At a temperature T a mass M of a substance occupies a volume V_{1} when on the saturated liquid curve and a volume V_{g} when on the saturated gas curve. Calculate the fractions by mass of the system which are in the gaseous and liquid phases, respectively, when the system has a total volume V intermediate between V_{1} and V_{g} and is at a temperature T (Fig. 1.3). *Clues:* It may help to draw a picture of the system showing the two phases. The total mass of the system is made up of the masses of . . . . These masses can be expressed in terms of volumes and densities. The densities are given by . . . .

1.5. Describe, in your own words, what happens to the phase of a substance as it is taken from state i to state f along the two routes shown in Fig. 1.18.

1.6. In what phases might the substance taken through the cycle *abcda* in Fig. 1.17 be found? Consider each section of the route separately.

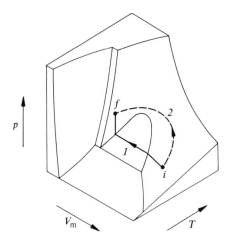

Fig. 1.18. Two different routes on the p–V_{m}–T surface.

24 The p–V_m–T surface

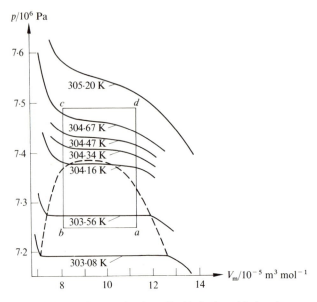

Fig. 1.19. Isotherms of carbon dioxide in the critical region.

1.7. Figure 1.19 shows some isotherms of carbon dioxide close to the critical point. (a) What is the equilibrium temperature of carbon dioxide at $p = 7\cdot4 \times 10^6$ Pa, $V_m = 11 \times 10^{-5}$ m^3 mol^{-1}? What phase is the substance in at this pressure and volume? (b) What volume would 10^{-2} mol of carbon dioxide occupy at $p = 7\cdot25 \times 10^6$ Pa, $T = 303\cdot56$ K? (c) Describe in words the phase changes which occur as the state of the system goes from a to b to c to d to a. (d) Graph isochores for $V_m = 8 \times 10^{-5}$ m^3 mol^{-1}, $V_m = 10 \times 10^{-5}$ m^3 mol^{-1}, and $V_m = 12 \times 10^{-5}$ m^3 mol^{-1} in the p–T plane. Indicate the phases which are present in different regions of the p–T projection. (e) Graph the isobar $p = 7\cdot43 \times 10^6$ Pa in the V_m–T plane. (f) At atmospheric pressure ($1\cdot0 \times 10^5$ Pa) carbon dioxide turns directly from the solid to the gaseous phase without passing through the liquid phase (i.e., it *sublimes*). What does this tell us about the triple point pressure of carbon dioxide?

1.8. (a) A common method of obtaining low temperatures is to reduce the pressure over a boiling liquid. What is the lowest temperature that can be attained by boiling nitrogen under reduced pressure, and what pressure is required to attain this temperature? Use the data given in Table 1.1 (p. 10). (b) One could construct a hydrogen liquefier which operates by increasing the pressure on the gas while holding its temperature constant. The hydrogen might be contained in a copper piston and cylinder arrangement which is immersed in a suitable liquid boiling under reduced pressure. Which of the substances listed in Table 1.1 could be used to cool the hydrogen? Explain.

1.9. Two moles of a substance are contained within an adiabatic enclosure like that shown in Fig. 1.14. (a) Assuming that the piston (of mass 10 kg) remains fixed while the string attached to the spindle of the paddle wheel is pulled out through a distance of 50 m with a force of 1·5 N, by how much will the molar internal energy of the system increase? (b) If the previous process is repeated but this time the piston (area 5×10^{-3} m²) rises through a distance of 6×10^{-2} m in pulling the string through the 50 m, what is the new value for the increase in the molar internal energy? Assume the atmosphere surrounding the cylinder is at a pressure of $1·0 \times 10^5$ Pa. (c) If the paddle wheel is replaced by a resistor of value 3 Ω carrying a current of 2 A, for how long should the current flow to ensure that the electrical work done on the system is the same as the mechanical work done via the paddle wheel in parts (a) and (b)? (d) If the 'electrical work' referred to in part (c) is now called 'heat' how does this alter your answers to parts (a) and (b)? Explain.

1.10. An electrical heater of resistance $R = 12 \, \Omega$ and carrying a current I which varies with time according to $I = I_0 \exp(-t/\tau)$, where I_0 and τ are constants, is placed in thermal contact with the diathermal wall of the apparatus shown in Fig. 1.1. Taking $I_0 = 5$ A and $\tau = 6$ s, how much heat will have passed through the wall during the time interval extending from $t = 0$ to $t = 9$ s?

1.11. When 10 J of heat is supplied to 0·2 mol of aluminium at s.t.p., its temperature is found to rise by 2·05 K. (a) What is $C_{p,m}$ for aluminium at s.t.p.? (b) Given the fact that the volume of aluminium increases by $7·5 \times 10^{-3}$ per cent for each 1 K rise in temperature (at s.t.p.), by how much will $C_{p,m}$ differ from $C_{V,m}$ at s.t.p.? One mole of aluminium occupies a volume of $1·0 \times 10^{-5}$ m³ at s.t.p. *Clue*: eqn (1.8).

1.12. One mole of carbon dioxide is contained within a piston and cylinder arrangement like that shown in Fig. 1.14. With the volume of the system held fixed at 8×10^{-5} m³ the work W_V, say, required to raise the temperature of the carbon dioxide from 304·16 to 304·34 K is measured. This work is done via the stirrer. The experiment is now repeated but this time the work W_p, say, required to raise the temperature of the carbon dioxide from 304·16 to 304·34 K is measured while the piston is subjected to a constant pressure of $7·38 \times 10^6$ Pa. Using the information contained in Fig. 1.19, calculate $W_p - W_V$.

1.13. We will shortly discover that the equation of state of all gases takes the form $pV_m = RT$ at sufficiently high molar volumes. Here R is a constant (of value 8·31 J K^{-1} mol^{-1}). Draw out the form of the p-V_m-T surface of such low-density gases. You will probably find it helpful to consider the form of an isotherm, an isobar, and an isochore.

1.14. Determine the value of $C_{p,m} - C_{V,m}$ for a low-density gas, whose equation of state has the form $pV_m = RT$, where R is a constant of value 8·31 J K^{-1} mol^{-1}. In answering this problem, you are to follow the same procedures as those adopted in answering problem 1.12 but with the isotherms given by the equation of state. You should therefore start by drawing out isotherms at, say, 300 and 301 K.

1.15. Deduce $C_{p,m} - C_{V,m}$ for a substance whose internal energy function is $U_m = \frac{3}{2}RT$ and whose equation of state is $pV_m = RT$.

1.16. Molar enthalpy, H_m, is defined as $H_m = U_m + pV_m$. Show that when one mole of a substance undergoes an isobaric change the heat added to the system is given by the change in molar enthalpy of the system.

1.17. Sketch out the general form of the U_m–p–T surface of a substance which contracts on melting. You will find it helpful to start with Fig. 1.15.

1.18. Columns 1 and 2 of Table 1.2 list the vapour pressure of argon vapour in equilibrium with solid argon. Columns 3 and 4 list the vapour pressure of argon vapour in equilibrium with liquid argon. Finally columns 5 and 6 give the pressures at which solid and liquid argon coexist in equilibrium. Using this data plot out on a p–T diagram the lines indicating phase equilibria. The following features should be clearly labelled; sublimation curve, vaporization curve, fusion curve, triple point, and critical point.

TABLE 1.2
Pressure–temperature data for argon

solid–gas		liquid–gas		solid–liquid	
T/K	$p/10^4$ Pa	T/K	$p/10^4$ Pa	T/K	$p/10^5$ Pa
66	0·35	84	7·08	84	0·91
68	0·52	86	8·81	85	4·86
70	0·77	88	10·91	90	25·33
72	1·12	90	13·37	100	76·00
74	1·59	100	32·52	110	125·6
76	2·20	110	66·77	120	175·3
78	3·01	120	121·4	130	223·0
80	4·07	130	202·4	140	276·6
82	5·40	140	317·1	150	334·4
83·78	6.88	150	474·2	160	390·1
		150·8	489·4	170	451·0
				180	506·6
				190	572·5

2. Characterizing atoms

IN this chapter we will look at how atoms may be characterized in terms of their mass and size, of how they interact with one another and in terms of the laws of motion that they obey. Throughout we shall look for generalizations that apply to many atoms rather than attempting to characterize a few atoms in depth. We shall, so to speak, establish the ground rules for all that lies ahead.

2.1. Atomic masses

The mass m of an object is defined operationally by Newton's second law

$$F = ma, \qquad (2.1)$$

where a is the acceleration produced by the force F. Assuming F to be constant and integrating eqn (2.1) with respect to distance gives the familiar result

$$Fs = \tfrac{1}{2}mv^2, \qquad (2.2)$$

where v is the final speed realized when the object, initially at rest, is acted on by force F through a distance s. The left-hand side of eqn (2.2) represents the energy transferred to the object.

Although there are many mass spectrometers which measure atomic masses via eqn (2.1), the one which we shall look at utilizes eqn (2.2).

The time-of-flight mass spectrometer

The most practical way to give an atom a known amount of energy is to *ionize* it—that is to remove one or more, say n, electrons—and then accelerate it through a known electrical potential difference. Fig. 2.1(a) shows a piece of apparatus where this approach may be realized. A short burst of ionized atoms (produced by bombarding the gas atoms with a transversely moving electron beam) each of mass m and each possessing a charge ne, where e is the charge of a proton ($+1.6 \times 10^{-19}$ C), is nudged through grid G_1 by momentarily making the backing plate B positive with respect to G_1. Once through G_1 each ion is accelerated by an electric field directed towards G_2. If the potential difference between G_1 and G_2 is V, the ion will possess an energy Vne on emerging through G_2. This assumes that the ion has suffered no collisions on the way across; a condition ensured by having a near vacuum in the apparatus. If the ion enters G_1 with effectively zero speed and leaves through G_2 with speed v eqn (2.2)

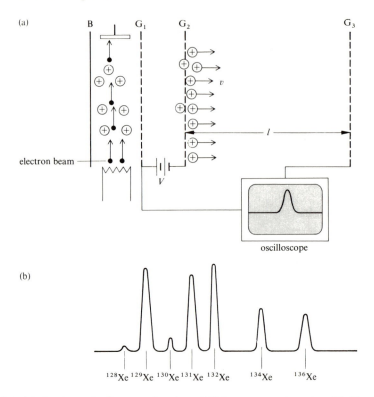

Fig. 2.1. (a) A schematic diagram of a time-of-flight mass spectrometer. (b) The trace obtained on the oscilloscope when naturally-occurring xenon is analysed in the spectrometer.

tells us that

$$Vne = \tfrac{1}{2}mv^2. \tag{2.3}$$

So, provided we can measure v and can guess n correctly, we can deduce the value of m.

To measure v we simply measure the time t taken by the ion to travel across a field-free region of length l. Since $v = l/t$, eqn (2.3) becomes

$$\boxed{Vne = \tfrac{1}{2}m(l/t)^2} . \tag{2.4}$$

The flight time t is measured using a cathode-ray oscilloscope. At the instant the burst of ions is nudged through G_1 the oscilloscope is triggered, tracing out a straight line until a blip is produced as the ions pass through grid G_3. Knowing the position of the blip and the sweep-rate of the oscilloscope we can readily deduce t, and hence m.

If ions of different masses are present each of these species will have different flight times (see eqn (2.4)) resulting in a series of blips. Fig. 2.1(b) shows the trace obtained with singly ionized atoms of xenon. The left-hand superscript, as for example in ^{131}Xe, tells us the total number of protons and neutrons present in a nucleus; this total is called the *mass number*. Since the mass of a proton $(1\cdot6726\times10^{-27}$ kg) and that of a neutron $(1\cdot6749\times10^{-27}$ kg) are practically identical it is usually a simple matter to deduce mass numbers—we divide m, as deduced from the position of the maximum of the blip, by $1\cdot67\times10^{-27}$ kg and round off the result. If we wish to record the number of protons present in a nucleus—that is the *atomic number*—we use the left subscript position. By writing $^{131}_{54}$Xe, for example, we are stating that we are dealing with an atomic species whose nucleus contains 54 protons and 131 protons-plus-neutrons (and therefore $131-54=77$ neutrons). The word *isotope* is used to describe atomic species having the same atomic number but with different mass numbers. Fig. 2.1(b) therefore shows seven isotopes of xenon. As another example, naturally-occurring carbon mainly consists of two stable isotopes; ^{12}C with an atomic mass of $1\cdot9926\times10^{-26}$ kg and ^{13}C with an atomic mass of $2\cdot1592\times10^{-26}$ kg. (We ignore the unstable ^{14}C isotope.) The ^{12}C isotope has a 98·89 per cent abundance (by mass); the ^{13}C isotope a 1·11 per cent abundance.

So far as the chemistry of an element is concerned all isotopes of that element behave in the same way, notwithstanding their different atomic masses. For this reason the mean atomic mass of an element as it occurs in nature is usually adequate. The mean mass follows immediately from the atomic masses of the isotopes and their relative abundances (proportional to the area under each blip on the oscilloscope trace). Thus, for example,

$$\begin{aligned}
\text{mean atomic mass} \atop \text{of natural carbon} &= \frac{98\cdot89}{100}(1\cdot9926\times10^{-26}\text{ kg})+\frac{1\cdot11}{100}(2\cdot1592\times10^{-26}\text{ kg}) \\
&= 1\cdot9944\times10^{-26}\text{ kg.}
\end{aligned}$$

2.2. Some definitions

Relative atomic mass

For many purposes, comparative values of atomic masses are quite adequate. By international agreement, the *relative atomic mass* A_r of an element is defined as

$$A_r = \frac{\text{mass of an atom of the substance}}{\text{mass of a }^{12}\text{C atom}}\times12, \qquad (2.5)$$

where 12 means precisely 12. Substituting the atomic masses as measured with a mass spectrometer gives, for example, $A_r(^{35}Cl) = 34 \cdot 9688$. When no left-hand superscript is shown (for example, $A_r(Cl)$) it indicates that the mean mass of the atom as found in the naturally-occurring form of the element has been used. Thus, $A_r(C) = 12 \times (1 \cdot 9944 \times 10^{-26} \text{ kg})/(1 \cdot 9926 \times 10^{-26} \text{ kg}) = 12 \cdot 0111$, whereas, of course, $A_r(^{12}C) = 12$ exactly.

Relative molecular mass

In an analogous way, the *relative molecular mass* M_r of a substance is defined as

$$M_r = \frac{\text{mass of a molecule of the substance}}{\text{mass of a } ^{12}C \text{ atom}} \times 12 \quad . \qquad (2.6)$$

Here the mass of the molecule is usually obtained by summing the constituent atomic masses. When the mean atomic masses are used this is indicated by the absence of the left-hand superscripts; for example, $M_r(CO_2) = 44 \cdot 009$. Had the molecule been composed of, say, ^{12}C and ^{16}O we would have written $M_r(^{12}C^{16}O_2)$. It is important to notice the word *relative* in our definitions of A_r and M_r. Without this word the phrases atomic mass and molecular mass will record, respectively, the mass of an atom of the element in question, and the mass of a molecule of the compound in question.

The mole

Because atomic masses range from about 10^{-25} to 10^{-27} kg there will be anything from 10^{25} to 10^{27} atoms in a kilogram of matter. Now in dealing with large numbers of anything—for example, seconds—it is convenient to have a multiple unit as it keeps the numerical values small. Such a quantity is provided by the *mole*. We say that there is one mole (written 1 mol) of a substance present in a system when it contains as many *named* entities (be they atoms, molecules, ions, electrons, or any other named particles) as there are atoms of ^{12}C in $0 \cdot 012$ kg of ^{12}C. This number of entities, known as the *Avogadro constant*, N_A is thus given by

$$N_A = \frac{0 \cdot 012 \text{ kg}}{\text{mass of } ^{12}C \text{ atom}} \text{ mol}^{-1} \quad . \qquad (2.7)$$

Substituting the measured mass of the ^{12}C atom gives $N_A = 6 \cdot 022 \times 10^{23} \text{ mol}^{-1}$.

When using the mole we *must* state to what entities it refers. A vessel does *not* contain '5 mol of water'; it does contain 5 mol of *water molecules* (i.e. 3.01×10^{24} molecules). Remembering that water is H_2O it could be said to contain 5 mol of oxygen atoms and 10 mol of hydrogen atoms. It so happens that if we measure out a mass *in grams* of an element X equal to the value of its relative atomic mass, $A_r(X)$, we obtain 1 mol of atoms of X. To prove this statement we ask how many atoms there are in $A_r(X)$ grams, that is in $A_r(X) \times 10^{-3}$ kg, of element X. The answer is

$$\frac{A_r(X) \times 10^{-3} \text{ kg}}{\text{mass of atom of X}}$$

which becomes, on substituting for $A_r(X)$ from eqn (2.5),

$$\frac{12 \times 10^{-3} \text{ kg}}{\text{mass of } ^{12}\text{C atom}}$$

and this (see eqn (2.7)) is equal to N_A. Likewise, if we measure out a mass in grams, of a compound Y equal to the value of $M_r(Y)$ we obtain 1 mol of molecules of Y.

Molar quantities

The *molar volume*, V_m, of a system is defined by

$$\boxed{V_m = \frac{\text{volume of the system}}{\text{amount of the substance present}}}, \qquad (2.8)$$

where *amount* means the number of moles of molecules present. In the case of an element the denominator means the number of moles of atoms present. As another illustration of a molar quantity, the *molar mass M* of a substance is defined by

$$M = \frac{\text{mass of the substance}}{\text{amount of the substance present}}. \qquad (2.9)$$

By way of example, if 3 mol of lead has a volume of $5 \cdot 48 \times 10^{-5}$ m³ and a mass of 0·622 kg the molar volume of lead is $1 \cdot 83 \times 10^{-5}$ m³ mol⁻¹ and its molar mass is 0·207 kg mol⁻¹.

2.3. Atomic sizes

Unlike the mass of an atom which is precisely defined by eqn (2.1), there is no such thing as *the* size of an atom. In a solid, for example, the atoms are in close proximity. But a solid *is* compressible, suggesting that the size of atoms depends on how hard they are squeezed together. An *order of magnitude* estimate of an atomic diameter—one correct to the nearest power of ten—will therefore suffice.

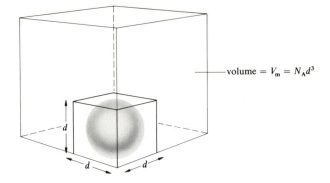

$$\text{volume} = V_m = N_A d^3$$

Fig. 2.2. If the atomic diameter is d, one mole of the solid may be pictured as made up of N_A atoms, each occupying a cube of volume d^3.

Roughly speaking, we may picture the solid phase of an element X as a collection of tightly-packed atoms (Fig. 2.2), not unlike lumps of sugar in a box. Each atom, of diameter d, will therefore occupy a volume d^3. If there is 1 mol of X, it follows that there are N_A atoms present and that the mass of the solid is $A_r(X) \times 10^{-3}$ kg. The total volume occupied by these N_A atoms is $(A_r(X) \times 10^{-3} \, \text{kg}/\rho) \, \text{mol}^{-1}$, where ρ, the density of the solid element, is given in units of kg m^{-3}. Equating this volume with the total volume of all N_A atoms, namely $N_A d^3$, gives

$$d = \left(\frac{A_r(X) \times 10^{-3}}{\rho N_A} \right)^{1/3}. \tag{2.10}$$

TABLE 2.1
Molecular data

Gas	M_r or A_r	Molecular diameter/10^{-10}m			$(\overline{u^2})^{1/2}/$ m s^{-1}	m.f.p.[d] $\lambda/10^{-10}$ m	de Broglie[e] $\lambda/10^{-10}$ m	$\gamma = \dfrac{C_{p,m}}{C_{V,m}}$
		From ρ[a]	From b[b]	From η[c]				
H_2	2·016	4·2	2·8	2·7	1838	800	1·14	1·41
He	4·003	4·2	2·7	2·2	1305	1010	0·81	1·66
CH_4	16·043	4·5	3·2	4·2	651	530	0·40	1·31
N_2	28·013	3·7	3·1	3·6	493	700	0·30	1·40
O_2	31·999	3·7	2·9	3·6	461	725	0·29	1·40
Ar	39·948	4·1	2·9	3·7	413	660	0·26	1·67
CO_2	44·010	4·0	3·2	4·6	393	540	0·24	1·30
Hg	200·59	3·3	2·4	6·3	184	520	0·11	—

Notes: All values are at s.t.p. The Hg data refers to the vapour phase. [a] Deduced from eqn (2.10) as modified for fcc lattice. [b] Deduced from eqn (6.8). [c] Deduced from eqn (7.20). [d] Deduced from eqn (5.61) with d as given by the mean of columns 3, 4, and 5. [e] Deduced from eqn (2.11) with the speed given by eqn (4.50).

As an example, solid lead has a density of $11 \cdot 3 \times 10^3 \, \text{kg m}^{-3}$. Since $A_r(\text{Pb}) = 207$ and $N_A = 6 \cdot 02 \times 10^{23} \, \text{mol}^{-1}$, it follows that $d = 3 \cdot 1 \times 10^{-10} \, \text{m}$.

Later on we shall meet other macroscopic properties, besides the density of a solid, which can be related to the diameter of the constituent atoms. The viscosity of a gas is one such property; its $p-V_m-T$ behaviour is another. Very often these different properties yield different answers for the atomic diameter, as can be seen by examining columns 3 to 5 of Table 2.1. (The other listed properties will be referred to later in the text.) Clearly, the whole concept of the 'size of an atom' will have to be looked at more carefully, and this we shall do in §2.5. But, as measured in most laboratory experiments, atomic diameters lie mainly in the range $(1-5) \times 10^{-10} \, \text{m}$.

2.4. How atoms behave

In discussing the time-of-flight mass spectrometer in §2.1 we assumed that Newtonian mechanics could be applied to atoms. But does an atom of mass about $10^{-26} \, \text{kg}$ obey the same laws of motion as, say, a motor car of mass $10^3 \, \text{kg}$?

A crucial experiment

Evidence that atoms do not obey the laws of Newtonian mechanics first came in the 1920s. In one experiment (developed by Stern from earlier work by Davisson and Germer) a beam of helium atoms all moving at effectively the same speed was directed at a lithium fluoride crystal and the angles at which the atoms were scattered for different settings of the crystal was studied (Fig. 2.3(a)). We shall look in some detail at the experimental procedures for we shall meet very similar ones in later chapters. Helium gas from a furnace—a container of heated helium—passes out through an orifice O and a channel S_1 to form a parallel beam of gas atoms. This beam next encounters two discs, D_1 and D_2, spinning about a common axis. Each of these discs has in its circumference 408 equivalent radial saw cuts. The discs are placed so that the slots on one disc exactly face the slots on the other. With the discs at rest or rotating very slowly, atoms of all speeds can pass through both discs. At a somewhat higher rate of rotation the only atoms which can get through are those which cover the distance d between the discs (about 3 cm) in the same time as it takes the next following slot of D_2 to travel up to the line of OS_1S_2, i.e., the time it takes for a point on the circumference of D_2 to travel a distance equal to the separation of adjacent radial slots. The speed of those atoms which are thereby allowed through S_2 to strike the crystal C is easily calculated from the rotational

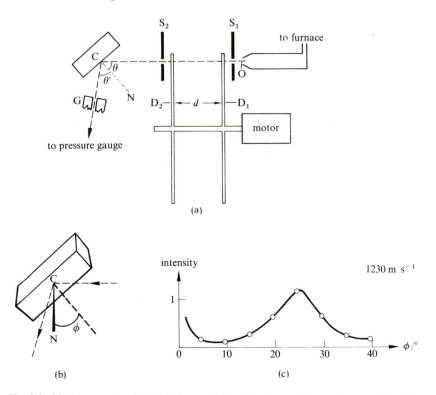

Fig. 2.3. (a) An apparatus in which the scattering of a beam of atoms by a crystal surface may be studied. (b) Showing the angle ϕ between the normal to the crystal surface and the plane of the incident and scattered beams (which lies in the plane of the paper, as in (a)). (c) Showing how the intensity of the scattered beam of helium atoms depends on ϕ when these atoms have a speed of 1230 m s^{-1}. (After Estermann, I., Frisch, R., and Stern, O. (1931). *Z. Phys.* **73**, 358, Fig. 11.)

speed of the disc and the separation of adjacent slots. Atoms scattered from the crystal surface are detected in a pressure gauge G which is set so that the angle of incidence (θ) of the beam equals the angle of reflection (θ'). What is changed in the experiment is the angle ϕ which the normal CN to the crystal surface makes with the plane containing the incident and the scattered beam (Fig. 2.3(b)).

Fig. 2.3(c) shows how the scattered intensity changed with ϕ for a beam of helium atoms moving with a speed of 1230 m s^{-1}. While Newtonian mechanics can readily explain the strong reflection at $\phi = 0$—this is how a ball thrown at a wall would behave—it cannot explain the maximum at 25°. This can only be explained by wave mechanics.

de Broglie's hypothesis

In 1924 de Broglie suggested that, at the microscopic level, matter might exhibit wave-like properties. He argued that instead of thinking of an atom as a well-defined particle of mass m moving in a straight line with a speed v (Fig. 2.4(a)) we should think of it rather as a packet of waves travelling as a group with a speed v (Fig. 2.4(b)). These waves he endowed with those properties, like momentum, which we have so far attributed to particles. In particular, he argued that the wavelength λ of these waves (Fig. 2.4(b)) might be related to the momentum of the 'particle' by

$$\lambda = \frac{h}{mv} , \qquad (2.11)$$

where h is a constant called Planck's constant $(6 \cdot 63 \times 10^{-34}\,\text{J s})$. The displacement of the wave at point x at time t is called the *wave function* $\psi(x, t)$ (Fig. 2.4(b)). Since the world is three-dimensional we should really write $\psi(x, y, z, t)$. The current interpretation says that $|\psi(x, y, z, t)|^2$ represents the relative probability *per unit volume* of finding the particle by an act of observation made at time t with a detector located at point (x, y, z). Therefore $|\psi(x, y, z, t)|^2 \,\delta V$ represents the relative probability of finding the particle within a volume δV at time t. If $|\psi(x, y, z, t)|^2$ is

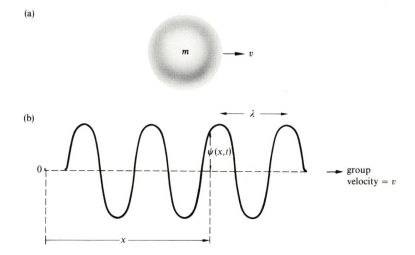

Fig. 2.4. (a) In Newtonian mechanics we consider a well-defined particle of mass m moving with a speed v in a straight line. (b) In wave mechanics we consider a packet of waves of wavelength $\lambda = h/mv$.

$10^{-2}\,\mathrm{m}^{-3}$ and $\delta V = 10^{-6}\,\mathrm{m}^3$, which might well be the 'active volume' of the detector, there is a one in a hundred million chance that the detector will 'click'.

Interpreting the results

To keep the discussion simple we will consider the case where $\theta(= \theta') = 0$ and $\phi \neq 0$ in Fig. 2.3. Fig. 2.5 shows the situation. It is clear that atoms arriving at the detector which have been scattered by atom Q will have travelled a greater distance than those scattered by atom P. This difference amounts to $2QR$ (being RQ in and QR out). Denoting the separation of atoms P and Q by a (see Fig. 2.5) we see that the

$$\text{path difference} = 2a \sin \phi. \tag{2.12}$$

If the signal at the detector is to be a maximum the de Broglie waves reaching the detector must be in phase. (When they are out of phase ψ, and therefore $|\psi|^2$, will be small, implying a small detector reading.) This demands that the path difference is an integral number n of wavelengths and thus, from eqn (2.12), that

$$2a \sin \phi = n\lambda$$

or, substituting for λ from eqn (2.11), that

$$\boxed{\sin \phi = \frac{nh}{2amv}} \ . \tag{2.13}$$

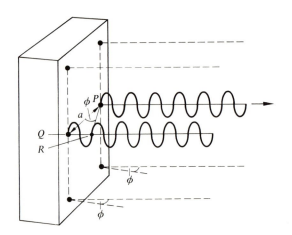

Fig. 2.5. The particular case in which the beam scattered from the crystal of Fig. 2.3(a) is along the line of the incident beam. (This particular geometry cannot be studied in the apparatus of Fig. 2.3(a).)

A single experiment, like that shown in Fig. 2.3(c), in which only ϕ is varied is a poor test of the correctness or otherwise of eqn (2.13). But Stern and his colleagues did study the effects of changing n (by looking at maxima at higher ϕ), a (by using a different crystal), and m (by using hydrogen in place of helium). In all cases their results were in accord with eqn (2.13) and thus with de Broglie's relation (eqn (2.11)).

The dilemma

It begins to look as if we are going to have to abandon Newtonian mechanics in favour of wave mechanics. Fortunately it so happens that we can employ Newtonian mechanics when the de Broglie wavelength of an object is significantly less than the dimension of anything with which that object interacts. This is manifestly true in the case of, say, a car of mass 10^3 kg moving at 20 m s^{-1}, where $\lambda = 3 \cdot 3 \times 10^{-38}$ m. On the other hand, as column 8 of Table 2.1 shows, the de Broglie wavelength $\lambda = h/mv$ of a molecule of a gas at s.t.p. (we shall see later how v can be found from the temperature of the gas) is around ten per cent of the molecular diameter. This suggests that Newtonian mechanics should not be used to discuss intermolecular collisions in a gas. However, we shall always attempt to apply Newtonian mechanics. Sometimes we shall get away with it; at other times we shall fail. The alternative is to adopt *wave mechanics* (or *quantum mechanics*) from now on!

Exercise 2.1

When naturally-occurring sodium (atomic number $= 11$) is examined in a time-of-flight mass spectrometer (Fig. 2.1(a)) a single blip is observed on the screen of the cathode-ray oscilloscope. A flight time of $5 \cdot 46 \times 10^{-6}$ s down a $0 \cdot 5$ m tube is recorded when the Na^+ ions have been accelerated through a potential difference of 1000 V.

(a) What is the measured mass of a sodium ion? (b) What is the de Broglie wavelength of a sodium ion in the field-free region between G_1 and G_2 in Fig. 2.1(a)? (c) What is the mass of a neutral sodium atom? (d) How many neutrons are present in the nucleus of the sodium isotope studied in this experiment? (e) Tables of relative atomic masses (formerly called 'atomic weights') list $A_r(Na) = 22 \cdot 9898$. How would you use this figure to deduce the atomic mass of a sodium atom?

Solutions. (a) Substituting $V = 1000$ V, $n = 1$, $e = 1 \cdot 6 \times 10^{-19}$ C, $l = 0 \cdot 5$ m, and $t = 5 \cdot 46 \times 10^{-6}$ s into eqn (2.4) gives $m = 3 \cdot 816 \times 10^{-26}$ kg. (b) It follows from eqn (2.3) that

$$mv = (2nVem)^{1/2}$$

and hence from de Broglie's relation (eqn (2.11)) that

$$\lambda = h/(2nVem)^{1/2}. \tag{2.14}$$

Substituting for the quantities on the right-hand side gives $\lambda = 1 \cdot 90 \times 10^{-13}$ m. (c) Since the mass of an electron is only $1/1836$ that of a proton we can ignore the small additional mass ($9 \cdot 1 \times 10^{-31}$ kg) to be added to that of the ion to give the mass of the neutral atom. (d) Dividing the mass of a sodium atom, $3 \cdot 816 \times 10^{-26}$ kg, by the (nearly equal) mass of a proton or neutron, $1 \cdot 67 \times 10^{-27}$ kg, and rounding the answer to the nearest integer, tells us that there are 23 protons plus neutrons in a neutral sodium atom. Since the atomic number is 11, there are thus 11 protons and hence $23 - 11 = 12$ neutrons present in the sodium nucleus. (e) As we proved in §2.2 the relative atomic mass expressed in grams, here $22 \cdot 9898 \text{ g} = 22 \cdot 9898 \times 10^{-3}$ kg, contains $N_A = 6 \cdot 022 \times 10^{23}$ atoms. Hence 1 atom has a mass of $(22 \cdot 9898 \times 10^{-3}/6 \cdot 022 \times 10^{23}) \text{ kg} = 3 \cdot 818 \times 10^{-26}$ kg.

Comments. We see that a sodium ion accelerated through a potential difference of 1000 V has a de Broglie wavelength of order 10^{-13} m; fairly insignificant compared with the diameter $d = 2 \times 10^{-10}$ m of a sodium ion. Sodium vapour at s.t.p. contains ions with a mean translational energy equal to the energy Ve that they would have received had they been accelerated through a potential difference $V = 0 \cdot 035$ V. Applying eqn (2.14) thus shows that $\lambda = 3 \cdot 2 \times 10^{-11}$ m at s.t.p. (some 16 per cent of the ion's diameter).

2.5. Atomic interactions; some general considerations

Without some form of interaction between atoms it would be impossible to account for the fact that some hundred-odd atomic species produce the enormous variety of materials of which we are all aware. We may find out quite a lot about this interaction by carrying out a *thought experiment* in which we interpret some large-scale or macroscopic properties of matter from a microscopic viewpoint.

The interatomic force

To begin with we shall consider interatomic forces between the atoms of an element (X, say). Our aim is to show how the force $F(r)$ which one atom of X exerts on another atom of X varies with their separation, measured by the distance r between their centres.

In our imagination we take one atom of X and fix its position. We now take another atom of X to which we have—again in our imagination—attached a Newton balance and we place this second atom at various values of r from the fixed atom (Fig. 2.6a). For each setting of r we try to decide whether the force $F(r)$ between the two atoms is attractive, is repulsive, or is zero. Attractive forces are given negative values (e.g. $-1 \cdot 5 \times 10^{-8}$ N), being directed in the opposite sense to that in which r is directed. Repulsive forces, being directed in the same sense as r, are given positive values. In Fig. 2.6(b) attractive forces are plotted along the upward ordinate and repulsive forces along the downward ordinate.[†]

> † The reason for not adopting the mathematically more usual convention of plotting positive forces along the upward ordinate and negative forces along the downward ordinate is that his leads to a force versus separation curve which superficially resembles the potential energy versus separation curve. The two curves are easily confused.

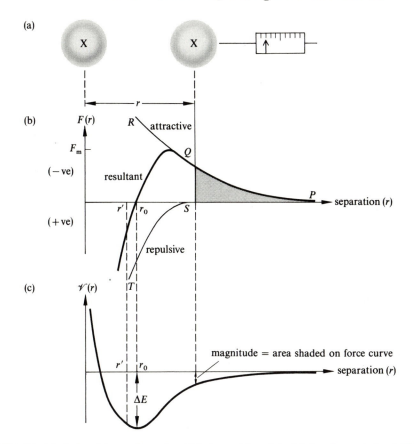

Fig. 2.6. (a) A thought experiment in which the position of the left-hand atom is fixed. The right-hand atom may be set various distances r (measured between centres) from the left-hand atom and the force of attraction (or repulsion) 'read' on a Newton balance. (b) The general form of the interatomic force between the two atoms of (a). Note that negative forces are plotted along the upward ordinate and positive forces along the downward ordinate. (c) The corresponding variation in potential energy.

In carrying out the thought experiment we will examine the following properties of matter from a microscopic viewpoint to see what they can tell us about the interaction between two atoms of X.

1. When they are well separated there is no significant force between two lumps of matter. This tells us that $F(r) \rightarrow 0$ as $r \rightarrow \infty$.
2. Two lumps of an element will only 'weld' together when they are in very close contact. This tells us that as r is reduced the force becomes attractive, as suggested by portion PQ of Fig. 2.6(b).
3. Discrete atoms do exist! Were curve PQ to continue ever upward

along QR (Fig. 2.6(b)) the two atoms would coalesce into a single entity. To prevent this happening repulsive forces, as suggested by curve ST, must come into play. The combined effect of the short-range attractive force followed, at smaller separations, by an even shorter-range repulsive force is shown as a heavy line in Fig. 2.6(b).
4. Under appropriate conditions of temperature and pressure *all* elements can exist in the solid phase. This demonstrates the universality of the hypothesized interaction.

Although we have only considered the interaction between two atoms of an element the arguments are equally true of the interaction between two molecules of a molecular solid—provided that the molecules do not interact 'chemically' in coming together to form the material. Were this to happen we would expect $F(r)$ to have different forms in the solid and gaseous phases.

The interatomic potential energy

The interaction between two atoms can also be expressed in terms of potential energy. We define the *potential energy* $\mathscr{V}(r)$ of two atoms a distance r apart as the energy *required* to bring one atom up from infinity to a distance r from the second one.† This definition labels $\mathscr{V}(r)$ as positive if energy is required in the operation of bringing the movable atom in from infinity to a distance r from the fixed atom; as negative if, on balance, energy is released in carrying out the operation.

The energy $\mathrm{d}\mathscr{V}(r)$ required to displace the movable atom of Fig. 2.6(b) through a distance $\mathrm{d}r$ is $-F(r)\,\mathrm{d}r$; the minus sign arises because the 'person' holding the movable atom must provide a force equal in magnitude, but opposite in direction, to the interatomic force $F(r)$. Thus

$$\mathrm{d}\mathscr{V}(r) = -F(r)\,\mathrm{d}r \tag{2.15}$$

and so

$$\mathscr{V}(r) = -\int_{\infty}^{r} F(r)\,\mathrm{d}r = \int_{r}^{\infty} F(r)\,\mathrm{d}r. \tag{2.16}$$

Hence the potential energy of the pair of atoms at a separation r is the area under the $F(r)$ curve between r and infinity. This operation of measuring an area and plotting the result as a point is shown in Fig. 2.6(b) and (c) (in examining these figures recall that we plotted negative forces

---

† The choice of where to take the zero of potential is, of course, arbitrary; our definition of potential energy takes the potential energy to be zero when the atoms are an infinite distance apart. In later chapters we shall sometimes find it more convenient to choose a different location for $\mathscr{V}(r) = 0$; for example, in §3.9 we shall take $\mathscr{V}(r) = 0$ at $r = r_0$.

upwards and positive forces downwards in Fig. 2.6(b)). Repeating this operation at all r leads to $\mathcal{V}(r)$ as shown in Fig. 2.6(c). Note that the minimum in $\mathcal{V}(r)$ occurs where $F(r) = 0$; the formal reason being that eqn (2.15), rewritten as

$$F(r) = -\frac{d\mathcal{V}(r)}{dr}, \qquad (2.17)$$

tells us that $F(r) = 0$ at a minimum in $\mathcal{V}(r)$ (where $d\mathcal{V}(r)/dr = 0$). More informally, the reason is that as the movable atom is brought in from infinity to r_0 all the energy is transferred to the person holding the atom (he gets pulled in). In coming in from infinity to $r' < r_0$ (Fig. 2.6) part of the energy gained between infinity and r_0 must be fed back as the person pushes against the repulsive forces existing between r_0 and r'. Hence $\mathcal{V}(r)$ is a minimum at $r = r_0$. Continuing in this informal vein we see that at some separation smaller than r' all the energy acquired from the attractive force will be spent in acting against the repulsive force and here the potential energy will be zero. At still-smaller separations the energy gained from the attractive force will be insufficient to push X against the repulsive force; the person must provide the missing energy, and thus the potential energy will be positive at such separations.

It is always good practice to invert arguments. So is Fig. 2.6 really consistent with the properties of solids? Fig. 2.6(b) shows stable equilibrium about $r = r_0$; increase the separation beyond r_0 and the attractive force will restore the separation towards r_0; decrease the separation below r_0 and the repulsive force will likewise restore the separation towards r_0. This accords with the essential stability of solids; compress them a little and they will return to their original form; extend them a little and they will likewise return. Fig. 2.6(b) also asserts that there is a maximum restoring force, F_m, between two atoms. Should a force greater than F_m be applied the atoms will be unable to match such a force and so will fly apart. This accounts—or could account—for the limited strength of materials. (We shall see later that materials are seldom as strong as we might expect from F_m.) Fig. 2.6 is therefore consistent with the behaviour of materials.

Exercise 2.2

Fig. 2.7 shows a somewhat fanciful model of the interatomic force curve between two atoms; fanciful because nature would not introduce sharp kinks! What, on this model, is (a) the equilibrium separation of the pair, (b) the force necessary to increase the separation of the pair of atoms by one per cent, (c) the minimum force necessary to dissociate the pair, and (d) the dissociation energy of the pair? Make a rough sketch of the interatomic potential energy $\mathcal{V}(r)$. (You will find it helpful to draw out $\mathcal{V}(r)$ to the same scale of r as in Fig. 2.7.)

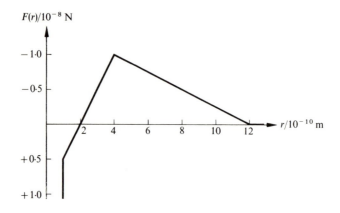

Fig. 2.7. A hypothetical interatomic force curve between two atoms.

Solutions. (a) The equilibrium separation (where $F(r)=0$) is $r=2\times10^{-10}$ m or $r\geqslant12\times10^{-10}$ m. Only at $r=2\times10^{-10}$ m is the equilibrium stable. (b) Since it takes 1×10^{-8} N to increase the interatomic separation from 2×10^{-10} to 4×10^{-10} m (that is, by 100 per cent) the force required for a one per cent increase in r is 1×10^{-10} N. Had $F(r)$ not been linear we would have drawn the tangent to $F(r)$ at the equilibrium separation r_0. (c) To completely separate the pair we must match the maximum restoring force $(1\cdot0\times10^{-8}$ N$)$ between the two atoms. (d) The dissociation energy is the magnitude of the area under the $F(r)$ plot between $r=2\times10^{-10}$ m and $r=12\times10^{-10}$ m, namely 5×10^{-18} J.

Comments. Your sketch of $V(r)$ will *not* be linear for 12×10^{-10} m$>r>1\times10^{-10}$ m (remember eqn (2.16)). It should have a minimum at $r=2\times10^{-2}$ m (where $F(r)=0$) and exhibit an infinitely-rapid rise at $r=1\times10^{-10}$ m, where the repulsive force becomes infinitely large. It should also have $V(r)=0$ at $r>12\times10^{-10}$ m.

2.6. Modelling the interaction

Although Fig. 2.6 describes graphically how two atoms (or molecules) interact it is often more convenient to have this information in mathematical form. Our task then is to find some mathematical expression which reproduces the main features of Fig. 2.6(b) *or* (c). The word *or* is deliberate; $F(r)$ and $V(r)$ are interrelated by eqn (2.17). Since differentiation is simpler to perform than integration it makes more sense to model $V(r)$, knowing that $F(r)$ follows via eqn (2.17), rather than to model $F(r)$, leaving $V(r)$ to be obtained by integration.

We can always view the potential energy curve of Fig. 2.6(c) as being the sum of the potential energy curve due to attractive forces (portion *PQR* of Fig. 2.6(b)) and the potential energy due to repulsive forces (portion *ST* of Fig. 2.6(b)). This is illustrated in Fig. 2.8 where the component potential energies are indicated in light line and the total

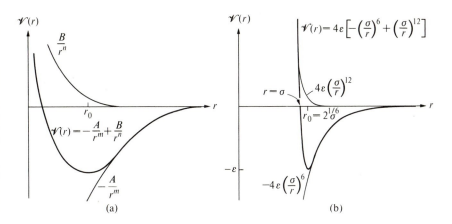

Fig. 2.8. Showing how $\mathcal{V}(r)$ can be represented as the sum of two separate curves. (a) In the Mie model the positive component (arising from interatomic repulsive forces) is represented by B/r^n; the negative component (arising from the interatomic attractive forces) by $-A/r^m$. Their combined effect is shown by the heavy line. (b) In the Lennard-Jones 6–12 model the two terms are given by $4\varepsilon(\sigma/r)^{12}$ and $-4\varepsilon(\sigma/r)^6$, where ε and σ are positive constants.

potential energy is indicated in heavy line. Our job is to find some suitable analytic functions for the two component energies and then to add them.

The Mie potential

Perhaps the simplest possible representation is that proposed by Mie in 1907. He represented the negative component by $-A/r^m$ and the positive component by B/r^n (see Fig. 2.8(a)). Here A and B are positive constants and m and n are positive integers with $n > m$. It is necessary to have $n > m$ so that the positive term will dominate as $r \to 0$ and will vanish more rapidly than the negative term as $r \to \infty$. Adding together the two components gives

$$V(r) = -\frac{A}{r^m} + \frac{B}{r^n} .$$ (2.18)

The Lennard-Jones potential

This potential has the form

$$V(r) = 4\varepsilon\left[-\left(\frac{\sigma}{r}\right)^p + \left(\frac{\sigma}{r}\right)^q \right] ,$$ (2.19)

where ε and σ are positive constants and p and q are positive integers. Like Mie's potential, Lennard-Jones' potential contains two continuously variable constants, ε and σ. Unlike Mie's constants A and B, the constant ε is solely associated with energy, the constant σ solely with the interatomic separation. (Eqn (2.19) shows that, say, doubling the value of ε doubles $V(r)$. Trebling σ, for example, stretches $V(r)$ along the abscissa; to obtain the same value $V(r)$ as before we now must go to $3r$.) This 'division of responsibility' between ε and σ can be a real asset in some situations. In practice the Lennard-Jones potential is employed almost exclusively in modelling van der Waals' interactions where, as we shall see later, $p = 6$. A value of $q = 12$ is usually also assumed, so eqn (2.19) becomes

$$V(r) = 4\varepsilon \left[-\left(\frac{\sigma}{r}\right)^6 + \left(\frac{\sigma}{r}\right)^{12} \right] .$$ (2.20)

For obvious reasons this is called the *Lennard-Jones 6–12 potential*. It is plotted graphically in Fig. 2.8(b). Some representative values of ε and σ are given in Table 2.2.

TABLE 2.2
Values of the Lennard-Jones constants ε and σ

Element	$\varepsilon/10^{-21}$ J	$\sigma/10^{-10}$ m
He	0·141	2·56
Ne	0·492	2·75
Ar	1·70	3·40
Kr	2·30	3·68
Xe	3·10	4·07
N_2	1·25	3·70
I_2	7·60	4·98
Hg	11·74	2·90
CCl_4	4·51	5·88

Getting rid of a constant

As they stand both the Mie and the Lennard-Jones potentials each contain four constants. It is however possible to eliminate either A or B from the Mie potential and σ from the Lennard-Jones potential—*provided* we agree to introduce the equilibrium separation r_0 of an isolated pair of atoms (a quantity which can be measured experimentally). We shall show how this reduction is carried out with a Lennard-Jones 6–12 potential, leaving you to deal with the Mie potential.

At the equilibrium separation r_0 we know that $F(r) = -dV(r)/dr = 0$.

Differentiating eqn (2.20) with respect to r gives

$$F(r) = -4\varepsilon\left[\frac{6\sigma^6}{r^7} - \frac{12\sigma^{12}}{r^{13}}\right]. \tag{2.21}$$

Setting $F(r_0) = 0$ gives

$$r_0 = (2)^{1/6}\sigma = 1\cdot122\sigma. \tag{2.22}$$

Substituting eqn (2.22) back into eqns (2.20) and (2.21) we obtain

$$V(r) = 4\varepsilon\left[-\frac{1}{2}\left(\frac{r_0}{r}\right)^6 + \frac{1}{4}\left(\frac{r_0}{r}\right)^{12}\right] \tag{2.23}$$

$$F(r) = -12\varepsilon\left[\frac{r_0^6}{r^7} - \frac{r_0^{12}}{r^{13}}\right]. \tag{2.24}$$

As an illustration of the use to which eqn (2.23) may be put let us work out the energy required to completely dissociate a pair of atoms which are initially at their equilibrium separation r_0. This is known as the *dissociation energy* ΔE. Since $V(r_0)$ is the energy required to *form* a pair so that they are at their equilibrium separation r_0 (Fig. 2.6(c)) it follows that

$$\Delta E = -V(r_0). \tag{2.25}$$

Substituting for $V(r)$ from eqn (2.23) with $r = r_0$ gives

$$\boxed{\Delta E = \varepsilon} \tag{2.26}$$

a result which is true for any values of p and q in eqn (2.19). Eqn (2.23), by contrast, is only true for a 6–12 potential.

Determining ε experimentally

In later chapters we shall see how the energy required to dissociate 1 mol of a solid into its component atoms (a quantity known as the *molar enthalpy of sublimation*, $H_{m,s}$) is related to the interatomic potential. However we can, at this stage, make a crude calculation of $H_{m,s}$ in terms of ΔE. Consider a single atom of the solid surrounded by its n nearest neighbours (known as the *first coordination number*). To pluck this atom out of the crystal and off to infinity calls for energy $n\Delta E$. Hence the total energy required to dissociate the 1 mol of crystal into N_A atoms is given by

$$\boxed{H_{m,s} = \tfrac{1}{2}nN_A\,\Delta E}, \tag{2.27}$$

where the factor of $\tfrac{1}{2}$ arises because in separating atom 1 from atom 2 we also separated atom 2 from atom 1. (Put differently, ΔE refers to a pair of atoms; we must therefore only count pairs.) In view of eqn (2.26) we may write eqn (2.27) as

$$H_{\mathrm{m,s}} = \tfrac{1}{2} n N_A \varepsilon. \tag{2.28}$$

Thus by measuring $H_{\mathrm{m,s}}$ we can evaluate ε provided that we know n. As we shall see in Chapter 8, $n = 12$ in closely-packed solids. By way of example, $H_{\mathrm{m,s}} = 1\cdot58 \times 10^4 \text{ J mol}^{-1}$ for xenon, from which we deduce (using eqn (2.28)) that $\varepsilon = 4\cdot37 \times 10^{-21}$ J. Since r_0, the equilibrium separation between centres of a pair of atoms, can be estimated from the density of the solid using eqn (2.10), eqns (2.23) and (2.24) are now determined.

Exercise 2.3

By following through the same procedures as those just adopted in eliminating one constant from the Lennard-Jones 6–12 potential, show that the Mie potential (eqn (2.17)) can be written as

$$V(r) = \frac{A}{r_0^m} \left[-\left(\frac{r_0}{r}\right)^m + \frac{m}{n}\left(\frac{r_0}{r}\right)^n \right], \tag{2.29}$$

where r_0 is the equilibrium separation of an isolated pair of atoms. Also show that the dissociation energy ΔE is

$$\Delta E = \frac{A}{r_0^m}\left(1 - \frac{m}{n}\right). \tag{2.30}$$

Proof. Evaluating $F(r)$ as follows

$$F(r) = -\frac{dV(r)}{dr} = -\frac{mA}{r^{m+1}} + \frac{nB}{r^{n+1}}$$

and setting $F(r_0) = 0$ gives

$$B = A\frac{m}{n} r_0^{n-m}.$$

Substituting this expression back into eqn (2.18) leads to eqn (2.29). Since $\Delta E = -V(r_0)$ we obtain the dissociation energy by multiplying the right-hand side of eqn (2.29) by -1 and letting $r = r_0$. This yields eqn (2.30).

Comment. In §9.1 we shall learn how the value of n is related to the compressibility of a solid. Hence compressibility measurements can be used to deduce the value of n.

2.7. Theories of the interaction

To go beyond the stage of modelling the interaction of two atoms in general calls for a grounding in quantum mechanics and a mathematical facility well beyond that assumed in this book. However, there are two types of interaction (ionic and van der Waals) which can be discussed without the need of quantum mechanics, but this is only possible because we know the electronic distributions in the interacting atoms.

Ionic bonding

In a crystal of sodium chloride there is an effectively complete transfer of one electron from each sodium atom to each chlorine atom: The sodium atoms exist as Na^+ ions (each of charge e) and the chlorine atoms as Cl^- ions (each of charge $-e$). The force $F(r)$ between an isolated Na^+–Cl^- ion pair is, from Coulomb's law, given by

$$F(r) = -\frac{e^2}{4\pi\varepsilon_0 r^2}, \tag{2.31}$$

where the constant $1/4\pi\varepsilon_0$ has a value of $8.99\times10^9\,\mathrm{N\,m^2\,C^{-2}}$. (The constant ε_0, not to be confused with ε in the Lennard-Jones potential, is known as the *permittivity* of free space.) Integrating eqn (2.31) via eqn (2.16) gives

$$\mathcal{V}(r) = -\frac{e^2}{4\pi\varepsilon_0 r} \tag{2.32}$$

as the potential energy due to the attractive part of the Na^+–Cl^- interaction. To obtain the Mie potential we would, of course, have to add the B/r^n term due to the repulsive part of the interaction. All we can simply say about the value of n is that since the repulsive forces are very short-range—they only come into play as the ions overlap—its value must be large. A value of $n = 9$ is often assumed although there is no theoretical basis for such a value. Rather, theory suggests that the potential energy due to the repulsive forces should be proportional to $\exp(-kr)$, where k is a constant. Fortunately, most calculations are very insensitive to the

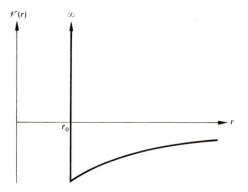

Fig. 2.9. The *hard-sphere* repulsion model in which the potential energy between two spheres increases towards infinity as they start to overlap. The hard-sphere repulsion has here been combined with a $-1/r$ potential (due to a $1/r^2$ attractive force).

exact value of n. For example, the dissociation energy of an isolated NaCl molecule (as may exist in the gas phase) is, from eqn (2.30) (with $m = 1$, $n = 9$, and $A = e^2/4\pi\varepsilon_0$),

$$\Delta E = \frac{e^2}{4\pi\varepsilon_0 r_0}\left(1 - \frac{1}{9}\right). \tag{2.33}$$

Substituting the values of $1/4\pi\varepsilon_0$, e, and taking $r_0 = 2.5 \times 10^{-10}$ m (the equilibrium separation of an isolated NaCl molecule) gives $\Delta E = 8.2 \times 10^{-19}$ J, which is within four per cent of the experimental value. Had we taken $n = \infty$ instead of $n = 9$ in eqn (2.30) it would only have changed ΔE by ten per cent. An infinite value of n corresponds to the *hard-sphere* repulsive model shown in Fig. 2.9.

van der Waals forces

This type of force represents the other extreme to ionic forces in that there is no transfer of charge from one atom (or molecule) to the other. The name actually encompasses three different types of interactions. Although present between all atoms, van der Waals forces are dominant in such diverse cases as the binding of certain anaesthetics (like ethylene) to cell membranes and in the interaction of inert-gas atoms.

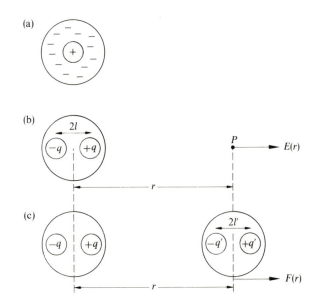

Fig. 2.10. (a) An isolated inert-gas atom. The centre of the electron cloud coincides with that of the nucleus and the atom appears to produce no external field. (b) The left-hand dipole produces an electric field. (c) This field polarizes the right-hand atom which leads to an attractive force between these atoms.

(a) *Dispersion forces.* We will start by considering a single isolated inert-gas atom, such as a neon atom. Because the electron cloud is spherically symmetric about the nucleus (Fig. 2.10(a)) the effective centre of the negative cloud coincides with that of the positive nucleus. As the cloud and the nucleus have equal, but opposite, charges no field should be produced outside the atom—making it hard to see why this atom should influence any other atom.

What we have said is true—but only true on average. To put it simply (if a little misleadingly) the electrons are in motion around the nucleus, so that at any instant the effective centre of the electron cloud may not coincide with the nucleus. The atom now consists of two charges, $+q$ and $-q$, separated by a distance $2l$ (Fig. 2.10(b)). Such an arrangement of charges is called an *electric dipole* and is said to possess an *electric dipole moment* $\mu = q(2l)$.

We shall now show that this dipole produces an electric field which so distorts the charge distribution in a second atom as to lead to a net attractive force between the two atoms.

Our first step is to calculate the electric field $E(r)$ produced at point P (Fig. 2.10(b)). This field, being the vector sum of the electric field due to charge $+q$ at distance $r-l$ from P and the field due to charge $-q$ at distance $r+l$ from P, is given by

$$E(r) = \frac{1}{4\pi\varepsilon_0}\left[\frac{q}{(r-l)^2} - \frac{q}{(r+l)^2}\right]$$

$$E(r) = \frac{q}{4\pi\varepsilon_0 r^2}\left[\frac{1}{[1-(l/r)]^2} - \frac{1}{[1+(l/r)]^2}\right].$$

Since $l \ll r$ this can be expanded via the binomial theorem to give, after ignoring second- and third-order terms,

$$E(r) = \frac{2\mu}{4\pi\varepsilon_0 r^3}. \tag{2.34}$$

The effect of this field on a second atom located at point P (Fig. 2.10(b)) will be to move its positive nucleus in the direction of $E(r)$ and its negative electron cloud in the opposite direction. This leads to the charge separation shown in Fig. 2.10(c) in which charges of $-q'$ and q' are separated by, say, $2l'$. The atom is said to be *polarized* with a dipole moment $\mu' = 2q'l'$ and the process is referred to as *induced polarization*. The field acting on the right-hand dipole varies from $E(r-l')$, acting on $-q'$, to $E(r+l')$ acting on $+q'$. This produces a resultant force on the right-hand dipole of

$$F(r) = [q'E(r+l') - q'E(r-l')]$$

or, since $E(r+l') - E(r-l') = (dE(r)/dr)2l'$ when $l' \ll r$,

$$F(r) = \mu' \frac{dE(r)}{dr}.$$

Substituting for $E(r)$ from eqn (2.34) gives

$$F(r) = -\frac{6\mu\mu'}{4\pi\varepsilon_0 r^4}. \tag{2.35}$$

The value of $\mu' = 2q'l'$ will depend both on the nature of the atom being polarized and on the strength of the polarizing field. Since $+q'$ and $-q'$ are the nuclear and electronic charges, respectively, these are fixed for a particular atomic species. Assuming that an atom obeys a form of 'Hooke's law' (extension is proportional to force) we can expect l', and hence $\mu' = 2q'l'$, to be proportional to $E(r)$, that is to $1/r^3$ (eqn (2.34)). Substituting this $1/r^3$ dependence of μ' into eqn (2.35) gives

$$F(r) \propto -\frac{1}{r^7} \tag{2.36}$$

which integrates (via eqn (2.16)) to yield

$$\mathcal{V}(r) \propto -\frac{1}{r^6}. \tag{2.37}$$

This justifies our earlier assertion that $p = 6$ when the Lennard-Jones potential is used to model the van der Waals interaction.

In arriving at eqn (2.36) we have advanced no adequate explanation as to the origin of the electric-charge fluctuations which produce the electric dipole moment of the left-hand atom in Fig. 2.10(b). To do so would require sophisticated quantum mechanics.

(b) *Dipole–dipole forces.* An individual molecule of HCl has a permanent electric dipole moment of 3.57×10^{-30} C m. This arises because there is a *partial* transfer of electronic charge from the hydrogen to the chlorine atom. (If the transfer were complete, the resulting HCl would have a dipole moment of $e \times$ interionic separation $(1.27 \times 10^{-10}$ m$) = 20.3 \times 10^{-30}$ C m.) The field produced by this permanent dipole will interact with the permanent dipole of a second HCl molecule producing an attractive force between the two dipoles. In view of eqn (2.35) we might guess that the force would have a $1/r^4$ dependence. However, unlike the case considered in Fig. 2.10, where the dipoles are always in line, the two HCl molecules may (at least in the gas phase) be orientated at any angle. When a suitable integration is performed over all angles the attractive force between the two permanent dipoles turns out to be proportional to $1/r^7$.

(c) *Dipole–induced dipole forces.* When a molecule possessing a permanent dipole (for example, HCl) approaches an atom (for example, Ne) the former induces a dipole moment in the latter. Now this essentially the same situation as that shown in Fig. 2.10 with the left-hand atom replaced by a permanent dipole. The same arguments hold true, leading to the conclusion that the attractive force between a permanent and an induced dipole varies as $1/r^7$.

As we have already remarked, the characteristic $1/r^7$ van der Waals attractive force is present between all atoms and molecules even though its presence is often masked by other stronger forces. Thus, for example, there are weak dispersion forces present in sodium chloride, but their contribution is masked by the $1/r^2$ ionic forces. Not unexpectedly, the relative contribution made by the three types of dipole–dipole interaction—all characterized by eqn (2.36)—will vary from species to species. In gaseous hydrogen bromide, for example, (in which only van der Waals forces are involved in the interaction between molecules) the dispersion, permanent–permanent, and permanent-induced forces occur in the ratio $44:1\cdot55:1$. In neon the dispersion force alone is present.

Covalent bonding

This type of bonding, by far the most prevalent in nature, is intermediate between ionic bonding, where there is an effectively complete transfer of an electron from one atom to the other, and the van der Waals interaction where there is no transfer at all of electrons. Put very simply, electrons are transferred to a 'halfway house' between the two nuclei. This is shown pictorially in Fig. 2.11 for the H_2 molecule. Although the two nuclei repel one another, each is also attracted to the electron cloud whose 'centre' is midway between them. At equilibrium, the attractive and the repulsive forces cancel one another.

Because the attractive force between two ions varies as $1/r^2$ and that between two atoms in the van der Waals interaction varies as $1/r^7$, you might guess that in covalent bonding the attractive force would be proportional to $1/r^n$, where $7 > n > 2$. In fact when the calculations are made a simple inverse power law of force is never found. This means, of course, that any conclusions arrived at via a Mie or Lennard-Jones type of potential need not necessarily apply to covalently bonded materials.

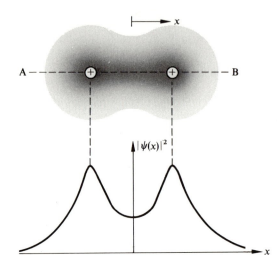

Fig. 2.11. A schematic representation of the electron cloud in a hydrogen molecule. This is suggested by shading and, more properly, by the probability density $|\psi(x)|^2$ at a point x along the line joining the centres of the two nuclei. The electron cloud is symmetrical about line AB. The important point to note is the high probability of finding an electron at points between the two nuclei.

2.8. Intermolecular forces in solids

So far we have only considered the force between an isolated pair of atoms, or molecules, such as might occur in the gas phase. *A priori*, there is no reason to expect the force to remain unaltered in the solid phase where there are, of course, many more atoms, or molecules, present. What actually happens varies from solid to solid. Roughly speaking, solids can be divided into four classes:

(a) *Molecular solids* (e.g. iodine, inert gases). Here the crystal consists of identifiable molecules (identifiable using X-ray diffraction techniques) whose structure closely resembles that found in the gaseous phase. The molecules are usually attracted to one another by weak van der Waals forces. This contrasts with the strong forces present *within* the molecule (covalent in the case of I_2).

(b) *Covalent solids* (e.g. silica, diamond). In contrast to molecular solids there are no identifiable molecular sub-units within the crystal. Such solids are best thought of as supermolecules in which atoms are joined to their neighbours by covalent bonds.

(c) *Ionic solids* (e.g. sodium chloride, calcite). As we shall see later (§8.4), these consist of a regular array of positive and negative ions.

(d) *Metals* (e.g. sodium, copper). Here positive metal ions are set in a sea of free (or nearly free) electrons. As with a covalent solid, a piece of metal is like a supermolecule.

In the case of covalent solids and metals the interaction between the component atoms will not have the same form as in the gas phase. Thus in sodium metal, for example, the atoms are ionized and surrounded by free electrons, whereas in the gaseous phase the sodium is mainly present as free atoms, causing $F(r)$ to be different in the two phases. Only in ionic and molecular solids can we assume $F(r)$ to have the same form in all three phases. For this reason we will, in future, be mainly concerned with substances like the alkali halides and the inert gases.

PROBLEMS

2.1. What is the flight-time for a Hg^+ ion in a time-of-flight mass spectrometer with an accelerating voltage of 60 V and a 0·5 m long drift tube? Take $A_r(Hg) = 200$.

2.2. Prove that if one measures out a mass in grams of a compound Y equal to $M_r(Y)$ one obtains 1 mol of molecules of Y.

2.3. (a) Make an order-of-magnitude correct estimate of the number of atoms in your body. (b) Remembering that organic material is composed mainly of carbon and hydrogen make a rough estimate of the number of electrons in your body. (c) If you could transfer 1 electron in 10^{10} of yours to a friend standing 1 m away what, roughly, would be the force of attraction between you? A 1 in 10^{10} loss of electrons could hardly seriously interfere with your biochemistry!

2.4. Make an order-of-magnitude estimate of the maximum area to which an oil drop could spread on water. Assume a drop volume of about 10^{-2} cm³. Also assume that a molecule of the oil contains roughly ten atoms.

2.5. On condensing a saturated vapour (a gas at $T < T_c$) to form a saturated liquid, the volume of the liquid is typically 10^{-3} times that of the vapour at the same pressure and temperature. What is the ratio of the mean separation between molecules in the vapour to the mean separation between molecules in the liquid?

2.6. Taking the density of solid sodium as $0·97 \times 10^3$ kg m⁻³, estimate the diameter of a sodium atom. Take $A_r(Na) = 23$.

2.7. Roughly how many moles of (a) water molecules, (b) neutrons, and (c) protons are there in the Mediterranean sea? Assume the hydrogen is present as $_1^1H$ and oxygen as $_8^{16}O$ and take $A_r(^1H) = 1$ and $A_r(^{16}O) = 16$. The density of water is 10^3 kg m⁻³. *Clue*: An atlas.

2.8. Table 2.3 records the results of a series of experiments performed by Stern in which he studied the dependence of the angle ϕ at which most helium atoms were scattered from a crystal as a function of the atoms' speed v. The apparatus used was as in Fig. 2.3. Check whether Stern's results are compatible with de Broglie's relation.

TABLE 2.3
The scattering of helium atoms

angle $\phi/°$	35	31	26·5	20
speed $v/\text{m s}^{-1}$	920	1065	1230	1590

2.9. What is the de Broglie wavelength of a nitrogen molecule moving at 300 m s^{-1}? Through what electrical potential difference would a N_2^+ ion have to be accelerated to acquire this speed? Take $A_r(N) = 14$.

2.10. Make accurate scale drawings of (a) the potential energy $V(r)$ corresponding to the interatomic force $F(r)$ shown in Fig. 2.7, and (b) the force $F(r)$ corresponding to $V(r)$ of Fig. 2.9.

2.11. Adopting a hard-sphere repulsion model write down an expression for $V(r)$ between the two ions in an NaCl molecule that is valid at separations greater than the equilibrium value r_0. Calculate the dissociation energy of such a pair taking $r_0 = 2 \cdot 5 \times 10^{-10}$ m.

2.12. Adopting a hard-sphere repulsion model, sketch $F(r)$ and calculate the maximum restoring force between the Na^+ and Cl^- ions of a NaCl molecule. Assume $r_0 = 2 \cdot 5 \times 10^{-10}$ m. Use your answer to calculate the tensional force required to break a grain of salt of cross-sectional area 10^{-6} m². Is this figure at all comparable with the sort of force which is required to smash a salt grain? (You surely must, at some stage, have crushed a grain of salt under a knife! In estimating the force you applied, remember that 1 newton is about equal to the weight of an average apple.)

2.13. Adopting a Lennard-Jones 6–12 potential and using the data given in Table 2.2 make accurate plots of the interatomic potential energy and the interatomic force characteristic of two isolated argon atoms. Using the data so plotted confirm the correctness of eqns (2.22) and (2.26).

2.14. Assuming a Lennard-Jones 6–12 potential and using the data given in Table 2.2, calculate the potential energy of two isolated helium atoms when their interatomic separation is (a) $4 \cdot 0 \times 10^{-10}$ m, (b) $2 \cdot 87 \times 10^{-10}$ m, and (c) $2 \cdot 0 \times 10^{-10}$ m.

2.15. Assuming a Lennard-Jones 6–12 potential and using the data given in Table 2.2, calculate the interatomic force between two isolated Xe atoms when their interatomic separation is (a) $5 \cdot 0 \times 10^{-10}$ m, (b) $4 \cdot 57 \times 10^{-10}$ m, and (c) $3 \cdot 0 \times 10^{-10}$ m.

2.16. The molar enthalpy of evaporation of water (density 10^3 kg m^{-3}) is $4 \cdot 1 \times 10^4 \text{ J mol}^{-1}$. Using these pieces of information and assuming that a water molecule is surrounded by approximately ten nearest-neighbours, deduce the form of the Lennard-Jones 6–12 potential between two water molecules. Take $M_r(H_2O) = 18$.

2.17. Show that the dissociation energy between a pair of atoms whose interaction is described by the general form of the Lennard-Jones potential model (eqn (2.19)) is equal to ε.

2.18. Another model of the interatomic potential is that of Morse:

$$V(r) = V_c[e^{-2(r-r_c)/a} - 2e^{-(r-r_c)/a}],$$

where V_c, r_c, and a are constants. (a) Show that the Morse potential does indeed fairly represent the interatomic interaction by plotting $V(r)/V_c$ against r/a for the case where $r_c = 2a$. (b) What is the equilibrium separation r_0? (c) Calculate the restoring force for a value of r slightly different from r_0 and hence calculate the restoring force per unit displacement from r_0 (the so-called *force constant*).

2.19. Using a hard-sphere repulsion model write down an expression for $V(r)$, valid at separations greater than the equilibrium separation r_0, for the potential energy of the two atoms in a NaCl molecule on the assumption that only gravitational (attractive) forces are present. Calculate the dissociation energy of the pair taking $r_0 = 2 \cdot 5 \times 10^{-10}$ m. Compare your answer with that you obtained in problem 2.11. (You will remember that the force $F(r)$ of gravitational attraction between two spheres of masses m_1 and m_2 a distance r apart between centres is $F(r) = -Gm_1m_2/r^2$, where G is a constant of value $6 \cdot 67 \times 10^{-11}$ N m^2 kg^{-2}.) Take $A_r(\mathrm{Na}) = 23$ and $A_r(\mathrm{Cl}) = 35 \cdot 5$.

2.20. Potassium fluoride (an ionic molecule like NaCl) has an electric dipole moment of $2 \cdot 87 \times 10^{-29}$ C m as measured in the gas phase.
(a) Assuming that the molecule consists of a $\mathrm{K}^+\text{--}\mathrm{F}^-$ ion pair what is the separation between the two charges in the molecule? (b) What field will a single KF molecule produce at a point 10^{-9} m along the axis of the dipole (measured from the centre of the dipole)? (c) What will be the force of attraction between two such dipoles when 10^{-9} m apart (between centres) and lying so that their axes are along a common line?

2.21. What will be the value (meaning magnitude and direction) of the electric field at a point 10^{-9} m from the centre of the dipole of problem 2.20, measured along a line perpendicular to the dipole's axis?

3. Thermal equilibrium

WE must now face up to a key question: Why, from a microscopic viewpoint, can one and the same substance sometimes exist as a solid, sometimes as a liquid, and sometimes as a gas? We shall see that the answer involves making comparisons between the kinetic energy of the substance's atoms and the pair dissociation energy ΔE. This will lead us to relate our arguments, based on the microscopic, to temperature—something macroscopic, read on a dial. In the process we will have to consider how atoms interchange energy with one another as they collide. This will aid our understanding of the nature of thermal equilibrium. Turning away from free atoms we will next evaluate the mean kinetic and potential energies of an atom bound within a solid. Finally we will show how results originally derived for individual atoms can be taken over into the macroscopic world to discuss several, so-called, fluctuation phenomena.

3.1. Why solids, liquids, and gases?

In attempting to answer this question we will direct our attention away from the isolated pair of atoms which have been our main concern till now to a whole *collection* of atoms or molecules. Furthermore, we will introduce the idea that atoms or molecules can have kinetic as well as potential energy. (For economy, we shall introduce the single word 'atom' to mean either an atom or a molecule—whichever is appropriate—of the substance under consideration.)

Since each atom in a collection of atoms can interact with any other atom in the manner summarized in Fig. 2.6, we can expect one or more bound pairs of atoms to form. To split up such a pair into its component atoms requires an energy ΔE, the dissociation energy of the pair. Let us therefore examine the possible fates of such bound pairs (like the pair shown in Fig. 3.1), looking in particular at what may happen when another atom with a kinetic energy $\frac{1}{2}mu^2$ collides with a bound pair.

If $\frac{1}{2}mu^2$ exceeds ΔE, there is enough energy available to dissociate the pair. Indeed if the average kinetic energy, written $\frac{1}{2}\overline{mu^2}$, greatly exceeds ΔE, virtually all collisions will lead to dissociation of the bound pairs. Under these conditions no bound pairs can persist; the system will consist mainly of individual atoms moving rapidly and with no apparent affinity for each other, along with a few pairs having a fleeting liaison. In fact, the atoms will have to be forced to occupy any given volume. Since these are just the properties displayed by gases, we can expect a gas to

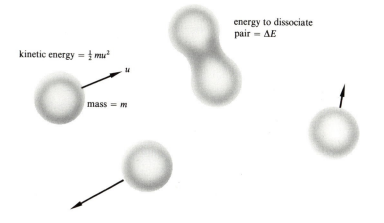

kinetic energy $= \frac{1}{2} mu^2$

energy to dissociate
pair $= \Delta E$

u

mass $= m$

Fig. 3.1. Showing the competing roles of kinetic and potential energies in a system of many atoms.

result when

$$\frac{1}{2}m\overline{u^2} \gg \Delta E. \tag{3.1}$$

At the other extreme, when the average kinetic energy $\frac{1}{2}m\overline{u^2}$ of a colliding atom is very much less than the dissociation energy of the bound pair, no dissociation will result for there is not the energy available. The result is that the colliding atom may be captured by the pair.† Before long, this process will be repeated many times over and a large conglomeration of atoms will form. Such a conglomeration—one that requires a significant amount of energy to dissociate it into a gas—is reminiscent of the solid phase. Therefore, on our microscopic model, the solid phase will occur when

$$\frac{1}{2}m\overline{u^2} \ll \Delta E. \tag{3.2}$$

Lastly, when $\frac{1}{2}m\overline{u^2}$ is of the same order of magnitude as ΔE, we can expect some 'clustering' to occur. Within the group of colliding atoms possessing an *average* energy $\frac{1}{2}m\overline{u^2} \approx \Delta E$ there will be some with a kinetic energy greater than ΔE and these will tend to dissociate any clusters

---

† To be captured, the kinetic energy of the colliding atom must be removed in the collision. If this kinetic energy is not removed the colliding atom will rebound with no loss of energy, that is, the collision will be elastic. A 'cold' surface is, as shall shortly see, a means of removing kinetic energy. In some types of atomic collisions, such as occur in hot flames, the collisions are sometimes inelastic; part of the translational kinetic energy of the colliding atoms goes to change their electronic configurations. These 'excited' atoms may subsequently lose all or part of their energy by emitting light. This type of process will not concern us further in this text.

which form. We can expect therefore to have a system in which a limited amount of association is possible: a fluctuating system, part-bound, part-free. The system will not possess the rigidity of a solid nor the unrestrained freedom of a gas. These are the conditions characteristic of a liquid. Therefore the liquid phase may well occur when

$$\tfrac{1}{2}m\overline{u^2} \approx \Delta E. \tag{3.3}$$

So by considering *both* the atoms' kinetic and potential energies we have been able to account for the solid, liquid, and gaseous phases. However, as it stands this theory is incomplete for it predicts that there should be a gradation of phases from solid through to gas. It does not explain the sudden nature of phase transitions. But perhaps it is not so surprising that a theory based on how a *single* pair of atoms behave— eqns (3.1) to (3.3) involve the dissociation energy of a *pair*—is unsuccessful in this respect. Theories which do seek to take account of the mutual interactions of large collections of atoms (known as *many-body* theories) are very difficult to develop, and even these theories have only had a limited success in explaining phase transitions.

There is a fourth phase of matter—the *plasma* phase—which consists of separated nuclei and electrons. This arises when the kinetic energy of the colliding atoms is so great that they knock each other asunder. The plasma phase is by far and away the most common—stars are plasmas— but it is also by far and away the most difficult to discuss quantitatively. For this reason we exclude it from further consideration.

3.2. A kinetic simulator

An economical way of studying how a system of many atoms will behave is to build a simulator employing components which interact in the same way as do atoms. To be successful the simulator must employ 'atoms' between which there is a short-ranged attractive force, followed by an even shorter-ranged repulsive force, and which can be given variable kinetic energies. For 'atoms' we may use steel balls (of the kind found in ball-bearings). The desired short-ranged attractive force may be simulated by coating the balls with oil. (Anyone who has assembled a bearing will know that oil-coated balls do stick together.) The required kinetic energy may be provided by placing the oil-coated balls on a tray whose base is made of rough-moulded glass and which is vibrated back and forth by means of a motor. (Because of energy dissipation in the oil films it is not sufficient to use a smooth-based tray and to feed in energy only at the boundaries.)

On pouring the oil-coated balls on to the centre of the stationary tray they 'crystallize' out into a regular two-dimensional pattern. This pattern,

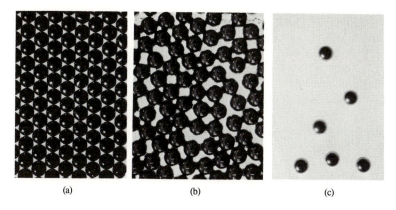

(a)   (b)   (c)

Fig. 3.2. A close-up of (a) a simulated solid, (b) a simulated liquid, and (c) a simulated gas.

reminiscent of how atoms are arranged in a solid, persists at low vibrational speeds of the tray; that is, at low $\frac{1}{2}\overline{mu^2}$ of the 'atoms' (Fig. 3.2(a)). As the speed of the tray, and therefore that of the balls, increases, the regular pattern remains essentially intact until, at a well-defined value of $\frac{1}{2}\overline{mu^2}$, the regularity disappears; the structure melts to give a liquid (Fig. 3.2(b)). On increasing the speed of the tray still further, the area occupied by the liquid shrinks (by evaporation around its perimeter) and the tray contains a single phase which is undoubtedly gas-like (Fig. 3.2(c)). However this only happens when there are few balls on the tray—that is when the 'molar area' (area per 'N_A' balls) is large. With many balls present— so that the 'molar area' is small—the liquid phase is seen to grow as $\frac{1}{2}\overline{mu^2}$ is increased until it fills the tray. This behaviour is entirely consistent with that of a real substance (recall routes 1 to 2 and 3 to 4 of Fig. 1.7). Thus the simulator reproduces conditions found in nature, and demonstrates the essential correctness of eqns (3.1) to (3.3).

3.3. A macroscopic view of temperature

Throughout this chapter we have been very hesitant about using the word 'temperature'. Yet in everyday speech we would have no such hesitancy; to us a substance is a gas when the temperature is 'high', is a solid when the temperature is 'low', and is a liquid when the temperature is 'middling'. Now the discussion summarized in eqns (3.1) to (3.3) has shown that a substance exists as a gas when $\frac{1}{2}\overline{mu^2}$ is 'high' (compared with ΔE), as a solid when $\frac{1}{2}\overline{mu^2}$ is 'low', and as a liquid when $\frac{1}{2}\overline{mu^2}$ is 'middling' (of order ΔE). These two viewpoints could be made consistent by writing

$$\text{the temperature of a substance} \propto \tfrac{1}{2}\overline{mu^2}. \qquad (3.4)$$

Before deciding to adopt such a definition we should remind ourselves that we have only looked at part of what is meant by a temperature scale—that part dealing with steady conditions. As commonly understood, a temperature scale also contains within it the idea that if two bodies have numerically different 'temperatures', energy will be transferred from the one of higher temperature to the one of lower temperature when the two are placed in thermal contact. We would probably wish to say, for example, that when a poker at room temperature is plunged into a fire of higher (meaning numerically higher) temperature it acquires energy from the fire. We know it acquires energy because it can now do 'useful jobs' (such as heating water) which it could not do previously.

3.4. A microscopic view of temperature

Once we have examined, at the atomic level, the conditions governing the transfer of energy from one body to another we will be better placed to attempt to tie together our two understandings of temperature.

A useful tactic to adopt in discussing any physical phenomenon is to try to break the discussion down into a series of discrete 'chunks'. We shall adopt this tactic here and in later chapters.

A single collision

Consider a hot poker (a solid) cooling in air (a gas). Excluding radiation losses, the only other way that energy can be transferred from the solid to the gas is through collisions between atoms of the gas and atoms of the solid. We therefore begin by considering what happens in a single such encounter.

Fig. 3.3(a) shows a gas atom of mass m moving up towards a solid composed of atoms of mass M. The ensuing collision is shown in Fig. 3.3(b). Before impact m has a velocity $\boldsymbol{u}$; after impact a velocity $\boldsymbol{v}$. Before impact M has a velocity $\boldsymbol{U}$, after impact a velocity $\boldsymbol{V}$. Each of these velocities may, of course, be resolved into components along the x-, y- and z-axes (in the manner indicated in Fig. 4.24(b)). The components are given in parentheses after the velocities in Fig. 3.3(b).

We start by applying the law of conservation of momentum. Since the interatomic force between M and m is directed along the line joining their centres (the x-axis in Fig. 3.3(b)) it follows that this is the only direction along which momentum can be exchanged between M and m. Thus the fact that the loss of momentum of M must equal the gain in momentum of m becomes

$$M(U_x - V_x) = -m(u_x - v_x). \tag{3.5}$$

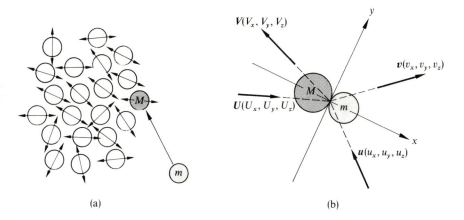

Fig. 3.3. (a) A gas atom of mass m approaches a solid composed of atoms of mass M. (b) The moment of impact between an atom of the gas and an atom of the solid. Here the x-axis is drawn along the line of centres of the atoms at impact. The y-axis lies in the tangent plane to each atom at impact. The z-axis (not shown) is mutually perpendicular to the x- and y-axes and comes out of the plane of the paper.

Because there can be no change in momentum along the y- and z-axes we have

$$U_y = V_y, \ U_z = V_z, \ u_y = v_y, \quad \text{and} \quad u_z = v_z. \tag{3.6}$$

When atoms collide they (normally) do so elastically; meaning that none of their translational kinetic energy gets converted into other forms (such as going to increase the potential energy of the atoms' electrons by moving them further from the nucleus). This being so, the equation of conservation of energy takes the simple form

$$\tfrac{1}{2}M(U_x^2 + U_y^2 + U_z^2) + \tfrac{1}{2}m(u_x^2 + u_y^2 + u_z^2)$$
$$= \tfrac{1}{2}M(V_x^2 + V_y^2 + V_z^2) + \tfrac{1}{2}m(v_x^2 + v_y^2 + v_z^2).$$

Using eqns (3.6) this reduces to

$$M(U_x - V_x)(U_x + V_x) = -m(u_x - v_x)(u_x + v_x)$$

or, substituting for $M(U_x - V_x)$ from eqn (3.5), to

$$U_x + V_x = u_x + v_x. \tag{3.7}$$

What interests us is the energy $\Delta\varepsilon$ transferred from M to m in the collision. Since the energy gained by m must equal the energy lost by M

we have†

$$\Delta \varepsilon = \tfrac{1}{2}M(U_x^2 + U_y^2 + U_z^2) - \tfrac{1}{2}M(V_x^2 + V_y^2 + V_z^2)$$

which becomes, on using eqns (3.6),

$$\Delta \varepsilon = \tfrac{1}{2}M(U_x - V_x)(U_x + V_x). \tag{3.8}$$

As it stands, eqn (3.8) involves the component speeds before and after impact. However, we can rewrite it so that it only involves the speeds U_x and u_x before impact. We do so by obtaining an expression for V_x in terms of U_x and u_x. This follows on substituting for v_x as given by eqn (3.7) into eqn (3.5), to yield

$$V_x = U_x \frac{(M-m)}{(M+m)} + \frac{2mu_x}{(M+m)}.$$

Putting this expression for V_x back into eqn (3.8) gives

$$\Delta \varepsilon = \frac{2mM}{(m+M)^2}[(MU_x^2 - mu_x^2) + (m-M)U_xu_x] \tag{3.9}$$

as the desired expression for the energy transfer from an atom of the solid to an atom of the gas in a single collision.

An average collision

To find the *average* energy $\overline{\Delta \varepsilon}$ which is transferred from the solid to the gas in a single collision we would look at, say, n collisions. Writing down n expressions like eqn (3.9), summing them, and dividing through by n, gives

$$\overline{\Delta \varepsilon} = \frac{2mM}{(m+M)^2}[(M\overline{U_x^2} - m\overline{u_x^2}) + (m-M)\overline{U_xu_x}]. \tag{3.10}$$

Now when a collision occurs, the component U_x of the atom of the solid may be either positive or negative. However, since this atom remains an atom of the solid it can have no continuous motion in the x-direction. Hence $\overline{U_x} = 0$, and so $\overline{U_xu_x} = 0$. On the other hand, u_x^2 and U_x^2 are always positive and so $\overline{u_x^2}$ and $\overline{U_x^2}$ will be non-zero. Eqn (3.10) therefore reduces to

$$\overline{\Delta \varepsilon} = \frac{2mM}{(m+M)^2}(M\overline{U_x^2} - m\overline{u_x^2}). \tag{3.11}$$

† We assume that the solid atom does not move significantly while it and the gas atom are in contact. Should M move significantly the forces present between it and the other atoms of the solid will 'do work' on m; so transferring some of the potential energy stored in the field between M and the other solid atoms to m. This would change its kinetic energy, of course. Put differently, we are assuming that the duration of the collision between m and M is very much less than the vibrational period of an atom (such as M) of the solid.

Looking back at Fig. 3.3(b) we are reminded that u_x is the x-component of the velocity u of the gas atom and that (see Fig. 4.24(b))

$$u^2 = u_x^2 + u_y^2 + u_z^2 \tag{3.12}$$

with a similar expression holding for U^2. Considering a sequence of n collisions we would write down n expressions like eqn (3.12), sum them, and divide through by n. This leads to

$$\overline{u^2} = \overline{u_x^2} + \overline{u_y^2} + \overline{u_z^2}. \tag{3.13}$$

Assuming, as seems reasonable, that $\overline{u_x^2} = \overline{u_y^2} = \overline{u_z^2}$ (were these unequal, a gas would have different properties in different directions) we have

$$\overline{u_x^2} = \tfrac{1}{3}\overline{u^2} \tag{3.14}$$

and likewise for $\overline{U_x^2}$. Hence eqn (3.11) becomes

$$\overline{\Delta\varepsilon} = \frac{4mM}{3(m+M)^2}(\tfrac{1}{2}M\overline{U^2} - \tfrac{1}{2}m\overline{u^2}).$$

Since m and M are constants this may be written as

$$\boxed{\overline{\Delta\varepsilon} \propto \tfrac{1}{2}M\overline{U^2} - \tfrac{1}{2}m\overline{u^2}} . \tag{3.15}$$

In other words the average energy $\overline{\Delta\varepsilon}$ transferred in a collision from an atom of the solid to an atom of the gas is proportional to the difference in the mean kinetic energy of the solid and gas atoms. If $\tfrac{1}{2}m\overline{u^2} < \tfrac{1}{2}M\overline{U^2}$ then $\overline{\Delta\varepsilon}$ is positive so that energy is transferred from the solid to the gas. If $\tfrac{1}{2}m\overline{u^2} > \tfrac{1}{2}M\overline{U^2}$ then $\overline{\Delta\varepsilon}$ is negative so that energy is transferred from the gas to the solid. Finally, if $\tfrac{1}{2}m\overline{u^2} = \tfrac{1}{2}M\overline{U^2}$ then $\overline{\Delta\varepsilon} = 0$ so that there is no net transfer of energy from solid to gas; under these conditions *thermal equilibrium* is said to exist.

3.5. The two viewpoints compared

We are now in a position to compare our macroscopic everyday usage of the word 'temperature' (which we shall denote by θ for short) with the description of a process as it occurs at the atomic level.

In so doing we will look at how these two viewpoints describe the act of placing a poker in a fire, leaving it until its state ceases to change, and then plunging it into water. The corresponding descriptions are summarized in Table 3.1, where the suffix g refers to the gas (the flame), s to the solid (the poker), and l to the liquid (water). The symbol m refers, of course, to the appropriate atomic (or molecular) mass and *not* to the masses of objects like the poker.

TABLE 3.1
Microscopic and macroscopic viewpoints compared

Observation	Description at the atomic level	Macroscopic description
Poker placed in fire; poker changes (gets 'hot')	Poker atoms acquire energy from gas atoms in flame: $\frac{1}{2}m_g\overline{u_g^2} > \frac{1}{2}m_s\overline{u_s^2}$	$\theta_g > \theta_s$
Poker ceases to change in fire	No net transfer of energy occurs between gas and poker atoms: $\frac{1}{2}m_g\overline{u_g^2} = \frac{1}{2}m_s\overline{u_s^2}$	$\theta_g = \theta_s$
Poker plunged into water; water changes	Water atoms acquire energy from the poker atoms: $\frac{1}{2}m_l\overline{u_l^2} < \frac{1}{2}m_s\overline{u_s^2}$	$\theta_l < \theta_s$

Comparing columns two and three of Table 3.1 we see that the microscopic and macroscopic viewpoints can be made compatible by writing

$$\boxed{\theta \propto \tfrac{1}{2}m\overline{u^2}} \quad . \tag{3.16}$$

This is also consistent with eqn (3.4), which was arrived at solely by considering the conditions under which a substance exists as a solid, as a liquid, or as a gas.

Although eqn (3.16) is never, in fact, used to define temperature it is instructive to see how it might be made operational. Clearly, one would have to devise a procedure for measuring $\frac{1}{2}m\overline{u^2}$; we shall shortly see how this may be done. One would also have to agree on the constant of proportionality in eqn (3.19), allowing it to be written as

$$\theta = C(\tfrac{1}{2}m\overline{u^2}), \tag{3.17}$$

where C could be given any value we fancy. Were we to set $C = 1$, for example, temperatures would be measured in the same units as energy; normally joules.

A time-of-flight thermometer

Even if we were to agree on a value for the constant C in eqn (3.17), there would still remain the problem of how to measure $\frac{1}{2}m\overline{u^2}$, or, assuming that we know the relative atomic (or molecular) mass of the substance, the problem of how to measure $\overline{u^2}$.

To find the temperature of, say, boiling water you might think it

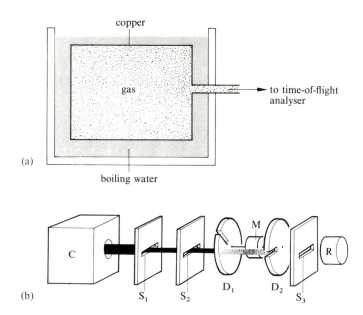

Fig. 3.4. A possible 'thermometer' for measuring the temperature of, say, boiling water. The copper vessel shown in (a) contains a gas, the mean square speed of whose atoms can be measured using the time-of-flight analyser shown in (b). The entire analyser is placed in an evacuated chamber (not shown). In practice, the discs D_1 and D_2 would contain many radial slots.

necessary to measure directly the speeds of the water molecules. However, if we immerse in the boiling water a cylinder made of a diathermal material like copper (Fig. 3.4(a)) and containing a known gas, say mercury vapour, then we know from our previous analysis that at thermal equilibrium the mean kinetic energies of the water molecules, the copper atoms, and the gas atoms are all equal.

The mean square speed $\overline{u^2}$—and hence the mean kinetic energy $\frac{1}{2}m\overline{u^2}$—of the gas atoms can be found using the apparatus shown in Fig. 3.4(b). Here the gas escaping through the hole in the copper container C is collimated into a fine beam by means of two slits S_1 and S_2. This beam next meets a pair of rotating discs D_1 and D_2 which, together, constitute a speed selector (§2.4). This allows through only those atoms which cover the distance between D_1 and D_2 in the time it takes for the slot in D_2 to move around to lie along the line S_1S_2. The number of those atoms which do get through D_2 is measured by allowing the beam, after it has passed through slit S_3, to strike a suitable recording detector R. Changing the speed of the motor M which drives the discs allows through atoms with a different, but again known, speed.

Exercise 3.1

Fig. 3.5 shows results obtained in an actual experiment of the type we have just described. The gas was mercury ($A_r(Hg) = 200$) and it was in thermal contact with boiling water. What is the mean kinetic energy of (a) the mercury atoms in the beam, (b) the mercury atoms in the oven, and (c) the water molecules in the bath? Because the faster gas atoms within C make more collisions per unit time with the hole in C (thereby escaping) than do the slower atoms, the emerging beam contains a disproportionate number of fast atoms. It turns out (see exercise 5.1) that the mean kinetic energy of the atoms in the beam is $\frac{4}{3}$ times the mean kinetic energy of the atoms within C.

Solutions. We first calculate the mass m of a mercury atom by recalling that there are N_A ($= 6{\cdot}022 \times 10^{23}$) atoms in $A_r(Hg)$ grams ($= 200 \times 10^{-3}$ kg) of mercury. This gives $m = 3{\cdot}32 \times 10^{-25}$ kg. If, out of a total of N gas atoms, n_i have speed u_i then, by definition, the mean square speed $\overline{u^2}$ of these atoms is

$$\overline{u^2} = \frac{\sum\limits_i n_i u_i^2}{\sum\limits_i n_i} = \frac{\sum\limits_i n_i u_i^2}{N} = \left(\frac{n_1}{N}\right)u_1^2 + \left(\frac{n_2}{N}\right)u_2^2 + \cdots . \tag{3.18}$$

Fig. 3.5 tells us the fractions (n_1/N), (n_2/N), etc., with speeds u_1, u_2, etc., which we may take to be the speeds at the mid-point of each block of the histogram, namely $1{\cdot}15 \times 10^2 \text{ m s}^{-1}$, $1{\cdot}65 \times 10^2 \text{ m s}^{-1}$, etc. The calculation of $(n_1/N)u_1^2$, $(n_2/N)u_2^2$, etc. is set out in Table 3.2. Adding up the numbers in the last column of this table tells us that (see eqn (3.18)) $\overline{u^2} = 5{\cdot}93 \times 10^4 \text{ m}^2 \text{ s}^{-2}$. (a) The mean kinetic energy of the mercury atoms in the beam is therefore given by $\frac{1}{2}m\overline{u^2} = \frac{1}{2}(3{\cdot}32 \times 10^{-25} \text{ kg})(5{\cdot}93 \times 10^4 \text{ m}^2 \text{ s}^{-2}) = 9{\cdot}84 \times 10^{-21}$ J. (b) The mean kinetic energy of the mercury atoms in the oven, which is $\frac{3}{4}$ of the mean kinetic energy of the mercury atoms in the beam, is $7{\cdot}38 \times 10^{-21}$ J. (c) The mean kinetic energy of the

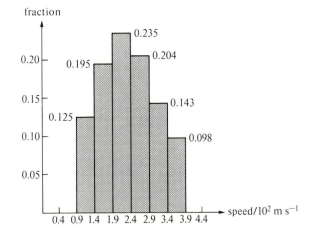

Fig. 3.5. Showing the distribution in speeds of gaseous mercury atoms in a beam emerging from a container which is in thermal equilibrium with boiling water. A time-of-flight instrument was used to measure the speed distribution. (Data from Lammert (1929). Z. Phys. **56**, 244.)

TABLE 3.2
Analysis of the data of Fig. 3.5

n_i/N	$u_i/10^2$ m s^{-1}	$u_i^2/10^4$ m^2 s^{-2}	$(n_i u_i^2/N)/10^4$ m^2 s^{-2}
0·125	1·15	1·32	0·165
0·195	1·65	2·72	0·530
0·235	2·15	4·62	1·086
0·204	2·65	7·02	1·432
0·143	3·15	9·92	1·418
0·098	3·65	13·30	1·303

water molecules in the bath is, at thermal equilibrium, equal to the mean kinetic energy of the mercury atoms in the oven, namely $7\cdot38 \times 10^{-21}$ J.

Comment. Problem 5.4 will consider a more recent study of the speed distribution of gas atoms.

3.6. The gas thermometer

There is in fact no agreed constant C in eqn (3.17) which would enable us to use it as a definition of temperature. What is agreed internationally is how to measure temperatures using a device called a gas thermometer. To fully describe a gas thermometer and to state precisely how a measurement is made would call for a chapter in itself. The essential steps can, however, be described quite concisely in the form of a prescription.

The basic article

Take a fixed mass of gas and measure its pressure p and volume V when the container holding the gas is in thermal contact with water at its triple point (Fig. 3.6(a)). Now evaluate the product of p and V, written $(pV)_{tr}$. Call the temperature of the gas $T_{tr} = 273\cdot16$ K exactly.

To find the temperature T of, say, boiling sulphur, immerse the bulb in the sulphur, measure the new values of p and V, and calculate the product pV afresh. The temperature T of the sulphur is (roughly) given by

$$\frac{T}{T_{tr}} = \frac{(pV)}{(pV)_{tr}}$$

and thus by

$$T = 273\cdot16 \text{ K} \frac{(pV)}{(pV)_{tr}}. \tag{3.19}$$

There is a major snag to the thermometer as we have described it.

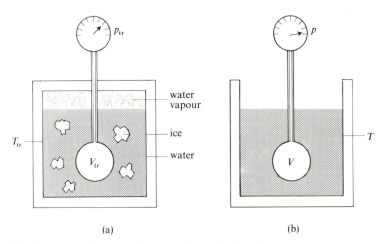

Fig. 3.6. A schematic diagram of the procedures involved in measuring a temperature on the gas scale. In (a) the bulb is immersed in water at its triple point. In (b) the bulb is immersed in the substance whose temperature is required.

Different observers would get different values of T for the same bath of boiling sulphur. The measured value depends on how the thermometer is constructed, on which gas is used, and on the amount of gas present in the thermometer's bulb. This might not matter much if one just wanted a rough-and-ready value for the temperature but discrepancies become very serious when we are trying to define operationally *the* temperature of the system.

Removing the snags

Over the years a great deal of effort has gone into refining gas thermometry. There is, however, no one 'approved' design of a gas thermometer. What exists is a consensus as to what constitutes good and bad practice. If you built a gas thermometer and obtained a value of 54·365 K for the triple point of oxygen, your entire experimental procedures would be scrutinized before that value was accepted as *the* triple point of oxygen on the gas scale.

As already mentioned, the temperature obtained on applying eqn (3.19) depends on which gas is used and on how much of it is present. Fortunately, the measured temperature gets less and less dependent on the gas the smaller is the amount of gas present. The procedure, or rather one procedure, that is followed is to measure $(pV_m)/(pV_m)_{tr}$ first at one fill. Next one removes, say, half the gas (thereby halving the amount of gas present) and measures $(pV_m)/(pV_m)_{tr}$ again. One removes still more of the gas and measures the ratio afresh. The limiting value of $(pV_m)/(pV_m)_{tr}$ as the amount of gas tends to zero is obtained by plotting $(pV_m)/(pV_m)_{tr}$

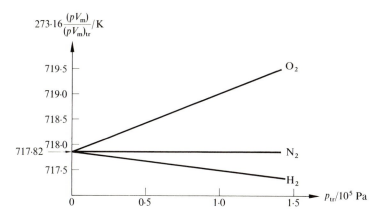

Fig. 3.7. Showing how the measured temperature at which sulphur boils (at standard atmospheric pressure) as deduced from the readings on a constant-volume gas thermometer depends on the thermometric gas and on the amount of gas present (indicated by its pressure p_{tr} at the triple point of water).

against p, the gas pressure at some convenient point, such as the triple point of water, and extrapolating the graph to where it crosses the $p = 0$ axis. This procedure is illustrated in Fig. 3.7 and shows that as p, and thus the amount of gas present, tends towards zero, the measured temperature (here the boiling point of sulphur at standard atmospheric pressure) ceases to depend on the type of gas used. Adopting some such procedure, eqn (3.19) becomes

$$T = 273 \cdot 16 \text{ K} \lim_{p \to 0} \left[\frac{(pV_m)}{(pV_m)_{tr}} \right]$$

which can be rewritten as†

$$T = 273 \cdot 16 \text{ K} \frac{\lim_{p \to 0}(pV_m)}{\lim_{p \to 0}(pV_m)_{tr}} . \qquad (3.20)$$

† As shown in practically any introductory text on calculus, if $f(x)$ and $g(x)$ are two functions of x such that $\lim_{x \to a} f(x)$ and $\lim_{x \to a} g(x)$ exist, then

$$\lim_{x \to a} \left[\frac{f(x)}{g(x)} \right] = \frac{\lim_{x \to a} f(x)}{\lim_{x \to a} g(x)}$$

provided $\lim_{x \to a} g(x) \neq 0$. These conditions hold true in our case.

This equation defines a temperature T on the *perfect gas scale* (also called the *ideal gas scale*). However it only defines T if all the 'rules' have been followed. Lurking behind eqn (3.20) is a very complex piece of equipment which must satisfy many different criteria!

The gas constant

Eqn (3.20) may be written as

$$\lim_{p \to 0}(pV_m) = RT, \tag{3.21}$$

where

$$R = \frac{\lim_{p \to 0}(pV_m)_{tr}}{273 \cdot 16\ \text{K}} \tag{3.22}$$

is known as the *gas constant*. The currently accepted value of R is $8 \cdot 31441(\pm 0 \cdot 00026)\ \text{J K}^{-1}\ \text{mol}^{-1}$. Because $\lim(pV_m)_{tr}$ as $p \to 0$ is an experimentally determined quantity (unlike the triple point temperature which is defined to be $273 \cdot 16\ \text{K}$ exactly) the value of R is subject to change as laboratory techniques are refined.

Boltzmann's constant

When we come to discuss the origins of gas pressure in Chapter 5 we shall show that (eqn (5.18))

$$pV_m = \tfrac{2}{3}N_A(\tfrac{1}{2}m\overline{u^2}) \tag{3.23}$$

provided that the gas is 'perfect' (meaning that the volume occupied by the gas atoms is negligible compared with the total volume and that the interatomic forces are effectively zero). Because these assumptions only hold true in a real gas as $p \to 0$, eqn (3.23) must be rewritten as

$$\lim_{p \to 0}(pV_m) = \tfrac{2}{3}N_A(\tfrac{1}{2}m\overline{u^2})$$

when it is to be applied to such a gas. Substituting for $\lim(pV_m)$ as $p \to 0$ from eqn (3.21) gives

$$T = \frac{2}{3k}(\tfrac{1}{2}m\overline{u^2}), \tag{3.24}$$

where

$$k = \frac{R}{N_A} \tag{3.25}$$

is called the *Boltzmann constant*. Since R and N_A are both determined experimentally the value of k depends on the accuracy to which these quantities are known; the currently accepted value of $k = 1\cdot380662(\pm 0\cdot000044)$ J K^{-1}.

If you compare eqns (3.17) and (3.24) you will see that both have fundamentally the same form, and that they become identical if we let $C = 2/3k$. You may therefore ask why we do not give k a fixed value in eqn (3.24) (perhaps rounding it off to $1\cdot38 \times 10^{-23}$ J K^{-1}) and then use that as our prime definition of temperature (knowing that $\frac{1}{2}m\overline{u^2}$ could be measured by a time-of-flight technique). Such a proposal must be rejected for it would remove temperature from the macroscopic (eqn (3.20)), where it properly belongs, to the realms of the microscopic (eqn (3.24)), where it is at the mercy of a particular model of the microscopic world.

Cross-multiplying eqn (3.24) gives

$$\frac{1}{2}m\overline{u^2} = \frac{3}{2}kT \tag{3.26}$$

or, since $\overline{u_x^2}(=\overline{u_y^2}=\overline{u_z^2}) = \frac{1}{3}\overline{u^2}$ (eqn (3.14)),

$$\frac{1}{2}m\overline{u_x^2} = \frac{1}{2}m\overline{u_y^2} = \frac{1}{2}m\overline{u_z^2} = \frac{1}{2}kT. \tag{3.27}$$

These results will prove to be of great value in later chapters.

Exercise 3.2

(a) Rewrite eqns (3.1) to (3.3) so that they involve the gas-scale temperature T. (b) Given that $\Delta E = 1\cdot4 \times 10^{-20}$ J (as estimated in problem 2.16) estimate the critical temperature of water.

Solutions. Substituting $\frac{1}{2}m\overline{u^2}$ as given by eqn (3.26) into eqns (3.1) to (3.3) tells us that a substance will exist as a gas when

$$kT \gg \Delta E, \tag{3.28}$$

as a solid when

$$kT \ll \Delta E, \tag{3.29}$$

and as a liquid when

$$\tfrac{3}{2}kT \approx \Delta E. \tag{3.30}$$

Because of the gross inequalities in eqns (3.1) and (3.2) we have approximated $\frac{3}{2}$ to unity in eqns (3.28) and (3.29). (b) The critical point is distinguished by the fact that at $T = T_c$ the liquid and gaseous phases become indistinguishable (Fig. 1.8). This suggests that at the critical point

$$\tfrac{3}{2}kT_c = \Delta E. \tag{3.31}$$

Substituting for k and ΔE predicts a critical temperature of 676 K for water. This is in close agreement with the measured value of 647 K, again vindicating the simple arguments employed in §3.1.

3.7. The International Practical Temperature Scale

The International Practical Temperature Scale (IPTS) is, as the name suggests, a scale of a somewhat more practical nature than the gas scale, yet one which is capable of very high precision. The scale *assigns* values to the temperatures of a number of reproducible equilibrium states (defining 'fixed points') and lays down the type of instrument that is to be used to measure temperatures between these fixed points. As examples of fixed points, the scale adopted in 1968, and currently in use (called the IPTS-68 scale), fixes the triple point of hydrogen as 13·81 K, the boiling point of neon at standard atmospheric pressure as 27·102 K, and the freezing point of zinc (equilibrium between solid and liquid phases) at standard pressure as 692·73 K. (Fig. 1.2 will remind you why it is necessary to specify the pressure of the boiling and freezing points.) These fixed point temperatures represent what are agreed to be the best determinations made on the gas scale. At regular intervals an international group of scientists assesses the 'state of the art' and may decide to change the value of any fixed point. It is important to note that when a numerical value is *assigned* to a fixed point no 'error' is attached to that value. Measurements made on the gas scale give the melting point of zinc as 692·73(±0·03) K—but IPTS-68 *fixes* the temperature as 692·73 K.

The second feature of the IPTS is that it lays down the instruments that are to be used in measuring temperatures between the fixed points. It says for example, that the standard instrument to be used from 13·81 K (the triple point of hydrogen) to 903·89 K (the freezing point of antimony at $1·01325 \times 10^5$ Pa) is the platinum resistance thermometer. It is an experimental fact that the resistance of a metal varies with temperature, so this property can clearly form the basis of a thermometer. On the IPTS the criteria which the instrument must satisfy are laid down, as is also the method of turning numbers—like the resistance of the platinum thermometer—into temperatures. Only when all these recommendations have been strictly adhered to can one claim to have measured a temperature on the IPTS.

Measuring a temperature on the IPTS is no easy matter. We therefore have a number of thoroughly practical thermometers, like the familiar mercury-in-glass thermometer, which can be calibrated against IPTS instruments. The usual mass-produced thermometer is several steps removed from the standard instruments.

3.8. Brownian motion

In §3.4 we discussed the *dynamics* of a single gas atom as it was buffeted about in collisions with other atoms. We now return to consider a single atom (or larger particle): in particular to ask how far we can expect such an atom to travel within a specified time. It is important to keep in mind that such movements are an entirely natural feature of a system which is in thermal equilibrium with its surroundings.

Experimental observations

In 1827 the English botanist Robert Brown noticed that grains of pollen suspended in water executed a highly erratic dance. This same erratic motion—known as *Brownian motion*—is also observed in particles (for example, smoke particles) suspended in air. If we regard each of these particles as being a rigid body of mass M, we would expect them to move at a mean-square speed $\overline{U^2}$ such that (see eqn (3.26))

$$\tfrac{1}{2}M\overline{U^2} = \tfrac{1}{2}m\overline{u^2} = \tfrac{3}{2}kT,$$

where m is the molecular mass of the medium in which the large particles are suspended. As an example, a 2×10^{-7} m radius grain of the resin gamboge will have a mass M of around 3×10^{-17} kg (which is some 10^9 times the mass of a water molecule) and hence a root-mean-square speed $(\overline{U^2})^{1/2} = (3kT/M)^{1/2} = 2 \cdot 0 \times 10^{-2}$ m s^{-1} at 300 K. The combination of such

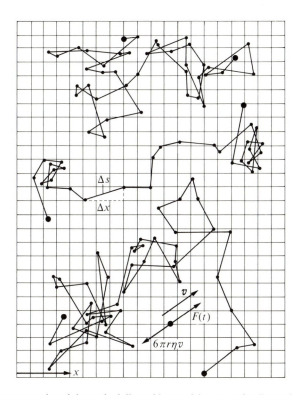

Fig. 3.8. Three examples of the paths followed by particles executing Brownian motion. The dots indicate the position of a particle (a gamboge grain suspended in water) at 30 s intervals. The size of the grid's divisions are not given by Perrin but are believed to be a few microns (1 micron = 10^{-6} m). Gamboge is a yellow resin used in water-colour paints. (From Perrin, J. (1909). *Ann. Chim. Physique* **18**, 1–114.)

slow speeds with such large sizes makes it possible to observe the motion of these suspended particles with a microscope. Fig. 3.8 shows the results of a study made by the French physicist Jean Perrin in 1909. What he did was to observe the position of a gamboge grain suspended in water at 30 s intervals. These positions, indicated by dots in Fig. 3.8, were then joined by straight lines. This means, of course, that the paths shown in Fig. 3.8 must not be interpreted too literally. To quote Perrin: 'If, in fact, one were to mark second by second, each of the straight line segments would be replaced by a polygon path of 30 sides, relatively as complex as the pattern here reproduced, and so on'.

Einstein's equation

Although the motion of a particle executing Brownian motion was first analysed by Einstein in 1906, the following discussion is that of P. Langevin (1908).

The random nature of the bombardment of a suspended particle (mass M) by the molecules of the medium means that the impulsive forces acting on M will not, in general, cancel. Hence, at any instant, there will be a resultant force $F(t)$ acting on M whose effect must be to accelerate it (Fig. 3.8). This motion will, however, be opposed by a frictional force whose value (according to Stokes' law) is $6\pi r\eta v$, where r is the radius of the particle, v is its speed, and η is the dynamic viscosity of the medium (a quantity to be defined later (p. 214). Resolving the motion of the particle along the x-direction (Fig. 3.8), applying Newton's second law, and multiplying the resulting expression through by x gives

$$X(t)x - 6\pi r\eta x \frac{\mathrm{d}x}{\mathrm{d}t} = Mx \frac{\mathrm{d}^2x}{\mathrm{d}t^2}, \tag{3.32}$$

where $X(t)$ is the x-component of $F(t)$. Now we can write

$$x \frac{\mathrm{d}x}{\mathrm{d}t} = \frac{1}{2} \frac{\mathrm{d}(x^2)}{\mathrm{d}t} \tag{3.33}$$

which becomes, when differentiated with respect to time,

$$x \frac{\mathrm{d}^2x}{\mathrm{d}t^2} = \frac{1}{2} \frac{\mathrm{d}}{\mathrm{d}t} \left[\frac{\mathrm{d}(x^2)}{\mathrm{d}t} \right] - \left(\frac{\mathrm{d}x}{\mathrm{d}t} \right)^2. \tag{3.34}$$

Substituting eqns (3.33) and (3.34) into eqn (3.32) and averaging the resulting expression over all the particles which are suspended in the liquid leads to

$$\overline{X(t)x} - 3\pi r\eta \frac{\overline{\mathrm{d}(x^2)}}{\mathrm{d}t} = \tfrac{1}{2}M \frac{\overline{\mathrm{d}}}{\mathrm{d}t}\left[\frac{\mathrm{d}(x^2)}{\mathrm{d}t} \right] - M\overline{\left(\frac{\mathrm{d}x}{\mathrm{d}t} \right)^2}. \tag{3.35}$$

Now at any fixed x (Fig. 3.8) we can expect a force $+X(t)$ acting on one particle to be matched by a force $-X(t)$ acting on another particle. Hence $\overline{X(t)x} = 0$. Furthermore, $M\overline{(dx/dt)^2} = kT$ (eqn (3.27)) and $\overline{d/dt[d(x^2)/dt]} = d/dt[\overline{d(x^2)}/dt]$. Eqn (3.35) therefore becomes

$$\tfrac{1}{2}M\frac{dw}{dt} + 3\pi r\eta w = kT, \tag{3.36}$$

where, for convenience, we have written

$$w = \frac{\overline{d(x^2)}}{dt}. \tag{3.37}$$

The general solution of the differential equation, eqn (3.36) is

$$w = \frac{kT}{3\pi r\eta} + C\exp\left(\frac{-6\pi r\eta t}{M}\right), \tag{3.38}$$

where C is a constant of integration. You can verify, by direct substitution, that eqn (3.38) is indeed a solution of eqn (3.36). Now $6\pi r\eta/M$ in the exponent of eqn (3.38) is a very large number (for example, it is around $2\times10^8\,\mathrm{s}^{-1}$ in Perrin's experiment). Thus the exponential term contributes so little to w at finite times t that it may usually be ignored, enabling us to write eqn (3.38) as

$$w = \frac{\overline{d(x^2)}}{dt} = \frac{kT}{3\pi r\eta}. \tag{3.39}$$

Integrating eqn (3.39) from $t = 0$ to $t = \tau$ yields

$$\overline{x^2} - \overline{x_0^2} = \left(\frac{kT}{3\pi r\eta}\right)\tau.$$

Finally, setting $x_0 = 0$ when $t = 0$ and writing Δx^2 for x^2 gives

$$\boxed{\overline{\Delta x^2} = \left(\frac{kT}{3\pi r\eta}\right)\tau}. \tag{3.40}$$

It is important to recall that this equation does *not* deal with the displacement Δs but with the *component* Δx of the displacement along a particular (but arbitrarily chosen) direction (Fig. 3.8). It tells us that we must take a 'snapshot' of the suspension at time $t = 0$ and again at time $t = \tau$, that we must measure Δx, and hence Δx^2 for each and every particle, and that we must then deduce $\overline{\Delta x^2}$ (by summing all the Δx^2 and dividing by the number of particles). With only three particles observed in Fig. 3.8 you might think that we could not reasonably deduce $\overline{\Delta x^2}$, and hence check eqn (3.40). However, we can assume that one particle followed for N successive intervals of time τ (where N is a large number) will have the same value of $\overline{\Delta x^2}$ as will all N particles observed during a single interval τ. The assumption that the time-averaged value of some property of a single particle has the same value as that found by averaging the property over all the particles at a single time is one which is frequently employed in physics. This assumption is ultimately an axiom, but not an unreasonable one. Of course, it only applies if the particles are in thermal equilibrium with one another.

Exercise 3.3

Deduce the radius of the particle whose Brownian motion is recorded by the lower zig-zag track in Fig. 3.8. You can assume that each of the dimensions of the grid is 1.0×10^{-6} m wide, and that the experiment was performed in water at 293 K (where $\eta = 1.0 \times 10^{-3}$ kg m^{-1} s^{-1}).

Calculation. Looking at the first twenty displacements (beginning at the right-hand end of the track) we see that Δx has the following values (all in units of 10^{-6} m): $+2.4$, $+1.2$, -1.6, -0.9, $+0.9$, -4.0, -1.5, $+1.7$, $+1.0$, $+0.3$, $+1.3$, -2.9, -3.1, -0.5, $+1.5$, $+0.7$, $+1.9$, -0.2, $+0.1$, -2.7. Squaring each of these twenty values of Δx, summing the Δx^2, and dividing by 20 gives $\overline{\Delta x^2} = 3.00 \times 10^{-12}$ m^2. It follows from eqn (3.40) that

$$r = \frac{kT\tau}{3\pi\eta \, \overline{\Delta x^2}}.$$

Substituting $k = 1.38 \times 10^{-23}$ J K^{-1}, $T = 293$ K, $\tau = 30$ s, $\eta = 1.0 \times 10^{-3}$ kg m^{-1} s^{-1}, and $\overline{\Delta x^2} = 3.00 \times 10^{-12}$ m^2 gives $r = 4.3 \times 10^{-6}$ m.

Comment. Perrin actually performed this experiment with (100 different) particles of known size. He therefore deduced the value of k, or since $N_A = R/k$ (eqn (3.25)), the value of N_A.

3.9. Bound atoms

We have already demonstrated that irrespective of the phase in which a substance exists, its constituent molecules have a mean *kinetic* energy $\frac{1}{2}m\overline{u^2} = \frac{3}{2}kT$ (eqn (3.26)); a result derived assuming that the laws of Newtonian mechanics hold true. In the case of a perfect gas ($\Delta E = 0$), or a real gas at low densities (where the interatomic separation is large), the kinetic energy is the *only* energy possessed by the molecules. This is manifestly untrue in the solid phase where each atom is bound by interatomic forces to its neighbours, as suggested pictorially by the springs of Fig. 3.9. We should therefore seek to calculate the total energy of a bound atom. However, we start by considering the equations of motion of such an atom.

When an atom like i in Fig. 3.9 is given kinetic energy it will be set moving around some complicated orbit. Despite its apparent complexity

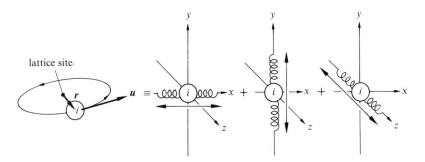

Fig. 3.9. Showing how the motion of an atom i of a solid may be resolved into three individual vibrational motions. The lattice site denotes the position the atom would occupy (classically) at $T = 0$.

this orbital motion can be resolved into the three individual motions shown in Fig. 3.9. (You have possibly seen a cathode ray tube demonstration of the so-called Lissajous' figures in which an elliptical orbit, albeit a two-dimensional one, is synthesized out of two perpendicular 'simple harmonic' motions.) We therefore start by considering how an atom behaves in a one-dimensional structure.

The simple-harmonic oscillator

Consider atom i in Fig. 3.10(a). At $T = 0$ it will be at rest $(\frac{1}{2}m\overline{u^2} = 0)$ midway between atoms j and k, at a distance r_0 from each neighbour. If atom i is now displaced so that it is at a distance r_1 from j and r_2 from k it will experience a restoring force $F(r_1)$ from atom j *and* a restoring force $F(r_2)$ from atom k. Approximating the interatomic force curve in the vicinity of r_0 by a straight line (Fig. 3.10(b)), we see that the overall restoring force acting on i per unit displacement from its equilibrium position—the so-called *force constant* k_f—is given by

$$k_f = 2k_s, \tag{3.41}$$

where k_s is the force constant between a pair of atoms. Hence a force of $-k_f x$ acts on i when it is displaced by x from its equilibrium position (Fig. 3.10(b)), and so its equation of motion is, from Newton's second law,

$$m\frac{d^2x}{dt^2} = -k_f x, \tag{3.42}$$

where m is the mass of atom i. This equation (the equation of motion of a *simple harmonic oscillator*) has a solution of the form

$$x = A \sin(\omega t + \varepsilon). \tag{3.43}$$

Here A represents the *amplitude* of vibration of i (eqn (3.43) tells us that

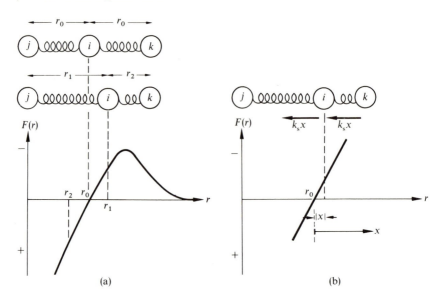

Fig. 3.10. (a) When atom i, initially at $r = r_0$, is displaced so that it is at a distance r_1 from j and r_2 from k, it will be subjected to a restoring force from each atom. (b) Showing the interatomic force curve approximated by a straight line in the vicinity of $r = r_0$. Displacements are now measured from $r = r_0$.

x lies between $\pm A$). To see the significance of ω we look at i at a time t and again at a time T_v (known as the *period* of the vibration) later such that the displacement of i is the same as it was at time t. Because the displacements are to be identical it follows from eqn (3.43) that

$$\sin(\omega t + \varepsilon) = \sin(\omega(t + T_v) + \varepsilon)$$

and hence that the *circular frequency* ω is given by

$$\omega = \frac{2\pi}{T_v}. \tag{3.44}$$

By defining $x = 0$ at $t = 0$ the *phase angle* ε is seen to be zero, so that eqn (3.43) becomes

$$x = A \sin \omega t. \tag{3.45}$$

Substituting eqn (3.45) into eqn (3.42) leads to

$$\boxed{\omega = \left(\frac{k_f}{m}\right)^{1/2} = \left(\frac{2k_s}{m}\right)^{1/2}.} \tag{3.46}$$

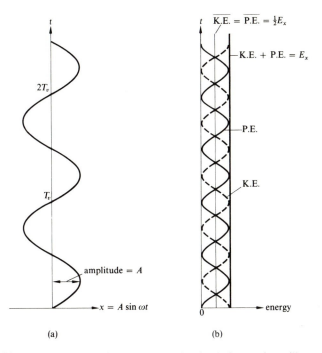

Fig. 3.11. (a) Showing how the displacement x of a simple harmonic oscillator varies with time t. (b) Showing how the kinetic, potential, and total energies of the oscillator vary with time.

Therefore eqn (3.45) can be written as

$$x = A \sin\left(\frac{k_f}{m}\right)^{1/2} t . \tag{3.47}$$

This equation is plotted graphically in Fig. 3.11(a).

The energy of a simple harmonic oscillator

Our main interest in atom i of Fig. 3.10(b) lies in its potential and kinetic energies. If we define the potential energy (P.E.) to be zero when $r = r_0$ in Fig. 3.10(b) then the P.E. of i at a displacement x is the energy *required* to move i to position x against the restoring force $-k_f x$. Hence

$$\text{P.E.} = \int_0^x k_f x \, dx = \tfrac{1}{2} k_f x^2. \tag{3.48}$$

Substituting for x from eqn (3.45) gives

$$\text{P.E.} = \tfrac{1}{2}k_f A^2 \sin^2\omega t \qquad (3.49)$$

and this is shown plotted in full line in Fig. 3.11(b).

The kinetic energy (K.E.) of atom i is, by definition, given by

$$\text{K.E.} = \tfrac{1}{2}m\left(\frac{\mathrm{d}x}{\mathrm{d}t}\right)^2 \qquad (3.50)$$

which becomes, on substituting for x from eqn (3.45) and appealing to eqn (3.46),

$$\text{K.E.} = \tfrac{1}{2}k_f A^2 \cos^2\omega t, \qquad (3.51)$$

a result which is shown plotted in dashed line in eqn (3.11(b)). Adding together eqns (3.49) and (3.51) tells us that the total energy E_x of atom i is

$$E_x = \tfrac{1}{2}k_f A^2 = \tfrac{1}{2}m\omega^2 A^2 \qquad (3.52)$$

which is independent of time (shown as the heavy line in Fig. 3.11(b)). Since the time-averaged values of $\sin^2\omega t$ and $\cos^2\omega t$ are each equal to $\tfrac{1}{2}$, it follows from eqns (3.49) and (3.51) that

$$\overline{\text{P.E.}} = \overline{\text{K.E.}} = \tfrac{1}{4}k_f A^2, \qquad (3.53)$$

where the bars indicate time-averaging for a *single* atom. Writing down similar expressions to eqns (3.49) and (3.51) for all the other atoms of the solid (we would expect each to have a different value of A), summing these expressions, and dividing through by the total number of atoms involved tells us that the average atom has a potential energy of $\tfrac{1}{2}k_f\overline{A^2}\sin^2\omega t$ and a kinetic energy of $\tfrac{1}{2}k_f\overline{A^2}\cos^2\omega t$. Time-averaging each of these quantities over a period T_v leads to the conclusion that

$$\boxed{\overline{\text{P.E.}} = \overline{\text{K.E.}}(=\tfrac{1}{4}k_f\overline{A^2}) = \tfrac{1}{2}kT} \,, \qquad (3.54)$$

where we have substituted $\overline{\text{K.E.}} = \tfrac{1}{2}kT$ from eqn (3.27). Another way of arriving at eqn (3.54) is to stay with a single atom for a sufficiently long time for it to have had its energy changed many times over as a result of interactions with the other atoms of the solid.

Finally, we may write the mean total energy of the one-dimensional oscillating atom as

$$\boxed{\bar{E}_x = \overline{\text{P.E.}} + \overline{\text{K.E.}} = kT} \,. \qquad (3.55)$$

The total energy of an atom

We now return to the atom i of Fig. 3.9, executing an elliptical orbit within the solid. Assuming that the restoring force k_f acting on i is spherically symmetric (meaning that it is independent of the orientation of the position vector R) we can write down that i has a P.E. given by

$$\text{P.E.} = \tfrac{1}{2}k_f R^2 = \tfrac{1}{2}k_f(x^2 + y^2 + z^2), \tag{3.56}$$

where we have, of course, applied Pythagoras' theorem. Likewise the K.E. of i can be written as

$$\text{K.E.} = \tfrac{1}{2}mu^2 = \tfrac{1}{2}m(u_x^2 + u_y^2 + u_z^2), \tag{3.57}$$

where u_x, u_y, and u_z are the component speeds of u along the x-, y-, and z-axes. Adding together eqns (3.56) and (3.57) gives the total energy of atom i as

$$E = E_x + E_y + E_z, \tag{3.58}$$

where $E_x = \tfrac{1}{2}k_f x^2 + \tfrac{1}{2}mu_x^2$, and similarly for E_y and E_z, are each equal to the instantaneous energy of a one-dimensional harmonic oscillator. Thus, on average,

$$\bar{E} = \bar{E}_x + \bar{E}_y + \bar{E}_z \tag{3.59}$$

which becomes, on substituting for $\bar{E}_x$ $(= \bar{E}_y = \bar{E}_z)$ from eqn (3.55),

$$\boxed{\bar{E} = 3kT} \tag{3.60}$$

as the desired total mean energy of a bound atom.

3.10. Fluctuations

In much the same way that we were able to take over the discussion of how two atoms behave in a collision (§3.4) and apply it to a macroscopic object like a smoke particle, so we can take over the discussion of the energy stored in a bound pair of atoms (eqn (3.54)) and apply it to macroscopic objects.

Length fluctuations in a spring

We will consider the mass–spring system shown in Fig. 3.12(a). Because of the thermal fluctuations in the number of atoms colliding with the mass, and because of thermal fluctuations in the spring itself, the mass will never be completely at rest. Instead it will oscillate back and forth in the erratic manner suggested in Fig. 3.12(b). When the displacement of the mass from its mean position is x the energy stored in the spring is

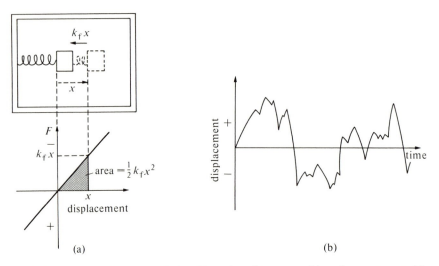

Fig. 3.12. (a) When the mass is displaced by x from its mean position, the energy stored in the spring is $\frac{1}{2}k_{f}x^2$; equal to the area shown shaded in the force–displacement graph. (b) Showing how the displacement might vary with time.

$\frac{1}{2}k_{f}x^2$ (eqn (3.48)); equal to the area shown shaded in Fig. 3.12(a). Thus, averaged over a long period of time, the mean energy stored in the spring is $\frac{1}{2}k_{f}\overline{x^2}$, where $\overline{x^2}$ is the mean square displacement of the mass. Setting this energy equal to $\frac{1}{2}kT$ (eqn (3.54)) leads to the conclusion that the root mean square (r.m.s.) displacement $x_{\text{r.m.s.}}$ $(=(\overline{x^2})^{1/2})$ of the mass is

$$x_{\text{r.m.s.}} = \left(\frac{kT}{k_{f}}\right)^{1/2}. \qquad (3.61)$$

In the case of a typical 'everyday' spring requiring a force of, say, 1 N to extend it by 1 cm (implying that $k_{f} = 10^2\,\text{N m}^{-1}$), $x_{\text{r.m.s.}}$ has the quite insignificant value of around 10^{-11} m. However, in the human eardrum—whose structure behaves like that of Fig. 3.12(a)—the value of $x_{\text{r.m.s.}}$ is about 10^{-10} m. Sounds which produce a lower amplitude than this in the eardrum cannot be 'heard' by the brain. In fact, the thermal motions of the eardrum determine the limits of audibility to frequencies in the range 10^3 Hz to 3×10^3 Hz.

Eqn (3.61) shows that $x_{\text{r.m.s.}}$ does not depend on the density of the gas surrounding the spring. If the density of the gas is high, there will be many more impacts per unit time with the mass than when the density is low. Consequently the movement of the mass is more erratic at high than at low densities. However, $x_{\text{r.m.s.}}$ is the same at all densities.

Density fluctuations in a gas

The thermal motions of the individual molecules of a gas lead to density fluctuations, and thus to pressure fluctuations, in the gas (Fig. 3.13(a)). As with the mass–spring system these fluctuations occur naturally in a system in which thermal equilibrium exists. In analysing the magnitude of these density fluctuations we start by calculating the energy required to produce a local increase in pressure from p_0 (the mean value) to $p_0 + \delta p$ within a small region of the gas. This can readily be done by picturing a portion of the gas, containing N molecules and originally occupying a volume V, as being compressed in the piston-and-cylinder arrangement shown in Fig. 3.13(b) until its volume is $V - \delta V$. The amount of work required to produce this compression—as produced by the 'person' holding the piston—is equal to the area shown shaded in the p–V diagram of Fig. 3.13(b). (Because V is very much less than the rest of the system p_0 remains essentially constant during the compression.) Assuming the gas to be perfect, its pressure p and volume V are related to the total number of atoms present N and the temperature T by eqn (5.20) (a relation we will later establish):

$$pV = NkT.$$

Thus, at constant temperature,

$$\frac{\mathrm{d}p}{\mathrm{d}V} = \frac{-NkT}{V^2}$$

which can, for small compressions, be set equal to $-\delta p/\delta V$. Using the

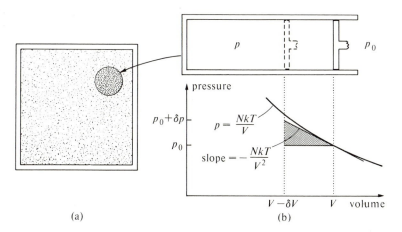

(a)

(b)

Fig. 3.13. (a) Local fluctuations in density occur naturally in any gas. (b) Showing how the work done in forming such a fluctuation can be calculated.

$\frac{1}{2}$(base)(height) rule for the area of a triangle we see that the area shown shaded in Fig. 3.13(b) has a value

$$\tfrac{1}{2}(\delta V)(\delta p)=\tfrac{1}{2}(\delta V)\left(\frac{NkT\,\delta V}{V^2}\right)=\tfrac{1}{2}NkT\left(\frac{\delta V^2}{V^2}\right).$$

Averaged over a long period of time, during which δV fluctuates erratically in a manner akin to the displacement in Fig. 3.12(b), the mean potential energy of the small volume of gas is $\frac{1}{2}NkT(\overline{\delta V^2}/V^2)$. Setting this equal to $\frac{1}{2}kT$ gives the root mean square fluctuation in volume $\delta V_{\text{r.m.s.}}$ $(=(\overline{\delta V^2})^{1/2})$ as

$$\boxed{\frac{\delta V_{\text{r.m.s.}}}{V}=\frac{1}{N^{1/2}}}. \tag{3.62}$$

Since the volume occupied by the N molecules fluctuates between $V+\delta V_{\text{r.m.s.}}$ and $V-\delta V_{\text{r.m.s.}}$, that is between $V(1+N^{-1/2})$ and $V(1-N^{-1/2})$, it follows that the number density of the molecules fluctuates between

$$\frac{N}{V(1+N^{-1/2})}\quad\text{and}\quad\frac{N}{V(1-N^{-1/2})}.$$

Writing n for the mean number density N/V and expanding $(1+N^{-1/2})^{-1}$ and $(1-N^{-1/2})^{-1}$ by the binomial theorem we conclude that the number density of the molecules fluctuates between

$$n\left(1-\frac{1}{N^{1/2}}\right)\quad\text{and}\quad n\left(1+\frac{1}{N^{1/2}}\right). \tag{3.63}$$

By way of example, there are, on average, around 3×10^{16} molecules in a volume of $10^{-9}\,\text{m}^3$ (1 mm^3) of air at s.t.p. Thus eqn (3.62) gives the fractional r.m.s. fluctuation in their volume as 6×10^{-9}. Consequently (see eqn (3.63)), the density fluctuates between $3\times10^{25}\,(1-6\times10^{-9})\,\text{m}^{-3}$ and $3\times10^{25}\,(1+6\times10^{-9})\,\text{m}^{-3}$. On the other hand a volume of $2\times10^{-19}\,\text{m}^3$ (the volume of a cube of side equal to the wavelength of yellow light) contains some 6×10^6 molecules, so that the fractional r.m.s. fluctuation in volume (or density) is now around 4×10^{-4}; that is 0·04 per cent. Such density fluctuations produce fluctuations in the index of refraction of the air and these in turn lead to scattering of light as it passes through a gas. The blueness of the sky arises because blue light is scattered more than red light.

PROBLEMS

3.1. An atom of mass m moving with velocity u makes a head-on elastic collision with another atom of mass m which is at rest. Show that the impact will result in the moving atom coming to rest and the atom which was at rest moving off with velocity $\mathbf{u}$.

3.2. An atom of mass m moving with velocity $\mathbf{u}$ makes an oblique collision with another atom of mass m which is at rest. (An oblique collision is one in which the path followed is not along the line joining the centres of the two atoms.) Show that if the collision is elastic, the paths of the two atoms after impact make an angle of $90°$ with each other. (*Clue:* Remember that momentum is a vector, so the momenta may be added vectorially.)

3.3. Air consists mainly of a mixture of some 75 per cent N_2 and some 23 per cent O_2. What is the ratio of the mean translational kinetic energy of a nitrogen molecule to that of an oxygen molecule? ($M_r(N_2) = 28$, $M_r(O_2) = 32$.)

3.4. A pollen grain of mass 10^{-21} kg is observed to move with an r.m.s. speed of $3 \cdot 5 \, \text{m s}^{-1}$ when it is suspended in water. What is (a) the mean translational kinetic energy of the water molecules? (b) the temperature of the water?

3.5. Fig. 3.14 show an instrument for measuring the speeds of gas atoms. Atoms from the oven O are collimated into a fine beam by being passed through two slits S_1 and S_2. As they travel the distance d they will be bent downwards by the force of gravity. The faster they are travelling the less they will be deflected. By moving the detector D in the vertical plane we can measure how many are deflected by a particular amount, and therefore how many atoms are moving with a particular speed. Derive an expression which relates the speed u of an atom arriving at the detector with the detector setting h and the flight path d.

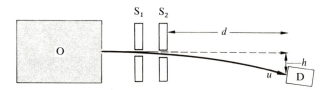

Fig. 3.14. An apparatus for determining the speeds of gas atoms.

3.6. What, roughly, is the r.m.s. speed of an oxygen molecule in the air surrounding you now?

3.7. The temperature reached during a hydrogen bomb explosion is some $10^8 \, \text{K}$. Free nuclei of ^1H and ^2H are present during the explosion. Assuming thermal equilibrium to exist, what is the ratio of the r.m.s. speed of a proton to that of a deuteron (a ^2H nucleus)? Assume that a proton and a neutron have the same mass.

3.8. (a) In one form of gas thermometer the volume of the gas is held constant—the pressure p of the gas being adjusted so as to maintain a constant volume. The pressure is recorded first with the bulb surrounded by water at its triple point, then with the bulb surrounded by the substance whose temperature is required. If the ratio of the second of these pressures to the first is $2 \cdot 00$, what is the temperature of the substance?
(b) In another, more unusual, form of gas thermometer the pressure is held constant while the volume is recorded with the bulb surrounded first by water at its triple point, and then by the substance whose temperature is

required. If the ratio of the second of these volumes to the first is 2·00 what is the temperature of the substance?

3.9. Assuming a hard-sphere repulsive model between a Na^+ and a Cl^- ion (as you did in problem 2.11) make a rough estimate of the melting (or fusion) temperature T_f of solid sodium chloride.

3.10. What, roughly, is de Broglie wavelength of a hydrogen molecule in gaseous hydrogen at a temperature of 50 K? $(A_r(H) = 1.)$

3.11. Repeat exercise 3.3 but using the top track of Fig. 3.8.

3.12. Using your everyday experiences of how long it takes for the smell of a freshly lit cigarette to travel the length of a room, assess whether Brownian motion of the smoke particles (average diameter of 2×10^{-8} m) could be responsible for the transmission of the smell. Air at 300 K has a dynamic viscosity of $1·8 \times 10^{-5}$ kg m^{-1} s^{-1}.

3.13. It is possible to boil a kettle by directing a high-speed stream of gas at the base of the kettle. Approximately what speed air blast (U) is necessary to ensure that water in the kettle will boil? *Clue:* The atoms in the air blast have a speed of approximately $U + (\overline{u^2})^{1/2}$ where $(\overline{u^2})^{1/2}$ is their r.m.s. thermal speed.

3.14. Put an upper limit to the temperature that the skin of a supersonic aircraft flying at twice the speed of sound might reach.

3.15. What is (a) the total kinetic energy of all the atoms present in 5 mol of a perfect gas at a temperature of 150 K, (b) the kinetic energy of all the atoms present in 5 mol of a monatomic solid at a temperature of 150 K, and (c) the potential energy of all the atoms present in 5 mol of a monatomic solid at 150 K (relative to their potential energy near 0 K)?

3.16 What is the vibrational period of atom i of Fig. 3.10(a) assuming that it has a mass of 2×10^{-26} kg and that the interatomic force between i and j (and between i and k) has the form shown in Fig. 2.7? Hence deduce the vibrational frequency $(1/T_v)$ and the circular frequency.

3.17. An object of mass 0·2 kg is attached to the end of a vertical spring which extends by 5×10^{-2} m as a consequence. The object is removed and replaced by another of mass 4 kg which is then pulled down by 3×10^{-2} m below its equilibrium position. What is (a) the period and (b) the circular frequency of the resulting oscillation? Assume $g = 10$ m s^{-2}.

3.18. An object of mass 2 kg, resting on a frictionless surface, is attached to the end of a horizontal spring with a force constant of 5×10^5 N m^{-1} which (like certain car suspension springs) can be both compressed and extended. The object is pulled sideways so as to extend the spring by 7×10^{-2} m and it is then released. What is (a) the period of vibration of the mass, (b) the maximum potential energy stored in the spring, (c) the mean potential energy stored in the spring, (d) the mean kinetic energy of the mass, and (e) the r.m.s. displacement of the mass?

3.19. An object of mass 0·3 kg is hung from the end of a vertical spring which thereby extends by 2×10^{-2} m. Assuming that the mass is at 'rest' what is its true r.m.s. displacement?

3.20. The moving part of an analogue current meter consists of a coil of wire (located in a magnetic field) to which is attached a pointer. The coil is able to rotate about a vertical axis against the opposing torque of a hair spring. Random motions of the surrounding air molecules produce torques which ensure that the pointer is never truly at rest. Assuming that the spring exerts a constant restoring torque of 10^{-11} N m rad^{-1}, what is the r.m.s.

displacement of the end of the pointer (length 5×10^{-2} m) if the instrument is at 300 K?

3.21. The elastic (or Young's) modulus of a bar is defined by $E = $ stress/strain $= (F/A)/(\delta l/l)$, where $\delta l/l$ is the fractional change in the length of the bar produced by a force F acting uniformly over its cross-sectional area A. Show that thermal fluctuations will cause the length of a bar (of mean length L) to have an r.m.s. fluctuation from L of $\Delta L_{\text{r.m.s.}} = (kTL/EA)^{1/2}$ at temperature T. Deduce the r.m.s. fluctuation at 300 K expressed as a fraction of L, for a bar of length 2×10^{-2} m with a cross-sectional area of 10^{-6} m^2 and made of a material with $E = 7 \times 10^{10}$ Pa.

3.22. Air at s.t.p. contains 2.7×10^{25} molecules m^{-3}. What is the r.m.s. fluctuation in the number of atoms contained within a cube of air at s.t.p. whose side is equal to the wavelength of blue light (around 4×10^{-7} m)? Also express your answer as a percentage of the mean number.

4. The Maxwell–Boltzmann distribution

So far we have been content to talk of the mean kinetic energy $\frac{1}{2}m\overline{u^2}$ of the atoms in a gas without asking how the energies are spread about the mean. In this chapter we will address ourselves to the problem of finding the precise form of the energy distribution function, that is, the function which tells us how many atoms in a gas have energies lying within unit range of any specified energy. We will also aim to find the related speed distribution function, which tells us the number of atoms with speeds lying within unit range of any specified speed.

We will start by looking at how the energy distribution function can be obtained by means of an experimental simulation. Since this is necessarily based on a small number of particles we will attempt to perform the simulation on paper with the hope that, once we have learnt the ground rules, it can be extended to deal with many particles. We will find that as the number of particles keeps increasing—there are, you will recall, 6×10^{23} in a mole—so the procedures become ever more time-consuming. We will be forced into adopting more formal procedures.

4.1. Distribution functions and mean values

The distribution function

We will start by considering what sort of graph should be used to show just how many atoms of a gas possess a particular speed. Our first hunch might be to plot a graph of *number* against *speed*. Such a plot would, however, be uninstructive as, at most, only one gas atom will have a speed of, say, $401 \cdot 392$ m s^{-1} precisely. What we must clearly do instead is to plot a histogram (as in Fig. 3.5) where we plot the fraction of the atoms possessing speeds lying within certain fixed intervals, say 50 m s^{-1}. The greater the amount of data available, the smaller can the intervals become. This is illustrated schematically in Fig. 4.1(a), where the fraction $\delta N/N$ of the population of gas atoms whose speeds lie within a range δu ($= 10$ m s^{-1}) is plotted against u. When dealing with large populations it actually becomes more convenient to plot not $\delta N/N$ against u, but $\delta N/(N \, \delta u)$ against u. By way of example, if out of a total population of 10^{20} gas atoms, 2×10^{18} have speeds lying between 300 and $300 \cdot 5$ m s^{-1} then $\delta N/(N \, \delta u) = 2 \times 10^{18}/(10^{20} \times 0 \cdot 5 \text{ m s}^{-1}) = 0 \cdot 04$ m^{-1} s. In the limit as $\delta u \rightarrow 0$

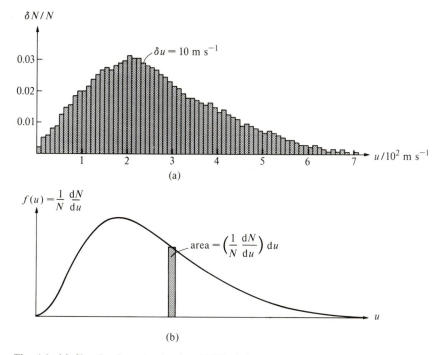

Fig. 4.1. (a) Showing how the fraction $\delta N/N$ of the atoms of a gas with speeds within a range δu (here $10\ \mathrm{m\,s^{-1}}$) of u varies with u. (b) The speed distribution function of a gas.

this ratio becomes

$$f(u) = \frac{1}{N}\frac{dN}{du}\ , \tag{4.1}$$

where $f(u)$ is called the *speed distribution function* of the gas. This is shown plotted in Fig. 4.1(b) in the case of a near-infinite population. Put into words, $f(u)$ tells us the fraction of the population per unit range of speed about u.† Cross-multiplying eqn (4.1) we see that the fraction of atoms with speeds between u and $u + du$ is

$$\frac{dN}{N} = f(u)\,du. \tag{4.2}$$

† Some authors define a speed distribution function as dN/du. Thus to them it defines the number, rather than the fraction, of atoms per unit range of speed at u. As such, its value depends on the total population (unlike eqn (4.1)). When we omit the word *function*, as in 'speed distribution', we will usually be referring to numbers (here dN/du), not fractions.

It follows from this that the area under Fig. 4.1(b) is given by

$$\int_0^\infty f(u)\,du = \frac{1}{N}\int_0^\infty dN = 1. \tag{4.3}$$

In an exactly similar manner the energy distribution function $f(E)$ of a gas can be defined as the fraction of the atoms per unit range of energy about E. That is

$$f(E) = \frac{1}{N}\frac{dN}{dE}. \tag{4.4}$$

Mean values

We will now look at how the concept of a distribution function may be used in practice to evaluate, say, the average speed of a gas atom. By average speed $\bar{u}$, we mean, of course, the sum of the speeds of all the gas atoms divided by the total population. Since the number of gas atoms with speeds between u and $u+du$ is, from eqn (4.2), given by $dN = Nf(u)\,du$, the combined speeds of this group of atoms is $u \times Nf(u)\,du$. Hence the sum of the speeds of all N atoms is $\int_0^\infty uNf(u)\,du$ and so

$$\bar{u} = \frac{N\int_0^\infty uf(u)\,du}{N}$$

$$\bar{u} = \int_0^\infty uf(u)\,du. \tag{4.5}$$

In general, we can evaluate the mean value $\bar{Q}$ of any function Q *provided* it can be expressed as a function of the variable (here u) in which the distribution function is given. Thus (cf. eqn (4.5))

$$\bar{Q} = \int_0^\infty Q(u)f(u)\,du = \frac{1}{N}\int_0^\infty Q(u)\frac{dN}{du}\,du. \tag{4.6}$$

Likewise, if Q is specified as a function of energy E and if the energy distribution function $f(E)$ is given;

$$\bar{Q} = \int_0^\infty Q(E)f(E)\,dE = \frac{1}{N}\int_0^\infty Q(E)\frac{dN}{dE}\,dE. \tag{4.7}$$

Eqns (4.6) and (4.7) will prove very useful later in the chapter.

Exercise 4.1

The height distribution of a (somewhat bizarre) population can be written as $f(h) = a(h - h_0)$ over the range $h_0 \le h \le h_1$, where $a = 0.7 \text{ m}^{-2}$ and $h_0 = 0.3 \text{ m}$. When $h_0 > h > h_1$, $f(h) = 0$.

(a) Sketch $f(h)$. (b) What is the value of h_1? (c) What is the mean height of the population? (d) Assuming that the mass m of each inhabitant is related to his or her height by $m = ch^2$, where $c = 30 \text{ kg m}^{-2}$, what is the mean mass of an inhabitant?

Solutions. (a) Your plot should be a straight line of gradient $a(= 0.7 \text{ m}^{-2})$ which meets the $f(h) = 0$ axis at $h = 0.3 \text{ m}$. The value of $f(h)$ must be zero at $h_0 > h > h_1$. (b) The value of h_1 is determined by using the fact that the total area under the height distribution curve is unity (c.f. eqn (4.3)). Hence

$$\int_0^\infty f(h) \, dh = \int_{h_0}^{h_1} a(h - h_0) \, dh = 1$$

$$\tfrac{1}{2}a(h_1^2 - h_0^2) - ah_0(h_1 - h_0) = 1$$

$$h_1 = h_0 + (2/a)^{1/2} = 1.99 \text{ m}.$$

(c) By analogy with eqn (4.5),

$$\bar{h} = \int_0^\infty hf(h) \, dh = \int_{h_0}^{h_1} ha(h - h_0) \, dh$$

$$\bar{h} = \tfrac{1}{3}a(h_1^3 - h_0^3) - \tfrac{1}{2}ah_0(h_1^2 - h_0^2).$$

Substituting the known values of a, h_0, and h_1 gives $\bar{h} = 1.43 \text{ m}$.
(d) By analogy with eqn (4.6),

$$\bar{m} = \int_0^\infty ch^2 f(h) \, dh = \int_{h_0}^{h_1} ach^2(h - h_0) \, dh$$

$$\bar{m} = \tfrac{1}{4}ac(h_1^4 - h_0^4) - \tfrac{1}{3}ach_0(h_1^3 - h_0^3).$$

Substituting the known values of a, c, h_1, and h_0 gives $m = 65.8 \text{ kg}$.
Comment. Any distribution function, such as $f(u)$, must satisfy eqn (4.3). We shall frequently make use of this condition in evaluating unknown constants which occur in the distribution function.

4.2. An air-table simulation

Simulating a gas

A particularly convenient simulator of a gas uses 'pucks' on an air-table, the boundary walls of which are vibrated back and forth by means of a motor. In this equipment the pucks—short cylindrical discs—float on a cushion of air which is forced up through holes in the table. A practical advantage of pucks over steel balls (as used in the simulator described in §3.2) is that they can easily be labelled A, B, C, etc. In so doing we are not, of course, saying that each atom in a gas actually carries a label. We label them because in classical physics each atom of a gas *is* unique; that is, identifiable.

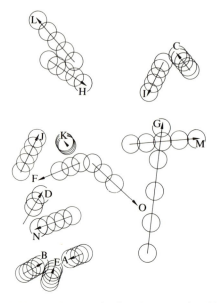

Fig. 4.2. A tracing made from a photograph of pucks on an air-table lit by a stroboscope. During the time exposure pucks F and O collided, as did F and D. The vectors indicate the velocity of each named puck.

To measure the speeds of the pucks—a necessary preliminary to measuring their energies—we may take a time-exposure photograph of the simulator while it is being lit by a stroboscope. Fig. 4.2 is a tracing made from an actual photograph obtained by this technique. During the time that the shutter was open, the strobe-light flashed five times. The separation of two consecutive images of a puck tells us how far that puck has travelled between flashes and so this separation, and therefore the total length of the track, is a measure of the speed of that puck. Knowing the speed u of each puck it would now be a simple matter to work out the energy $\frac{1}{2}mu^2$ of each puck. However, just to measure up the speeds, to work out the energies, and to plot an appropriate histogram would, at this stage, throw away all the information we have about the *direction* in which the pucks are moving. Before long we will wish to distinguish between particles having the same speed but different velocities.

Recording the information

To preserve the full information contained within the photograph of the air-table we can draw in velocity vectors, as shown in Fig. 4.2. (The sense of movement of a puck may be obtained by reducing the camera's aperture during the time exposure; the puck then moves in the sense in which the image intensity on the negative decreases.) Where a collision has appeared in the photograph the velocities shown in Fig. 4.2 have been recorded before or after the event; not both. On bringing the velocity vectors of Fig. 4.2 to a common origin we obtain Fig. 4.3(a). To find the number of pucks possessing a kinetic energy within a particular range of energies a series of rings is superimposed on the vector representation as shown in Fig. 4.3(b). If any one ring has a radius of u, say, and the next one out has a radius of $u + \delta u$ then all those named pucks whose vector tips lie in the annulus between radii u and $u + \delta u$ will have kinetic energies within a range of energy

$$\delta E \approx \left(\frac{\mathrm{d}E}{\mathrm{d}u}\right) \delta u = mu\,\delta u \tag{4.8}$$

about the energy

$$E = \tfrac{1}{2}mu^2 \tag{4.9}$$

in the case of small δu. In constructing the ring system of Fig. 4.3(b) δE has been made the same in each annulus. Thus the number of named tips in the inner circle gives the number of pucks with energies within the

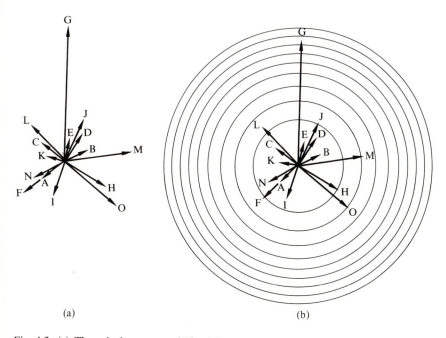

(a)   (b)

Fig. 4.3. (a) The velocity vectors of Fig. 4.2 transposed to a common origin. (b) Showing rings of constant energy superimposed on the vectors.

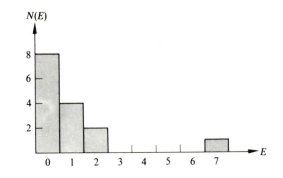

Fig. 4.4. Shows the number $N(E)$ of pucks of Fig. 4.2 with an energy E.

range 0 to 1 δE; the number of tips falling within the first annulus gives the number of particles with energies between 1 δE and 2 δE, etc. Counting up the number of distinct vector tips which fall within each annulus gives the energy distribution shown in Fig. 4.4. Here the energy of those pucks whose velocity vector tips fall within the inner circle is taken as 0; the energy of those whose vector tips fall within the first annulus as 1, and so on.

The average distribution

So far we have only learnt about the particular energy distribution that was present when one particular photograph—that of Fig. 4.2—was taken.

To obtain a feel for the average energy distribution the simulated gas must be photographed many more times and the results obtained from individual photographs then averaged. Figs. 4.5(a)–(c) shows the results obtained on carrying out the same procedures which led to Fig. 4.3(b) on three more randomly-selected photographs of the table. Here, instead of drawing in the velocity vectors in full, only the terminating points of the vector heads are shown. To obtain the average number of named pucks lying within any energy interval we simply count the total number of tips (i.e. the number of labels A, B, C, etc.) falling within the appropriate annulus in all the photographs' representations and then divide by the number of representations concerned. On carrying out this procedure with the four velocity distributions contained in Fig. 4.3(b) and in Figs. 4.5(a)–(c), the average energy distribution shown in Fig. 4.6 was obtained. This indicates that in a two-dimensional gas there would be a steady fall-off in the number of atoms possessing a specified energy (within a unit range of energy) as the energy is increased. Real gases are, however, three-dimensional!

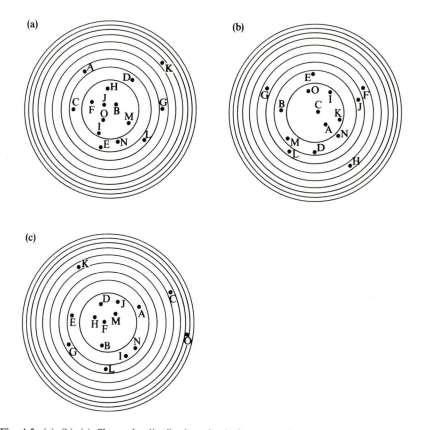

Fig. 4.5. (a), (b), (c). Shows the distribution of velocity vectors from three randomly-chosen photographs of the air-table. The dot adjacent to each letter indicates that the velocity of that puck is represented by the vector joining the centre of the ring system to the dot in question.

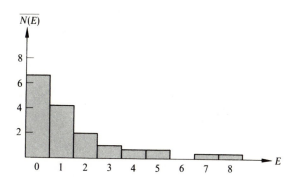

Fig. 4.6. Showing the average number $\overline{N(E)}$ of pucks with an energy E. The average is taken over the representations of Fig. 4.3(b) and of Figs. 4.5(a), (b), and (c).

Simulating a three-dimensional gas

A randomly-shaken (transparent) box of steel balls could simulate the gas. Stereoscopic photographs of the box taken with stroboscopic illumination would allow the velocity of each ball to be determined. As in two dimensions, vectors erected along the tracks would then be transposed to a common origin. This time, however, instead of drawing in annuli we would draw in spheres, as in Fig. 4.7. If the radii of a spherical shell are u and $u + \delta u$ then all those named vector tips which fall within the shell of thickness δu will have energies in the range $\delta E = mu\,\delta u$ about $E = \frac{1}{2}mu^2$.

Technical problems aside, gravity makes the simulation very difficult to achieve. If the box has a height h then a ball of mass m will lose energy mgh in climbing through a height h, where g is the local acceleration due to gravity. To ensure that $mgh \ll \frac{1}{2}m\overline{u^2}$ the box must be shaken very violently. Our three-dimensional simulation remains a thought experiment.

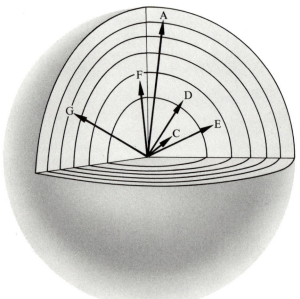

Fig. 4.7. A hypothetical distribution of velocity vectors for a three-dimensional simulated gas.

4.3. Predicting the photographs

Even with the highly practical two-dimensional simulator all we can hope to do is to *sample* the behaviour of the gas. However, if pictures like those of Fig. 4.5 could be generated on paper we would be able to do more than sample the gas at random moments; we could generate every conceivable picture. Furthermore, if the technique works in two dimensions we may be able to extend it to three dimensions. We will therefore attempt to generate the pictures without further resort to experiment.

The ground rules

The sequence of pictures shown in Fig. 4.3(b) and Fig. 4.5 is characterized by a constant number of pucks sharing out a constant *total* energy. What changes from picture to picture is the *way* that the energy is shared out. We can therefore assert that the ground rules to be used in predicting the photographs are:

    (1) Each atom is unique (and so can be labelled A, B, C, etc.).
    (2) The total number of atoms is constant.
    (3) The total energy of the atoms is constant.

An additional rule

Were we to attempt to use these three rules alone we would very soon run into problems. We can see how the problems arise by considering the ways that, say, four atoms (A, B, C, and D) might share out five units of energy. In so doing we will use the ring system shown in Fig. 4.8. As before, atoms whose velocity vector tips fall within the innermost circle are supposed to have zero energy, those whose vector tips fall within the first annulus have unit energy, and so on.

Clearly, the arrangements shown in Fig. 4.8(a), (b) are allowed ways of sharing out the five units of energy. Equally clearly, they represent

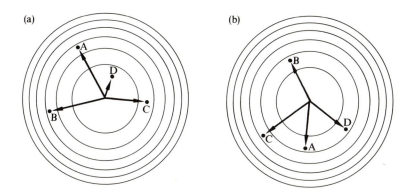

Fig. 4.8. (a) and (b). Two different arrangements of velocity vectors for four atoms of a two-dimensional gas sharing out five units of energy.

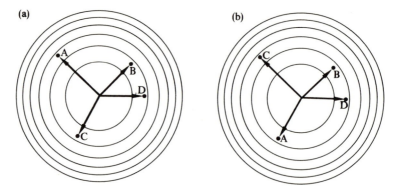

Fig. 4.9. (a) and (b). Two different arrangements of velocity vectors of four atoms of a two-dimensional gas sharing our five units of energy.

physically distinct situations. Likewise, because the atoms are unique, Figs. 4.9(a) and (b) represent different situations; atoms A and C are each behaving differently in the two cases. Now look at the two pictures shown in Fig. 4.10. Careful scrutiny will reveal that although the speed u of A, and therefore its energy $\frac{1}{2}mu^2$, is the same in both cases, its direction of motion, and therefore its momentum mu, is different in the two pictures. Through just how small a distance can the point of the vector tip be moved to generate a new picture? If the answer is through an infinitesmal distance, there will be an infinity of pictures. To keep the number of pictures finite we must introduce yet another rule; one that will allow us to decide whether two arrangements of vectors, two pictures, are different. Operationally, here is what this new (fourth) rule states:

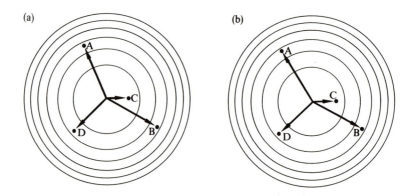

Fig. 4.10. The velocity (but not the speed) of atom A is slightly different in (a) and (b).

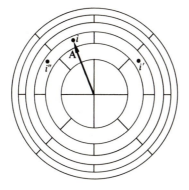

Fig. 4.11. Showing the division of two-dimensional velocity space into cells of equal area. Moving the tip of the velocity vector of atom A from point i to point i' does generate a new picture; moving it from i to i'' does not generate a new picture.

(4) The space in which the velocity vectors are drawn (*velocity space*) is to be carved up into *cells* of equal 'size'; in two dimensions the cells are to have equal area; in three dimensions the cells are to have equal volume. Changing the position of a vector tip while keeping the tip within a cell does *not* generate a new picture. Changing the position of the vector tip so as to move it into a new cell *does* generate a new picture.

There is nothing in the rule about the shape of the cells, provided that they have equal size throughout velocity space and do fill the space. Fig. 4.11 shows the carving-up procedure applied in two dimensions. By way of illustration of the rule, if the vector representing the velocity of particle A, which at present has its tip at point i in Fig. 4.11, is moved to i' we *do* obtain a new picture. If, however, the tip only moves to point i'' within the same cell as i, while the other vectors are kept fixed, we *do not* obtain a new picture. Although the rule may appear somewhat arbitrary at this stage our subsequent development will show that the final form of the energy distribution does not depend on the size of the cell which may therefore (at least according to classical physics) be made as small as we wish.

Having established the rules we may now generate the complete set of distinct pictures in which the total energy of all the named particles is a constant.

4.4. Manual sorting in two dimensions

We can readily show that if the energy rings in two dimensions are divided up into cells of equal area, each energy ring contains the same number of cells. If two adjacent circles drawn in velocity space have radii u and $u + \delta u$, corresponding to energies E and $E + \delta E$, then the area of the enclosed ring is $2\pi u\, \delta u$ (that is, the length $2\pi u$ multiplied by the thickness δu). If each cell has constant area A_0 then the number $\mathcal{N}_2(E, A_0)$ of cells in the ring whose radii correspond to energies E and $E + \delta E$ is given by

$$\mathcal{N}_2(E, A_0) = \frac{2\pi u\, \delta u}{A_0}$$

or, substituting $u\, \delta u = \delta E/m$ from eqn (4.8),

$$\mathcal{N}_2(E, A_0) = \frac{1}{A_0}\left(\frac{2\pi}{m}\right)\delta E. \tag{4.10}$$

Therefore if δE is constant throughout the ring system each ring will contain the same number of cells (four in the case of Fig. 4.11).

Before attempting to generate the complete set of distinct pictures it is convenient to transform the velocity-space diagram of Fig. 4.11 into the rectangular diagram of Fig. 4.12. Thus if atom A, for example, has two units of energy (which means, of course, that the tip of its velocity vector falls in the second ring in Fig. 4.11) it is shown in row 2 of Fig. 4.12. The actual cell in which A resides depends, of course, on the cell in which its velocity vector tip ends in Fig. 4.11.

We will now attempt to answer the following questions (the first of which has already been posed on p. 97):

1. How many distinct arrangements of a given number of atoms (say 4; A, B, C, and D) can be made such that the total energy of each arrangement sums to a given number (say, 5)?
2. Averaged over all the distinct arrangements or 'pictures' (equally weighted), what is the mean number of atoms in a two-dimensional gas possessing a given energy?

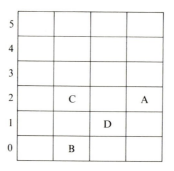

Fig. 4.12. An alternative representation of the cells of Fig. 4.11.

If you were to try to answer the first question, you would discover that even with only four cells per level there are a daunting number of distinct arrangements. We first try a simpler problem.

Exercise 4.2

Considering only a single vertical column of cells (such as one of the columns of Fig. 4.12) draw out all the distinct ways that four atoms A, B, C, and D can share out a total of five units of energy, and hence work out the mean number of atoms with a particular energy. You are to assume that (see rule (4)) rearranging labels within a given cell does not produce a new picture; CA, for example, is the same as AC. In carrying through this exercise you may just find it helpful to think of A, B, C, and D as books being sorted among various levels of a bookshelf so that their total (gravitational potential) energy remains constant. Here the cells would become 'modules' of a given shelf.

Solution. There are 56 distinct ways for four atoms to share out five units of energy. These are shown in Fig. 4.13. Here the arrangement of type (5, 0, 0, 0)

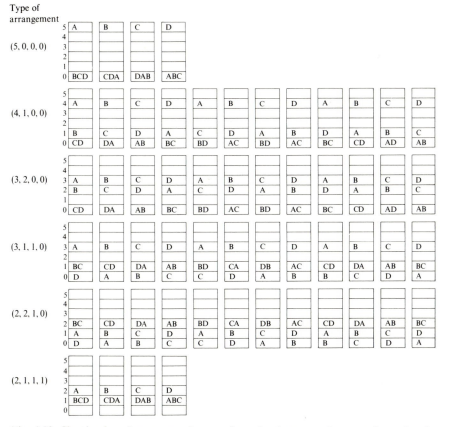

Fig. 4.13. Showing how four atoms of a two-dimensional gas can share out five units of energy.

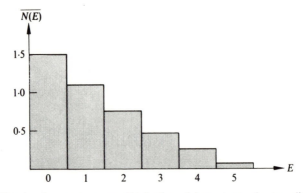

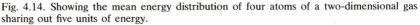

Fig. 4.14. Showing the mean energy distribution of four atoms of a two-dimensional gas sharing out five units of energy.

means that one atom has five units of energy while the other three have zero energy (hence the three zeros). Similarly, (4, 1, 0, 0) means one atom has four units, one has one unit, and two have zero energy, and so on for the other types of arrangement. To find the mean number of atoms with a particular energy we merely count the total number of appearances of an atom along a particular row and divide by the total number of pictures (56). As an example, the total number of atoms which appear in level 2 is given by 4 atoms appearing in the (2, 1, 1, 1) arrangement, by 24 atoms in the (2, 2, 1, 0) arrangement, and by 12 atoms in the (3, 2, 0, 0) arrangement (see Fig. 4.13). So the mean number of atoms with 2 units of energy, averaged over all 56 pictures, is $(4 + 24 + 4)/56 = 0.71$. Working out the corresponding numbers at all the other energies and plotting the results as a histogram leads to Fig. 4.14.

We must now return to the original problem posed by Fig. 4.12 (that is, by Fig. 4.11) with its four cells at each level. We shall discover that while the complete set of arrangements is much larger than 56, a histogram *identical* to Fig. 4.14 will emerge from the calculation. For *each* of the original 56 arrangements (an example of which is given in Fig. 4.15(a)), 3 *new* arrangements can be generated by moving atom A horizontally, as shown in Fig. 4.15(b), (c), and (d), giving 4 arrangements in all. By moving both atoms A and B we get 16 possible arrangements, and by moving all four atoms we get $4 \times 4 \times 4 \times 4 = 256$ arrangements for *each* arrangement generated with a single column of cells. This means that when we evaluate the number of atoms along a particular row—say, row three—the original number of appearances (24) is multiplied by 256. However, in evaluating the mean number of atoms in row three we divide the total number of appearances (24×256) by the total number of arrangements (now 56×256). The result is that the histogram of Fig. 4.14 remains unchanged. Since the form of the distribution is not altered by scaling up the number of cells per ring by a common factor it is clear that the fineness of division of the energy rings is irrelevant to the form of the

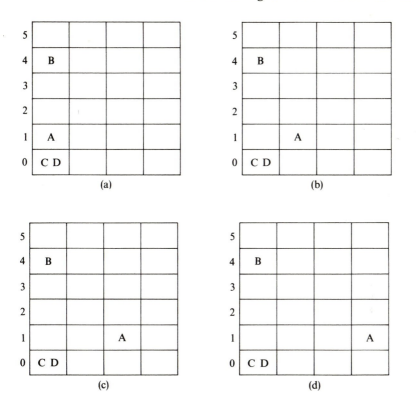

Fig. 4.15. (a) One of the original arrangements of Fig. 4.13. The three new arrangements shown in (b), (c), (d) result from the movement of book A in Fig. 4.15(a).

energy distribution. The distribution of Fig. 4.14 obtained with each energy annulus as a single cell will therefore be the same as would be obtained with the finest subdivision imaginable. This justifies our original assertion (p. 99) that the form of the energy distribution does not depend on the size of the cell.

4.5. Manual sorting in three dimensions

We will now attempt to extend the technique so that it can be used to predict the complete set of arrangements of the velocity vectors in a three-dimensional gas (one sample set of which is shown in Fig. 4.7). By way of example, we will concentrate on the problem of four atoms of a three-dimensional gas sharing out five units of energy. The question to be answered is this: On average, how many atoms have 0, 1, 2, 3, 4, or 5 units of energy?

The first step is to ascertain how many cells of equal volume V_0 can be put into a shell in velocity space between two spheres of radii u and $u + \delta u$ (such as between two spheres shown schematically in Fig. 4.7). This is, of course, analogous to finding how many cells can be put into an annulus in two-dimensional velocity space. The volume δV of a shell contained between two spheres of radii u and $u + \delta u$ is, to a close approximation, given by the surface area of the inner sphere $(4\pi u^2)$ multiplied by the thickness of the shell (δu):

$$\delta V = 4\pi u^2 \, \delta u. \tag{4.11}$$

Since $E = \frac{1}{2}mu^2$, it follows that $u^2 = 2E/m$ and that $\delta E = mu \, \delta u = (2mE)^{1/2} \, \delta u$. So

$$\delta u = \frac{\delta E}{(2mE)^{1/2}}. \tag{4.12}$$

Thus eqn (4.11) becomes

$$\delta V = \left(\frac{32\pi^2}{m^3}\right)^{1/2} E^{1/2} \, \delta E. \tag{4.13}$$

Dividing δV by the cell volume V_0 gives the number $\mathcal{N}_3(E, V_0)$ of cells of volume V_0 in the spherical shell between energies E and $E + \delta E$ as

$$\mathcal{N}_3(E, V_0) = \frac{1}{V_0}\left(\frac{32\pi^2}{m^3}\right)^{1/2} E^{1/2} \, \delta E \tag{4.14}$$

or as

$$\mathcal{N}_3(E, V_0) = CE^{1/2} \, \delta E, \tag{4.15}$$

where C is a constant.

Notice how in three dimensions $\mathcal{N}_3(E, V_0)$ is not constant in all shells of constant δE but varies as $E^{1/2}$. This means that in setting up the

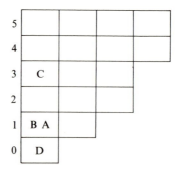

Fig. 4.16. The energy-level diagram of a three-dimensional gas.

corresponding cell diagram (akin to that shown in Fig. 4.15 for a two-dimensional gas) the number of cells at each energy must be proportional to $E^{1/2}$. Such a diagram is shown in Fig. 4.16. (This was obtained by letting $C = 2$ and $\delta E = 1$ in eqn (4.15) and rounding $\mathcal{N}_3(E, V_0)$ down to the the nearest integer: An exception is made at $E = 0$ where $\mathcal{N}_3(0, V_0)$ is taken as unity.)

Exercise 4.3

Using the energy-level diagram shown in Fig. 4.16, find the total number of distinct ways that four atoms, A, B, C, and D, can share out five units of energy. Hence deduce the mean number of atoms with energies of 0, 1, 2, 3, etc. units.

Procedures. Rather than drawing out all the distinct arrangements in a haphazard fashion (there are 676 such arrangements!) it is easier to proceed—as you did in exercise 4.2—by using the first column of cells alone to find the permitted *types* of arrangement and the associated number of arrangements of each type. Finally you can make use of the known number of cells at each level to deduce the total number of distinct arrangements of the atoms.

Solution. The only permitted types of arrangement (those totalling to five units of energy) are those shown in Fig. 4.13. This figure also shows the required number of arrangements of each type obtained using only the first column of cells in Fig. 4.16. For example, for the $(3, 1, 1, 0)$ type there are 12 distinct arrangements.

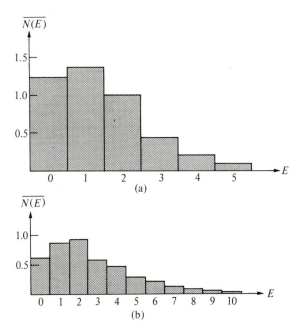

Fig. 4.17. Showing the mean energy distribution of four atoms of a three-dimensional gas sharing out (a) five units of energy, (b) ten half-units of energy.

We must next work out how many arrangements ('pictures', if you like) may be formed by sorting the atoms among the cells existing at each level. We see that there are now 3 cells for an atom at level 3 instead of the original single cell. This means that the twelve original arrangements of the $(3, 1, 1, 0)$ type shown in Fig. 4.13 becomes 12×3 arrangements as a result of adding the extra cells at level 3. Adding the extra cell at level 1 multiplies the 12 original arrangements by 4 since atom A, say, may be in either of two cells, as may atom B. There is no additional freedom at level 0 as there is still only one cell there. The total number of pictures of the $(3, 1, 1, 0)$ type therefore comes to $(3 \times 4 \times 1)$ multiplied by the original number of arrangements (12) obtained using only a single column of cells, giving 144 distinct pictures of the $(3, 1, 1, 0)$ type in all. The total number of pictures for each of the other types can be worked out similarly.

As in the two-dimensional problem, we obtain the mean number of atoms appearing at a particular level, that is the mean number of atoms with a particular energy, by totalling the number of atom appearances at this level and then dividing by the total number of pictures involved. Fig. 4.17(a) shows the resulting energy-distribution histogram.

Extending the procedures

We are starting to get some intimations that there is a peak in the energy distribution of a three-dimensional gas that is absent in the energy distribution of a two-dimensional gas (compare Figs. 4.17(a) and 4.14). In view of the rough-and-ready way in which velocity space was carved up into cells in Fig. 4.16, it would be unwise to be too dogmatic about their being a peak in Fig. 4.17(a). However, when the calculations are repeated using an energy-level diagram in which successive levels differ by half a unit of energy the peak in the histogram is still there (Fig. 4.17(b)).

As the energy levels are divided ever more and more finely the number of *types* of arrangement increases. If, for example, three atoms share out 4 units of energy, the *types* of arrangement are; $(2, 1, 1)$, $(3, 1, 0)$, $(2, 2, 0)$, and $(4, 0, 0)$. If the three atoms share out 8 half-units of energy (again totalling to 4 units) the *types* of arrangement are; $(4, 3, 1)$, $(5, 2, 1)$, $(3, 3, 2)$, $(4, 2, 2)$, $(5, 3, 0)$, $(6, 1, 1)$, $(6, 2, 0)$, $(7, 1, 0)$, $(4, 4, 0)$, and $(8, 0, 0)$. What is more important than the increase in the number of types of arrangement is the fact that *one type* of arrangement becomes relatively ever more and more important as the energy levels become ever more and more finely divided. In the example just cited, where the three atoms share out eight half-units of energy, it is the $(4, 3, 1)$ type of arrangement which makes the single biggest contribution in evaluating the mean energy distribution. In a sufficiently finely divided system containing very many atoms there is a *single* type of arrangement which contains an overwhelming number of pictures—so much so that all other types of arrangement can be disregarded in evaluating the mean number of atoms with a particular energy.

Using these manual sorting procedures with ever increasing numbers of atoms and with ever increasing fineness of energy levels becomes ever more and more tedious. We will therefore try to predict theoretically the mean energy distribution of a large population in which the energy levels are (virtually) infinitely finely divided. This will introduce you to the ideas of statistical mechanics.

4.6. The energy distribution function

The density of states

The first step is essentially to redraw the rings in two dimensions (Fig. 4.5) and the shells in three dimensions (Fig. 4.7) so as to reduce the energy spread δE encompassed within a single ring or shell. Reducing δE means that the difference δu in radii of adjacent rings or shells will also shrink according to $\delta u = \delta E/mu$ (eqn (4.8)). Proceeding then to the limit $\delta u \to 0$, we see that eqn (4.10) may be written as

$$\mathcal{N}_2(E, A_0) = g_2(E) \, dE, \tag{4.16}$$

where

$$g_2(E) = \frac{1}{A_0} \left(\frac{2\pi}{m}\right) \tag{4.17}$$

and that eqn (4.14) may be written as

$$\mathcal{N}_3(E, V_0) = g_3(E) \, dE, \tag{4.18}$$

where

$$g_3(E) = \frac{1}{V_0} \left(\frac{32\pi^2}{m^3}\right)^{1/2} E^{1/2}. \tag{4.19}$$

It follows from eqn (4.16) that $g_2(E)$ ($= \mathcal{N}_2(E, A_0)/dE$) is equal to the number of cells per unit range of energy at energy E in a two-dimensional gas. Likewise, $g_3(E)$ is the number of cells per unit range of energy at energy E in a three-dimensional gas. In formal statistical mechanics $g(E)$ is called the *density-of-states* function, the word *density* arising because $g(E)$ is the number of cells per unit range of energy.

The full line in Fig. 4.18 is a graphical representation of eqn (4.19) with $g_3(E)$ plotted (somewhat unorthodoxly) along the abscissa and E along the ordinate. We can immediately see similarities between this figure and our crude energy-level diagram for a three-dimensional gas (Fig. 4.16). In particular, if we arbitrarily set $dE = 1$ in eqn (4.18) (this unit of energy must be such that $dE/E \ll 1$) we see that the abscissa of Fig. 4.18 tells us the number of cells to be drawn in at energy E. For future convenience the subscript i is used to denote the conditions pertaining to this level.

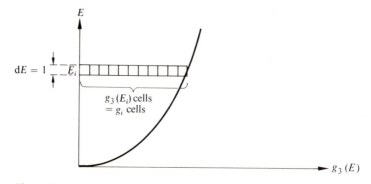

Fig. 4.18. The density-of-states function $g_3(E)$ for a three-dimensional gas.

Arranging the atoms

Bearing in mind the procedures we adopted in discussing how four atoms of a three-dimensional gas can share out five units of energy we must somehow now write down or 'draw out' the full range of pictures in which the energy of the gas atoms totals to some specified value. We must then compute the average number of atoms at each level; that is, the average number of atoms possessing energies within unit range of E.

We begin by first calculating the number of pictures, say W, which are obtained for a given *type* of arrangement; one with n_i atoms at level i (each with an energy E_i), n_j atoms at level j (each with an energy E_j), etc. Adopting the language of statistical mechanics we shall use the word *macrostate* for 'type of arrangement' and *microstate* for 'picture'. (Thus in this new language we can say that the top row of Fig. 4.13 shows all four microstates corresponding to the $(5, 0, 0, 0)$ macrostate.) So let us therefore set up a particular macrostate, where the number n_i of atoms at each level i is going to be specified, and deduce the associated number of microstates, W.

Fig. 4.19 shows a blank row of cells into which we are going to place the specified number n_i of atoms. These atoms must come from the total number of atoms N in the gas. The number of ways that n_i atoms can be

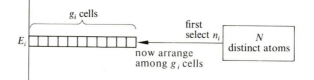

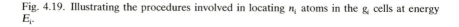

Fig. 4.19. Illustrating the procedures involved in locating n_i atoms in the g_i cells at energy E_i.

selected as *would-be occupants* for level i—atoms still to be placed at this level—is given by $_NC_{n_i}$ where

$$_NC_{n_i} = \frac{N!}{n_i!(N-n_i)!}.$$ (4.20)

Having selected the n_i would-be occupants these must now be distributed among the available cells at level E_i. The first of these n_i can go into any of the g_i (where $g_i = g_3(E_i)$) cells at this level. The same is true of the second atom, of the third atom, and so on for all n_i atoms. Therefore the number of distinct ways of arranging the n_i atoms selected for level i is $(g_i)^{n_i}$. But there are, as we have seen, $_NC_{n_i}$ ways of selecting these n_i atoms from our population. So the total number of possible distinct arrangements of atoms on level i is given by

$$W_i = {_NC_{n_i}} \times (g_i)^{n_i}$$

$$W_i = \frac{N!}{n_i!(N-n_i)!}(g_i)^{N_i}.$$ (4.21)

In selecting the n_j atoms for level j we have only $(N-n_i)$ atoms left from which to select n_j. This selection may be made in $_{(N-n_i)}C_{n_j}$ ways and each of these selected atoms can be placed in any of the g_j cells in level j, giving $(g_j)^{n_j}$ distinct ways of arranging the n_j atoms selected for this level. Therefore the number W_j of distinct arrangements of atoms on level j is given by

$$W_j = {_{(N-n_i)}C_{n_j}} \times (g_j)^{n_j}.$$ (4.22)

For level k the n_k atoms must be selected from the $(N-n_i-n_j)$ unallocated atoms remaining. This leads to W_k distinct arrangements where

$$W_k = {_{(N-n_i-n_j)}C_{n_k}} \times (g_k)^{n_k}.$$ (4.23)

Since any one of the W_i distinct arrangements of atoms on level i may coexist with any one of the W_j distinct arrangements on level j and with any one of the W_k distinct arrangements on level k, and so on, it follows that the total overall number W of microstates (pictures) corresponding to the macrostate characterized by n_i atoms in level i, n_j atoms in level j, and so on, is given by

$$W = W_i W_j W_k \cdots .$$ (4.24)

Substituting from eqns (4.21), (4.22), and (4.23) this becomes

$$W = \frac{N!}{n_i!(N-n_i)!} \cdot \frac{(N-n_i)!}{n_j!(N-n_i-n_j)!} \cdot \frac{(N-n_i-n_j)!}{n_k!(N-n_i-n_j-n_k)!} \cdots$$

$$\times (g_i)^{n_i}(g_j)^{n_j}(g_k)^{n_k} \cdots$$

or, on cancelling terms,

$$W = \frac{N!}{n_i!n_j!n_k!\ldots} (g_i)^{n_i}(g_j)^{n_j}(g_k)^{n_k}\cdots. \tag{4.25}$$

Thus far we have only evaluated W for *one* particular macrostate; the one with n_i atoms at level i, n_j atoms at level j, and so on. Were we to follow the procedures we adopted in manual sorting we would recalculate W for a new macrostate; one with, say, n_i' atoms at level i, n_j' atoms at level j, and so on. On the supposition that each microstate occurs with equal probability—this is a fundamental hypothesis of statistical mechanics—we would then calculate the average number of atoms at each level. In fact we can simplify the operation. The simplification arises because in a large population (recall there 6×10^{23} molecules in a mole of gas) there is one particular *macrostate* whose number of microstates, W, is overwhelmingly greater than the value of W for any other macrostate; so much so that all the other microstates arising from all the other macrostates may be ignored in calculating the mean energy distribution. The problem facing us is therefore to find this *one* macrostate which possesses the greatest number of microstates. Once this type is known the energy distribution follows immediately.

Finding the dominant macrostate

We must now find the type of distribution which maximizes W. In practice it is easier to maximize $\ln W$ (this, of course, also maximizes W). Taking the logarithm of eqn (4.25) gives

$$\ln W = \ln N! - \sum_i \ln n_i! + \sum_i n_i \ln g_i. \tag{4.26}$$

By using a result, known as Stirling's formula, for the logarithm of the factorial of a large number, namely

$$\ln (n!) \approx n \ln n - n, \tag{4.27}$$

and by recalling that the total number of atoms, N, is constant, eqn (4.26) becomes

$$\ln W = \text{constant} + \sum_i (n_i \ln g_i - n_i \ln n_i + n_i).$$

The problem then is to find the set of numbers $(n_0, n_1, n_2,$ etc.) which maximizes $\ln W$, and hence W. When $\ln W$ is a maximum, small changes in the value of n_i (meaning $n_0, n_1, n_2,$ etc.) will not change $\ln W$. Thus

$$\delta(\ln W) = \sum_i (\delta n_i \ln g_i - \delta n_i - \delta n_i \ln n_i + \delta n_i) = 0$$

$$\sum_i \ln (g_i/n_i)\, \delta n_i = 0. \tag{4.28}$$

The quantities δn_1, δn_2, etc. represent the changes in the numbers of atoms at energy levels E_1, E_2, etc. These changes occur naturally in a gas as a result of collisions during which energy may be exchanged between atoms (§3.4). However, the various δn_i s are subject to two constraints. Firstly, the total number of atoms, N, namely

$$N = \sum_i n_i, \tag{4.29}$$

must remain constant. This implies

$$\delta N = \sum_i \delta n_i = 0 \tag{4.30}$$

or, in other words, any increases in the population at some levels must be counterbalanced by decreases in the population at other levels. Secondly the total (internal) energy of the gas, namely

$$E = \sum_i n_i E_i \tag{4.31}$$

is fixed (the gas is isolated). This implies

$$\delta E = \sum_i E_i \, \delta n_i = 0 \tag{4.32}$$

and therefore any changes that give extra energy to certain atoms must be balanced by other changes that give lower energies to the other atoms. Eqns (4.30) and (4.32) represent the conditions which are imposed on the δn_i in eqn (4.28).

Following a technique known as Lagrange's method of undetermined multipliers, we now multiply eqn (4.30) through by a constant, α, (giving $\sum_i \alpha \, \delta n_i = 0$) and eqn (4.32) through by a constant, β, (giving $\sum_i \beta E_i \, \delta n_i = 0$). We next subtract these modified forms of eqns (4.30) and (4.32) from eqn (4.28), to give

$$\sum_i (\ln(g_i/n_i) - \alpha - \beta E_i) \, \delta n_i = 0. \tag{4.33}$$

Since the δn_i are now in effect independent, the coefficient of each must vanish (otherwise eqn (4.33) cannot be satisfied for finite values of δn_1, δn_2, etc.). So for all i

$$\ln(g_i/n_i) - \alpha - \beta E_i = 0$$
$$n_i = g_i \, e^{-\alpha} \, e^{-\beta E_i}. \tag{4.34}$$

These n_i describe the macrostate which possesses the largest number of microstates. According to our assumptions this macrostate describes the 'average' conditions existing in our gas. Now what we have been calculating is the 'picture-averaged' distribution ('ensemble averaged' to give it the formal name). In experimental practice we would, however, be interested in the time-averaged distribution. It is an axiom of elementary statistical mechanics that the time-averaged and the ensemble-averaged distributions are the same.

You will recall how, to keep the notation simple, we deliberately set $dE = 1$ in eqn (4.18), allowing us to write the number of cells at level i, $\mathcal{N}_3(E_i, V_0)$, as $g_3(E_i) = g_i$. Introducing the correct number of cells (given by eqns (4.16) and (4.18)) into eqn (4.34) and writing dN for the actual number of atoms within the energy range dE at energy $E = E_i$, gives

$$dN = \text{constant } e^{-\beta E}\, dE \qquad (4.35)$$

for a two-dimensional gas, and

$$dN = \text{constant } e^{-\beta E} E^{1/2}\, dE \qquad (4.36)$$

for a three-dimensional gas. These constants include not only the various constants occurring in eqns (4.17) and (4.19) but also $e^{-\alpha}$ which is, of course, also constant as α is a constant. Remembering that the total number of atoms N is constant we may rewrite eqn (4.35) to give the

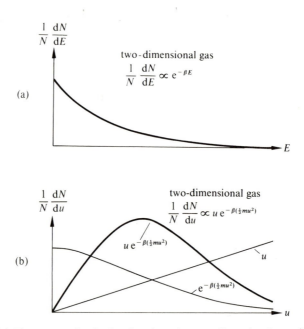

Fig. 4.20. (a) The energy distribution function of a two-dimensional gas. (b) The speed distribution function of a two-dimensional gas.

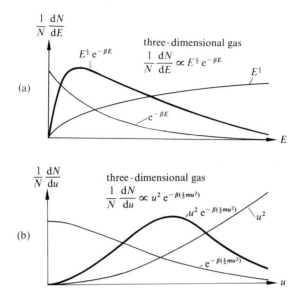

Fig. 4.21. (a) The energy distribution function of a three-dimensional gas. (b) The speed distribution function of a three-dimensional gas.

energy distribution function of a two-dimensional gas as

$$f_2(E) = \frac{1}{N}\frac{dN}{dE} = \text{constant } e^{-\beta E}. \tag{4.37}$$

Likewise, eqn (4.36) gives the energy distribution function of a three-dimensional gas as

$$f_3(E) = \frac{1}{N}\frac{dN}{dE} = \text{constant } E^{1/2} e^{-\beta E}. \tag{4.38}$$

Eqn (4.37) is shown plotted in Fig. 4.20(a) and eqn (4.38) in Fig. 4.21(a). Comparing these plots with the results obtained by manual sorting (Figs. 4.14 and 4.17) we see that, as expected, the two techniques do yield broadly similar results.

4.7. The speed distribution function

Since the energy E of a gas atom is related to its speed u by $E = \frac{1}{2}mu^2$, it is a simple matter to convert an energy distribution function into a speed distribution function, measuring the fraction of atoms per unit range of speed about a particular speed u. Substituting $E = \frac{1}{2}mu^2$ and

$dE = mu\, du$ into eqns (4.37) and (4.38) gives the speed distribution function as

$$f_2(u) = \frac{1}{N}\frac{dN}{du} = A_2 u\, e^{-\beta(\frac{1}{2}mu^2)} \tag{4.39}$$

for a two-dimensional gas, and as

$$f_3(u) = \frac{1}{N}\frac{dN}{du} = A_3 u^2\, e^{-\beta(\frac{1}{2}mu^2)} \tag{4.40}$$

for a three-dimensional gas. Here A_2 and A_3 are, as yet, undetermined constants. Eqn (4.39) is shown plotted in Fig. 4.20(b) and eqn (4.40) in Fig. 4.21(b). We see that the speed distribution function of a two-dimensional gas has a peak that is absent in the corresponding energy distribution function (Fig. 4.20(a)). We also see that there is a peak in both the speed and the energy distribution functions of a three-dimensional gas (Fig. 4.21), although the detailed form of the two functions is, of course, different.

Evaluating the constants

There are still unknown constants to be determined in eqns (4.39) and (4.40) (and therefore in eqns (4.37) and (4.38)). In fact we will only evaluate the constants for a three-dimensional gas; their evaluation for a two-dimensional gas will be left as an exercise.

We proceed by using additional pieces of information about the gas. The first piece of information is summarized in eqn (4.3); applied to eqn (4.40) it tells us that

$$A_3 \int_0^\infty u^2\, e^{-\beta(\frac{1}{2}mu^2)}\, du = 1. \tag{4.41}$$

(The same result follows on using the fact that the total number of atoms—namely $\int_0^\infty (dN/du)\, du$—is equal to N.) The second piece of information is that the average energy $\bar{E}$ of the gas atoms is $\frac{3}{2}kT$ (eqn (3.26)). Now it follows from eqn (4.6) that the mean value $\bar{E}$ of any quantity E which can be expressed as a function of u is given by

$$\bar{E} = \int_0^\infty E f_3(u)\, du$$

or, substituting for $f_3(u)$ from eqn (4.40) with A_3 as given by eqn (4.41),

$$\bar{E} = \frac{\displaystyle\int_0^\infty E u^2\, e^{-\beta(\frac{1}{2}mu^2)}\, du}{\displaystyle\int_0^\infty u^2\, e^{-\beta(\frac{1}{2}mu^2)}\, du}. \tag{4.42}$$

Writing $E = \frac{1}{2}mu^2$ and $\bar{E} = \frac{3}{2}kT$ gives

$$\frac{3}{2}kT = \frac{\frac{1}{2}m \int_0^\infty u^4 \, e^{-\beta(\frac{1}{2}mu^2)} \, du}{\int_0^\infty u^2 \, e^{-\beta(\frac{1}{2}mu^2)} \, du}. \tag{4.43}$$

To evaluate the integrals occurring in the numerator and denominator of eqn (4.43) we use the standard integrals listed in Table 4.1 with $\lambda = \frac{1}{2}\beta m$. This table tells us that the integral occurring in the numerator is $I_4 = \frac{3}{8}(\pi/\lambda^5)^{1/2} = \frac{3}{2}(2\pi/\beta^5 m^5)^{1/2}$ and that the integral occurring in the denominator is $I_2 = \frac{1}{4}(\pi/\lambda^3)^{1/2} = \frac{1}{2}(2\pi/\beta^3 m^3)^{1/2}$. Substituting these results into eqn (4.43) gives

$$\beta = \frac{1}{kT}. \tag{4.44}$$

Inserting this expression for β into eqn (4.41) and using I_2 of Table 4.1 with $\lambda = m/2kT$ gives

$$A_3 = \left(\frac{2}{\pi}\right)^{1/2} \left(\frac{m}{kT}\right)^{3/2}. \tag{4.45}$$

TABLE 4.1

Integrals   of   the   form
$I_n = \int_0^\infty u^n \, e^{-\lambda u^2} \, du$[a]

n	I_n	n	I_n
0	$\frac{1}{2}\left(\frac{\pi}{\lambda}\right)^{1/2}$	1	$\frac{1}{2\lambda}$
2	$\frac{1}{4}\left(\frac{\pi}{\lambda^3}\right)^{1/2}$	3	$\frac{1}{2\lambda^2}$
4	$\frac{3}{8}\left(\frac{\pi}{\lambda^5}\right)^{1/2}$	5	$\frac{1}{\lambda^3}$
6	$\frac{15}{16}\left(\frac{\pi}{\lambda^7}\right)^{1/2}$	7	$\frac{3}{\lambda^4}$

[a] This integral can only be evaluated from $u = 0$ to $u = \infty$. If interested, you will find it and other like integrals discussed in more detail, in for example, Born, M. (1957). *Atomic Physics*. 6th edn. Blackie, London, (Appendix 1). When $n = 1, 3, 5, 7$, etc. (Table 4.1) the integral can be evaluated 'by parts'.

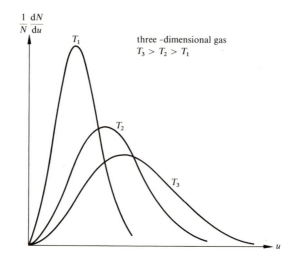

Fig. 4.22. The speed distribution function of a three-dimensional gas at temperatures $T_3 > T_2 > T_1$.

Substituting eqns (4.44) and (4.45) back into eqn (4.40) gives

$$f_3(u) = \frac{1}{N}\frac{dN}{du} = 4\pi \left(\frac{m}{2\pi kT}\right)^{3/2} u^2\, e^{-mu^2/2kT} \quad . \tag{4.46}$$

This equation, known as the *Maxwell–Boltzmann speed distribution function*, is plotted out graphically in Fig. 4.22 at three different temperatures. Its truth has been confirmed experimentally by time-of-flight measurements of the type already described in §3.5 (see also problem 5.4). To convert eqn (4.46) into an energy distribution function we merely substitute $E = \frac{1}{2}mu^2$ and $du = dE/mu$. This leads to the *Maxwell–Boltzmann energy distribution function*:

$$f_3(E) = \frac{1}{N}\frac{dN}{dE} = \frac{2}{\pi^{1/2}}\left(\frac{1}{kT}\right)^{3/2} E^{1/2}\, e^{-E/kT} \quad . \tag{4.47}$$

Instead of interpreting eqn (4.46) as telling us the fraction of the gas atoms whose speeds lie within unit range of u we can, of course, interpret it as telling us the probability that any *one* atom has a speed within unit range of u. A similar remark, framed in terms of energy, applies to eqn (4.47).

The mean and the r.m.s. speeds compared

Before leaving the speed distribution it is worthwhile using it to evaluate the mean speed $\bar{u}$ and the root mean square speed $(\overline{u^2})^{1/2}$ in a gas. According to the averaging procedure summarized in eqn (4.6),

$$\bar{u} = \int_0^\infty u f_3(u)\, du \qquad (4.48)$$

$$(\overline{u^2})^{1/2} = \left(\int_0^\infty u^2 f_3(u)\, du \right)^{1/2}. \qquad (4.49)$$

Substituting for $f_3(u)$ from eqn (4.46) and making use of the standard integrals in Table 4.1 gives

$$\boxed{\bar{u} = \left(\frac{8kT}{\pi m} \right)^{1/2}} \qquad (4.50)$$

$$\boxed{(\overline{u^2})^{1/2} = \left(\frac{3kT}{m} \right)^{1/2}}. \qquad (4.51)$$

Therefore

$$(\overline{u^2})^{1/2}/\bar{u} = (3\pi/8)^{1/2} = 1\cdot 08. \qquad (4.52)$$

The fact that the r.m.s. speed and the mean speed only differ by eight per cent will allow us, in later chapters, to roughly equate these two speeds.

Exercise 4.4

Evaluate the unknown constants occurring in the speed distribution function of a two-dimensional gas (eqn (4.39)). Transform the speed distribution function into the corresponding energy distribution function. Note that in a two-dimensional gas

$$\tfrac{1}{2}m\overline{u^2} = \tfrac{1}{2}m\overline{u_x^2} + \tfrac{1}{2}m\overline{u_y^2}$$

and so, from eqn (3.27),

$$\tfrac{1}{2}m\overline{u^2} = kT. \qquad (4.53)$$

Solution. The procedures are essentially those followed in evaluating the constants in the speed distribution function of a three-dimensional gas (eqn (4.40)). Eqn (4.3) as applied to eqn (4.39) tells us that

$$A_2 \int_0^\infty u\, e^{-\beta(\frac{1}{2}mu^2)}\, du = 1. \qquad (4.54)$$

In the case of a two-dimensional gas eqn (4.42) takes the form

$$\bar{E} = \frac{\displaystyle\int_0^\infty Eu\, e^{-\beta(\frac{1}{2}mu^2)}\, du}{\displaystyle\int_0^\infty u\, e^{-\beta(\frac{1}{2}mu^2)}\, du}.$$

Writing $E = \frac{1}{2}mu^2$ and $\bar{E} = kT$,

$$kT = \frac{\frac{1}{2}m \displaystyle\int_0^\infty u^3\, e^{-\beta(\frac{1}{2}mu^2)}\, du}{\displaystyle\int_0^\infty u\, e^{-\beta(\frac{1}{2}mu^2)}\, du}. \tag{4.55}$$

The integrals occurring in the numerator and denominator of eqn (4.55) are I_3 and I_1, respectively, in Table 4.1 with $\lambda = \frac{1}{2}\beta m$. This leads to $\beta = 1/kT$. Substituting this expression for β into eqn (4.54) and using I_1 of Table 4.1 gives

$$A_2 = (m/kT).$$

Hence the speed distribution function, eqn (4.39), becomes

$$\boxed{f_2(u) = \frac{1}{N}\frac{dN}{du} = 2\pi\left(\frac{m}{2\pi kT}\right)u\, e^{-mu^2/2kT}}. \tag{4.56}$$

The corresponding energy distribution function follows on writing $E = \frac{1}{2}mu^2$ and $du = dE/mu$:

$$\boxed{f_2(E) = \frac{1}{N}\frac{dN}{dE} = \left(\frac{1}{kT}\right)e^{-E/kT}}. \tag{4.57}$$

Comment. Although the discussion of a two-dimensional gas may appear to have been little more than a cerebral exercise there are situations where gases actually do behave in this manner. An example is a gas loosely bound to the surface of a solid.

4.8. The one-dimensional gas

The behaviour of a hypothetical one-dimensional gas is the subject of the structured problem 4.6. Following through the arguments outlined in this problem leads to

$$\boxed{f_1(u) = \frac{1}{N}\frac{dN}{du} = 2\left(\frac{m}{2\pi kT}\right)^{1/2}e^{-mu^2/2kT}} \tag{4.58}$$

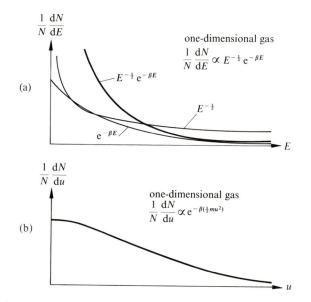

Fig. 4.23. (a) The energy distribution function of a one-dimensional gas. (b) The speed distribution function of a one-dimensional gas.

as the speed distribution function, and to

$$f_1(E) = \frac{1}{N}\frac{\mathrm{d}N}{\mathrm{d}E} = \frac{1}{\pi^{1/2}}\left(\frac{1}{kT}\right)^{1/2} E^{-1/2}\, e^{-E/kT} \qquad (4.59)$$

as the energy distribution function. These distribution functions are plotted graphically in Fig. 4.23. The importance of eqn (4.58) lies mainly in the fact that when u is replaced by u_x it gives the speed distribution function of the x-components of velocity of the atoms in a three-dimensional gas; a statement we shall presently justify.

4.9. The velocity distribution function

In later chapters we will need to known the fraction of atoms in a gas which possess not only speeds that lie within specified limits but which also move within a defined range of angles. We will therefore seek to find the *velocity distribution function*, by which we mean the fraction of the gas atoms per unit range of speed per unit solid angle about $\boldsymbol{u}$.

We start by noting that the speed distribution function $f_3(u)$ (eqn (4.46)) can be written as

$$\mathrm{d}N = \rho(4\pi u^2\, \mathrm{d}u), \qquad (4.60)$$

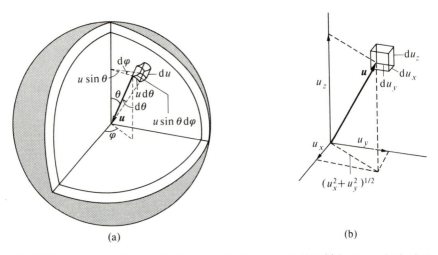

Fig. 4.24. Showing an element of volume in velocity space specified (a) in terms of spherical polar coordinates, and (b) in terms of cartesian coordinates.

where

$$\rho = \frac{dN}{4\pi u^2 \, du} = \frac{Nf_3(u)}{4\pi u^2} = N\left(\frac{m}{2\pi kT}\right)^{3/2} e^{-mu^2/2kT}. \tag{4.61}$$

It is clear from Figs. 4.7 and 4.24(a) that $4\pi u^2 \, du$ is the volume contained between two adjacent shells in velocity space of radii u and $u + du$. Eqn (4.61) therefore tells us the number of points per unit volume of velocity space (at radius u); points at which vectors terminate (in the case of Fig. 4.7) or from which vectors set off to the origin (in the case of Fig. 4.24(a)). (Because vectors can always be transposed parallel to themselves Fig. 4.7 is equivalent to Fig. 4.24(a) with the vectors 'slid along themselves' until their tails, rather than their heads, are at the origin.) If we now draw in the cell shown in Fig. 4.24(a), then all those vectors which begin at points lying within this cell and which terminate at the origin (or the other way round) will represent atoms which have speeds between u and $u + du$, and which are moving in directions between θ and $\theta + d\theta$, ϕ and $\phi + d\phi$. The triplet (u, θ, ϕ) define the *spherical polar coordinates* (here in velocity space); θ is usually called the *polar angle*; ϕ the *azimuth*. Since the cell shown in Fig. 4.24(a) has a volume $u \, d\theta \times u \sin \theta \, d\phi \times du = u^2 \sin \theta \, d\theta \, d\phi \, du$ it follows that the number of atoms $d^3N_{u\theta\phi}$ with speeds between u and $u + du$ and moving in directions between θ and $\theta + d\theta$, ϕ and $\phi + d\phi$ is given by

$$d^3N_{u\theta\phi} = \rho u^2 \sin \theta \, d\theta \, d\phi \, du \tag{4.62}$$

or, substituting $\rho = Nf_3(u)/4\pi u^2$ (eqn (4.61)) and dividing through by N,

$$\frac{d^3 N_{u\theta\phi}}{N} = \frac{1}{4\pi} f_3(u) \sin \theta \, d\theta \, d\phi \, du \quad . \tag{4.63}$$

Alternatively, we may substitute $\rho = N(m/2\pi kT)^{3/2}\exp(-mu^2/2kT)$ (eqn (4.61)) into eqn (4.62) and obtain

$$\frac{d^3 N_{u\theta\phi}}{N} = \left(\frac{m}{2\pi kT}\right)^{3/2} e^{-mu^2/2kT} u^2 \sin \theta \, d\theta \, d\phi \, du.$$

Now the spherical lower surface of the cell shown in Fig. 4.24(a) is at a distance u from the origin and has an area $dA = u \sin \theta \, d\phi \times u \, d\theta$. It therefore (by definition) subtends a solid angle $d\omega$ at the origin given by

$$d\omega \left(= \frac{dA}{u^2} \right) = \sin \theta \, d\theta \, d\phi. \tag{4.64}$$

Eqn (4.63) can thus be rewritten as

$$f_3(u) = \frac{1}{N} \frac{d^2 N}{du \, d\omega} = \frac{1}{4\pi} f_3(u) = \left(\frac{m}{2\pi kT}\right)^{3/2} u^2 \, e^{-mu^2/2kT} \tag{4.65}$$

which is the required velocity distribution function of a three-dimensional gas. The fact that it is $f_3(u)/4\pi$ is to be expected since a sphere subtends a solid angle of $4\pi u^2/u^2 = 4\pi$ steradian (written sr) at its centre.

The distribution of velocity components

Instead of specifying velocities by means of spherical polar coordinates, as in Fig. 4.24(a), we may, of course, use cartesian coordinates, as in Fig. 4.24(b), where a velocity u is specified by its three components u_x, u_y, and u_z. In this coordinate system a cell in velocity space will have a volume $du_x \, du_y \, du_z$ (see Fig. 4.24(b)) and contain

$$d^3 N_{u_x u_y u_z} = \rho \, du_x \, du_y \, du_z = N\left(\frac{m}{2\pi kT}\right)^{3/2} \exp[-m(u_x^2 + u_y^2 + u_z^2)/2kT] \, du_x \, du_y \, du_z \tag{4.66}$$

points at which vector tips from the origin terminate (or leave for the origin). Eqn (4.66) may be written as

$$\frac{d^3 N_{u_x u_y u_z}}{N} = \left[\left(\frac{m}{2\pi kT}\right)^{1/2} e^{-mu_x^2/2kT} du_x\right]\left[\left(\frac{m}{2\pi kT}\right)^{1/2} e^{-mu_y^2/2kT} du_y\right]$$
$$\times \left[\left(\frac{m}{2\pi kT}\right)^{1/2} e^{-mu_z^2/2kT} du_z\right]. \tag{4.67}$$

This tells us either the fraction of the gas atoms whose velocity components lie in the range u_x to $u_x + du_x$, u_y to $u_y + du_y$, and u_z to $u_z + du_z$ or the probability that any *one* atom has velocity components within this range. Assuming that the probability of a gas atom having a velocity component between, say, u_x and $u_x + du_x$ is independent of the values of its other two velocity components (here u_y and u_z) we know that the overall probability that an atom has velocity components in the range u_x to $u_x + du_x$, u_y to $u_y + du_y$, and u_z to $u_z + du_z$ will be equal to the probability that its x-component of velocity lies in the range u_x to $u_x + du_x$, multiplied by the probability that its y-component of velocity lies in the range u_y to $u_y + du_y$, multiplied by the probability that its z-component of velocity lies in the range u_z to $u_z + du_z$. Comparing this statement with the right-hand side of eqn (4.67) we therefore see that the probability that an atom has a velocity component between u_x and $u_x + du_x$ can be written as

$$f_3(u_x)\, du_x = \left(\frac{m}{2\pi kT}\right)^{1/2} e^{-mu_x^2/2kT}\, du_x$$

$$f_3(u_x) = \left(\frac{m}{2\pi kT}\right)^{1/2} e^{-mu_x^2/2kT}, \qquad (4.68)$$

and similarly for the other two component velocities.

To convert eqn (4.68) into a speed distribution function the right-hand side must be multiplied by two. This is because u_x and $-u_x$ (which are different velocities) represent the same speed. Therefore

$$\boxed{f_3(u_x) = 2 \left(\frac{m}{2\pi kT}\right)^{1/2} e^{-mu_x^2/2kT}} \qquad (4.69)$$

which has exactly the same form as the speed distribution in a hypothetical one-dimensional gas (eqn (4.58)).

As an example of the application of eqn (4.69) we will use it to calculate the mean value $\bar{u}_x$ of the x-component speeds. According to eqn (4.6),

$$\bar{u}_x = \int_0^\infty u_x f_3(u_x)\, du_x = -2 \left(\frac{kT}{2\pi m}\right)^{1/2} \int_0^\infty \frac{-mu_x}{kT} e^{-mu_x^2/2kT}\, du_x$$

$$\bar{u}_x = \left(\frac{2kT}{\pi m}\right)^{1/2}. \qquad (4.70)$$

Had we wished to find $\overline{u_x}$ we would have used $f_3(u_x)$ (eqn (4.68)) in place of $f_3(u_x)$ and would have integrated from $u_x = -\infty$ to $u_x = +\infty$. As expected, this would have given $\overline{u_x} = 0$.

4.10. The speed distribution function in liquids and solids

In writing the total energy E of a gas atom as $\frac{1}{2}mu^2$—a substitution we made on many occasions throughout this chapter—we were implicitly assuming that $\Delta E = 0$. Such a step is not possible in liquids (where $\Delta E \approx \frac{1}{2}\overline{mu^2}$) or in solids (where $\Delta E \gg \frac{1}{2}\overline{mu^2}$). Nevertheless it turns out that the Maxwell–Boltzmann speed distribution function (eqn (4.46)) is equally applicable to all three phases—provided that the atoms' motions can be described by the laws of Newtonian mechanics (a big qualification in the case of solids). Thus if, in imagination, we take a given mass of a substance and increase its pressure while holding the temperature fixed (thereby following the isotherm shown in Fig. 1.4) the substance passes from the gas phase through the liquid phase to the solid phase. Yet throughout this transition eqn (4.46) accurately describes the speed distribution of the atoms. Likewise, eqn (4.47) accurately describes their *kinetic energy* distribution. However, except in the gas phase, where the interatomic potential energy may normally be ignored, eqn (4.47) does *not* describe the way that the total energy is distributed.

Perhaps the simplest way of confirming the correctness or otherwise of these assertions is to use the kinetic simulator described in §3.2 to simulate the three phases of matter. A cine film made of such a simulator in action—Fig. 3.2 shows representative frames from such a film—allows the speed distribution of the 'atoms' to be determined in all three phases. The results agree closely with the Maxwell–Boltzmann speed distribution function (in the form (eqn (4.56)) appropriate to a two-dimensional gas). Real experiments in liquids confirm that the velocity distribution is indeed Maxwell–Boltzmann in character. In these experiments neutrons from a reactor are allowed to come into thermal equilibrium with a liquid, after which their speed distribution is measured using time-of-flight techniques. Since this distribution is Maxwell–Boltzmann in form it can be argued that the liquid's atoms must also have a Maxwell–Boltzmann speed distribution. Evidence to be presented in Chapter 10 shows that this is generally not the case in solids (except at high temperatures).

4.11. The Boltzmann distribution

Having seen how useful a distribution function can be—particularly in working out mean values—we will now attempt to change the form of the speed distribution function (eqn (4.46)) so that it can be applied to systems other than (near perfect) gases.

Atoms in a gravitational field

We start by recalling that the probability of a gas atom having component velocities in the range u_x to $u_x + du_x$, u_y to $u_y + du_y$, and u_z to

$u_z + du_z$ is, from eqn (4.66), given by

$$\frac{d^3 N_{u_x u_y u_z}}{N} = \left(\frac{m}{2\pi kT}\right)^{3/2} \exp[-m(u_x^2 + u_y^2 + u_z^2)/2kT]\, du_x\, du_y\, du_z.$$

$$(4.71)$$

Now in §5.6 we will show that when the atoms of a gas are subjected to a constant gravitational field (as happens in the Earth's atmosphere) the Maxwell–Boltzmann speed distribution still applies although the number density $n(h)$ of the atoms decreases as $n(h) = n_0 \exp(-mgh/kT)$. Here n_0 is a constant (equals to the number density at $h = 0$), m is the mass of an atom, and g is the acceleration due to gravity (so that mgh is the potential energy $V(h)$ of an atom at height h). Hence the fraction of the gas atoms to be found at heights between h and $h + dh$—and therefore the probability that any one named atom will be found between these heights—is given by

$$\frac{dN_h}{N} = \frac{n(h)\, dh}{\displaystyle\int_0^\infty n(h)\, dh} = \frac{n_0\, e^{-mgh/kT}\, dh}{n_0 \displaystyle\int_0^\infty e^{-mgh/kT}\, dh} = \left(\frac{mg}{kT}\right) e^{-V(h)/kT}\, dh. \quad (4.72)$$

Thus the probability that any particular molecule will have velocity components in the range u_x to $u_x + du_x$, u_y to $u_y + du_y$, u_z to $u_z + du_z$ *and* be located between h and $h + dh$ is equal to the product of the two probabilities given by eqns (4.71) and (4.72). This joint probability can be therefore be written as

$$\frac{d^4 N_{u_x u_y u_z h}}{N} = C \exp[-(\tfrac{1}{2}mu_x^2 + \tfrac{1}{2}mu_y^2 + \tfrac{1}{2}mu_z^2 + V(h))/kT]\, du_x\, du_y\, du_z\, dh,$$

$$(4.73)$$

where we have written C for $(mg/kT)(m/2\pi kT)^{3/2}$. Hence we can define the distribution function $f_3(u_x, u_y, u_z, h)$ as

$$f_3(u_x, u_y, u_z, h) = \frac{d^4 N_{u_x u_y u_z h}}{N\, du_x\, du_y\, du_z\, dh}$$

$$f_3(u_x, u_y, u_z, h) = C \exp[-(\tfrac{1}{2}mu_x^2 + \tfrac{1}{2}mu_y^2 + \tfrac{1}{2}mu_z^2 + V(h))/kT] \quad (4.74)$$

Having established the form of the distribution function for a gas in a gravitational field we can now use it to evaluate, say, the mean energy $\bar{E}$ of a gas atom. By analogy with eqns (4.3) and (4.6):

$$C \int_0^\infty \int_{-\infty}^\infty \int_{-\infty}^\infty \int_{-\infty}^\infty \exp[-(\tfrac{1}{2}mu_x^2 + \tfrac{1}{2}mu_y^2 + \tfrac{1}{2}mu_z^2 + V(h))/kT]$$

$$\times du_x\, du_y\, du_z\, dh = 1 \quad (4.75)$$

$$\bar{E} = C\int_0^\infty \int_{-\infty}^\infty \int_{-\infty}^\infty \int_{-\infty}^\infty E \exp[-(\tfrac{1}{2}mu_x^2 + \tfrac{1}{2}mu_y^2 + \tfrac{1}{2}mu_z^2 + V(h))/kT]$$

$$\times du_x \, du_y \, du_z \, dh. \quad (4.76)$$

Substituting for C from eqn (4.75) into eqn (4.76) and making the limits to the integrals implicit gives

$$\bar{E} = \frac{\int \cdots \int E \exp[-(\tfrac{1}{2}mu_x^2 + \tfrac{1}{2}mu_y^2 + \tfrac{1}{2}mu_z^2 + V(h))/kT] \, du_x \, du_y \, du_z \, dh}{\int \cdots \int \exp[-(\tfrac{1}{2}mu_x^2 + \tfrac{1}{2}mu_y^2 + \tfrac{1}{2}mu_z^2 + V(h))/kT] \, du_x \, du_y \, du_z \, dh}$$

$$(4.77)$$

where E is given by

$$E = \tfrac{1}{2}mu_x^2 + \tfrac{1}{2}mu_y^2 + \tfrac{1}{2}mu_z^2 + V(h).$$

Rigid diatomic molecules

It is not difficult to guess how to extend eqn (4.77) so that it applies to more general systems than the Earth's atmosphere. Consider, for example, a gas composed of rigid diatomic molecules (like that shown in Fig. 5.18(a)). Ignoring the effects of gravity—permissible when the gas is held in a laboratory vessel—we can write the energy E of the molecule as

$$E = \tfrac{1}{2}mu_x^2 + \tfrac{1}{2}mu_y^2 + \tfrac{1}{2}mu_z^2 + \tfrac{1}{2}I_1\omega_1^2 + \tfrac{1}{2}I_2\omega_2^2 + \tfrac{1}{2}I_3\omega_3^2, \quad (4.78)$$

where u_x, u_y, and u_z are the x-, y-, and z-components of the velocity $\boldsymbol{u}$ of the mass centre, where I_1, I_2, and I_3 are the principal moments of inertia of the molecule at its mass centre, and where ω_1, ω_2, and ω_3 are the components of the angular velocity $\boldsymbol{\omega}$ in the directions of the principal axes of inertia. We might therefore guess (correctly as it happens) that eqn (4.77) can be applied to give the mean energy $\bar{E}$ of a molecule of the gas as

$$\bar{E} = \frac{\int \cdots \int E \exp[-(\tfrac{1}{2}mu_x^2 + \tfrac{1}{2}mu_y^2 + \tfrac{1}{2}mu_z^2 + \tfrac{1}{2}I_1\omega_1^2 + \tfrac{1}{2}I_2\omega_2^2 + \tfrac{1}{2}I_3\omega_3^2)/kT] \times du_x \, du_y \, du_z \, d\omega_1 \, d\omega_2 \, d\omega_3}{\int \cdots \int \exp[-(\tfrac{1}{2}mu_x^2 + \tfrac{1}{2}mu_y^2 + \tfrac{1}{2}mu_z^2 + \tfrac{1}{2}I_1\omega_1^2 + \tfrac{1}{2}I_2\omega_2^2 + \tfrac{1}{2}I_3\omega_3^2)/kT] \times du_x \, du_y \, du_z \, d\omega_1 \, d\omega_2 \, d\omega_3}$$

$$(4.79)$$

Substituting for E from eqn (4.78) leads to the result that the right-hand side of eqn (4.79) is the *sum* of six integrals, each of the form

$$\frac{\int_{-\infty}^{\infty} \tfrac{1}{2}mu_x^2 \exp[-(\tfrac{1}{2}mu_x^2)/kT]\,du_x \int \cdots \int \exp[-(\tfrac{1}{2}mu_y^2+\tfrac{1}{2}mu_z^2+\tfrac{1}{2}I_1\omega_1^2 + \tfrac{1}{2}I_2\omega_2^2+\tfrac{1}{2}I_3\omega_3^2)/kT]\,du_y\,du_z\,d\omega_1\,d\omega_2\,d\omega_3}{\int_{-\infty}^{\infty} \exp[-(\tfrac{1}{2}mu_x^2)/kT]\,du_x \int \cdots \int \exp[-(\tfrac{1}{2}mu_y^2+\tfrac{1}{2}mu_z^2+\tfrac{1}{2}I_1\omega_1^2 + \tfrac{1}{2}I_2\omega_2^2+\tfrac{1}{2}I_3\omega_3^2)/kT]\,du_y\,du_z\,d\omega_1\,d\omega_2\,d\omega_3}$$

$$(4.80)$$

Since they are identical, the integrals over u_y, u_z, ω_1, ω_2, and ω_z can be cancelled in the numerator and denominator, allowing eqn (4.80) to be written as

$$\frac{\tfrac{1}{2}m \int_{-\infty}^{\infty} u_x^2 e^{-(m/2kT)u_x^2}\,du_x}{\int_{-\infty}^{\infty} e^{-(m/2kT)u_x^2}\,du_x} = \frac{2(\tfrac{1}{2}m)\int_0^{\infty} \xi^2 e^{-\lambda\xi^2}\,d\xi}{2\int_0^{\infty} e^{-\lambda\xi^2}\,d\xi} = \frac{m}{2}\left(\frac{1}{2\lambda}\right) = \tfrac{1}{2}kT,$$

$$(4.81)$$

where the integrals have been evaluated using the standard forms I_0 and I_2 of Table 4.1 with $\lambda = (m/2kT)$. Hence eqn (4.79) becomes

$$\bar{E} = 6 \times \tfrac{1}{2}kT = 3kT. \qquad (4.82)$$

For later use it is convenient to set out eqn (4.81) afresh as

$$\boxed{\frac{\int_0^{\infty} \xi^2 e^{-\lambda\xi^2}\,d\xi}{\int_0^{\infty} e^{-\lambda\xi^2}\,d\xi} = \frac{1}{2\lambda} = \tfrac{1}{2}kT} \,. \qquad (4.83)$$

The important conclusion to be drawn from the discussion is that *each independent quadratic term* that features in the expression for the energy of an *individual* molecule (terms like $\tfrac{1}{2}mu_x^2$ or $\tfrac{1}{2}I_1\omega_1^2$) contributes an average energy of $\tfrac{1}{2}kT$ towards the total energy of the system. This conclusion only applies to quadratic terms. Linear terms (like mgh, for example) contribute an average energy of kT (see problem 4.13).

PROBLEMS

4.1.   A little-known tribe has a mass distribution of the form $f(m) = p(m - m_0)^2$ over the range $m_0 \leq m \leq m_1$, where $p = 3 \times 10^{-3}\,\mathrm{kg}^{-3}$ and $m_0 = 4\,\mathrm{kg}$.

(a) Sketch $f(m)$ on the assumption that $f(m)=0$ when $m_0>m>m_1$. (b) What is the value of m_1? (c) What is the mean mass of a member of the tribe? (d) Assuming that the annual income of a member of the tribe is related to his or her mass by $I=cm^3$, where $c=2$ marks kg^{-3} (the mark being the local unit of currency), what is mean annual income of a tribe member?

4.2. Using a single column of cells, as in Fig. 4.13, sketch out all the way that three atoms, A, B, and C of a two-dimensional gas can share out four units of energy. Hence calculate the mean number of atoms with 0, 1, 2, etc. units of energy and plot the results as a histogram. *Clue*: Start by listing the permitted *types* of arrangement.

4.3. Using a cell diagram similar to that shown in Fig. 4.16, sketch all the way that three atoms A, B, and C of a three-dimensional gas can share out 4 units of energy. Hence calculate the mean number of atoms with 0, 1, 2, etc. units of energy and plot your results as a histogram.

4.4. Repeat problem 4.3 but this time with two atoms A and B sharing out (a) three units of energy, and (b) four units of energy. By raising the total energy from three to four units we are increasing the mean kinetic energy per atom from $\frac{3}{2}$ to 2 units. This is equivalent to raising the temperature of the gas by a factor of $2/(\frac{3}{2})=4/3$. By comparing the histogram you obtain in (b) with the histogram you obtain in (a) you should be able to see, in a rough sort of way, how raising the temperature affects the energy distribution in a three-dimensional gas.

4.5. Show that Fig. 4.6 does indeed represent the energy distribution of the pucks featured in Figs 4.3(b) and Fig. 4.5.

4.6. The aim of this problem is to derive the speed distribution function (eqn (4.58)) and the energy distribution function (eqn (4.59)) of a one-dimensional gas. We start with Fig. 4.25 which shows a 'picture' of velocity vectors for a one-dimensional gas; for clarity these have been displaced sideways. This is the equivalent of Fig. 4.3(b) for a two-dimensional gas and of Fig. 4.7 for a three-dimensional gas. In a one-dimensional gas the 'cells of constant size' in velocity space are segments of constant length l_0 (shown marked off by dotted lines in Fig. 4.25). Your first task is to show that the number of such cells within the interval in velocity space corresponding to energies E and $E+\delta E$ is given by

$$\mathcal{N}_1(E, l_0) = \frac{1}{l_0}\left(\frac{2}{m}\right)^{1/2} E^{-1/2}\, \delta E$$

or, in the limit $\delta u \to 0$, by

$$\mathcal{N}_1(E, l_0) = g_1(E)\, dE,$$

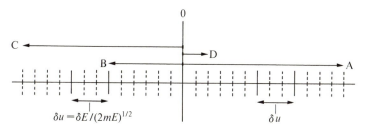

Fig. 4.25. Velocity vectors for a one-dimensional gas.

where

$$g_1(E) = \frac{1}{l_0} \left(\frac{2}{m}\right)^{1/2} E^{-1/2}.$$

Following the procedures adopted with two and three-dimensional gases, this value of $g_1(E)\,dE$ is substituted for g_i in eqn (4.34) with n_i replaced by dN. Hence

$$dN = \text{constant } E^{-1/2}\, e^{-\beta E}\, dE.$$

The constants are evaluated by first converting this into a speed distribution function and then applying eqns (4.3) and (4.6). The average energy of an atom of a one-dimensional gas is $\frac{1}{2}kT$ (eqn (3.27)).

4.7. The *most probable speed* of atoms in a gas is that at which the Maxwell–Boltzmann speed distribution function is a maximum. Show that in a three-dimensional gas the most probable speed is $(2kT/m)^{1/2}$. By how much is the r.m.s. speed in a gas greater than the most probable speed of the gas atoms?

4.8. Prove that $\bar{u}$ and $(\overline{u^2})^{1/2}$ are as given by eqns (4.50) and (4.51) for a three-dimensional gas.

4.9. Starting with the relevant speed distribution functions, show that the mean speed of an atom in one-dimensional gas is $(2kT/\pi m)^{1/2}$ and that the mean speed of an atom in a two-dimensional gas is $(\pi kT/2m)^{1/2}$. What is the ratio of the r.m.s. to the mean speed in (a) a one-dimensional gas and (b) a two-dimensional gas?

4.10. Plot out graphically the Maxwell–Boltzmann speed distribution function, eqn (4.46), for molecular oxygen at a temperature of 300 K. Take $M_r(O_2) = 32$.

4.11. In a two-dimensional gas at a temperature of 100 K; (a) what fraction of the atoms have an energy lying between 2.0×10^{-21} J and 2.1×10^{-21} J?, (b) what fraction have an energy greater than 1.38×10^{-21} J?, and (c) what fraction have an energy greater than 3.76×10^{-21} J?

4.12. What is $\overline{(1/u)}/(1/\bar{u})$ in a three-dimensional gas?

4.13. What is the mean energy (kinetic energy plus gravitational potential energy) of a molecule in Earth's atmosphere? Assume the atmosphere to be isothermal at 290 K.

4.14. Show that each independent cubic term that features in the expression for the energy of an individual particle in a system (meaning $E = \alpha x^3$, where α is a constant) contributes an average energy $\bar{E} = \frac{1}{3}kT$ to the energy of the system.

4.15. Show that a diatomic molecule composed of atoms of mass m_1 and m_2, and between which the force constant is k, vibrate about their centre of mass with a period $T_v = 2\pi(\mu/k)^{1/2}$, where $\mu = m_1 m_2/(m_1 + m_2)$. What is the average interatomic potential energy of such a molecule, relative to its potential energy when the atoms are at rest at their equilibrium positions? *Clue:* eqn (4.83).

4.16. Give the underlying physical reasons for the following facts: (a) A three-dimensional gas has a peak in its energy distribution function whereas a two-dimensional gas does not. (b) The density of   states is constant, independent of energy, in a two-dimensional gas. (c) The r.m.s. speed of a gas molecule is greater than its mean speed.

5. Perfect gases

In this chapter we will attempt to predict some of the key properties of a so-called *perfect* (or *ideal*) gas. As has already been said, such a gas is supposed to contain atoms whose pair dissociation energy $\Delta E = 0$ (implying a total absence of any *attractive* interatomic forces) and whose diameter is effectively zero. We can anticipate that the predicted properties of such a gas will be in approximate agreement with the properties of real gases (whose atoms have, of course, a finite ΔE and a finite diameter) when the gas is at a sufficiently high temperature to ensure that $\frac{1}{2}m\overline{u^2} \gg \Delta E$ and at a sufficiently high molar volume to ensure that the volume occupied by the atoms *per se* is very much less than V_m. We can also expect the near agreement to break down as $\frac{1}{2}m\overline{u^2}$ becomes comparable with ΔE and/or at low molar volumes. Imperfect gases are the subject of the next chapter.

5.1. The interaction of gas atoms with solid surfaces

Before discussing why a pressure must be applied to contain a gas within a cylinder—one of the main concerns of this chapter—it makes good sense to start by asking how gas atoms behave in practice when they collide with a solid surface, such as a piston face. There are two main aspects to consider. The first is the temporal behaviour: does a gas atom rebound, in tennis ball fashion, immediately after being brought to rest at the piston, on does it 'dwell' for some time on the solid surface? The second is the spatial behaviour; how many atoms leave the surface moving along a particular direction?

Temporal behaviour

Evidence that gas atoms do 'stick' (or 'dwell', or 'linger', or 'sojourn') on a surface for some time before leaving first came from experiments performed in the 1920s. In one such study (illustrated in Fig. 5.1), a collimated beam of gas atoms strikes a rotating disc at a point S. If these atoms stick they will be carried along on the rotating surface. A movable detector enables one to measure the number which survive the journey to the point D on the disc, opposite the detector entrance, and which are desorbed at this point. By varying the speed of rotation of the disc the number of atoms that stick on its surface for various known times can be measured. These experiments, and their later developments, showed that the average dwell times of an inert gas atom on a steel or glass surface, held at room temperature, were in the microsecond to nanosecond range.

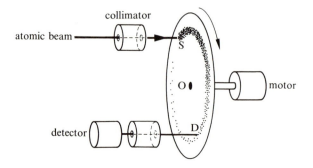

Fig. 5.1. An apparatus for measuring the time a gas atom dwells on a surface. A beam of gas atoms strikes a rotating disc at S. The number coming off at position D is measured with a suitable detector. The entire apparatus is contained within a highly evacuated chamber.

Such dwell times are long compared with the 10^{-12} to 10^{-14} s vibrational period of the constituent atoms of the absorbing surface and allow the colliding gas atoms to exchange momentum and energy with the atoms of the surface. Were a gas atom to behave in tennis ball fashion—squashing itself against the surface and then dilating itself as it leaves—we might expect to find a dwell time of order $d/\bar{u}$, where d is the diameter of the gas atom and $\bar{u}$ is its mean speed. Adopting typical values of $d = 5 \times 10^{-10}$ m and $\bar{u} = 500$ m s^{-1} would suggest that immediate rebounds could be characterized by dwell times of less than 10^{-12} s. Such short times are only observed with particularly clean surfaces carefully prepared under high vacuum conditions. We may note, in passing, that certain processes, like the removal of poisonous or odorous gases from air by passing it through charcoal, require long dwell times if they are to succeed. Here dwell times of centuries are not uncommon.

Spatial behaviour

The apparatus shown in Fig. 5.2 allows one to study how gas atoms are scattered from a solid surface. A beam of collimated atoms from oven O strikes the solid, while a movable detector D (usually a bulb with a small hole in it, connected to a sensitive pressure gauge) records the number of atoms scattered into a fixed solid angle $\delta\omega$ (the acceptance angle of the gauge) at an angle θ.

Fig. 5.3(a) shows the results obtained when a beam of nitrogen atoms incident along a direction IS strikes a polished glass surface. The length of the line SR is drawn proportional to the number of atoms scattered per unit time along direction SR. Plotted in this way the results lie on a circle. By simple geometry, the intensity $I(\theta)$ of the beam scattered through an

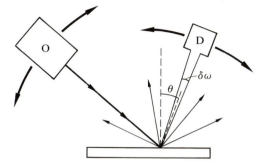

Fig. 5.2. Atoms from an oven O strike a surface. The number scattered in different directions is measured with a detector D. The entire apparatus is contained within a highly evacuated chamber.

angle θ with the normal to the surface (see Fig. 5.3(a)) is

$$I(\theta) = I_0 \cos \theta ,\qquad (5.1)$$

where I_0 is the intensity of the beam scattered normal to the surface. Although the beam is shown incident along a particular direction in Fig. 5.3(a) this *cosine-law* behaviour is true, or nearly true, for all other angles of incidence. The law is closely obeyed by a wide variety of materials (e.g. sheet aluminium, steel, teflon) which have not been specially prepared, other than being 'workshop' polished.

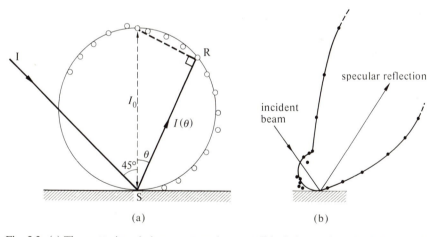

Fig. 5.3. (a) The scattering of nitrogen atoms from a polished glass surface. Both beam and solid are at 300 K. (Data from Hurlbut, F. C. (1957). *J. appl. Phys.* **28**, 844.) (b) The scattering of hydrogen molecules from a single crystal of tungsten ((100) surface) covered with a strongly-bound monolayer of hydrogen. Both beam and crystal are at 300 K. (Data from Hayward, D. O. and Walters, M. R. (1974). *Jap. J. appl. Phys.*, Suppl. 2, Pt. 2, 587.)

The cosine-law of scattering is not obeyed with certain carefully-prepared single crystals. We have already met such a case in Stern's studies of the scattering of helium atoms from a lithium fluoride crystal (Fig. 2.3(c)). Fig. 5.3(b) shows how a beam of hydrogen molecules is scattered from a single crystal of tungsten (covered with a strongly-bound monolayer of hydrogen). Only 35 per cent of the hydrogen atoms are diffusely scattered; the remaining 65 per cent are in the broad specularly-directed lobe. By analogy with the behaviour of a light beam striking a solid surface, we usually speak of *diffuse scattering* if the cosine-law is obeyed and of *specular reflection* if the angle of scattering is equal to the angle of incidence.

As a rough-and-ready rule, diffuse scattering will always occur if the size of any surface irregularities on the solid is greater than the de Broglie wavelength $\lambda = h/m\bar{u}$ of the gas atoms. Under these circumstances the surface will appear rough to the incoming atoms. Substituting for $\bar{u}$ from eqn (4.50) gives $\lambda = h(\pi/8kTm)^{1/2}$, which has a value (see Table 2.1) of $1\cdot1\times10^{-10}$ m for H_2 at 273 K and a value of $0\cdot30\times10^{-10}$ m for N_2 at 273 K. Since most mechanically polished surfaces have irregularities greatly in excess of 10^{-10} m we see that diffuse scattering is indeed to be expected from such materials. Only a cleaved crystal will be free of these gross irregularities.

Adsorption forces

Although we can frequently ignore the attractive force existing between atoms in the gas phase the fact that even inert-gas atoms do dwell on solid surfaces shows that it is usually *not* possible to ignore the attractive force that exists betwen a gas atom and a solid. This difference arises because, unlike atom i of Fig. 5.4, which lies within the body of the

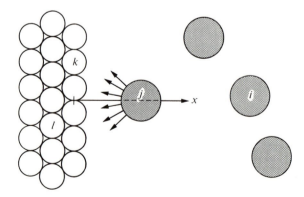

Fig. 5.4. Unlike atom i within the body of the gas, atom j experiences a significant attractive force directed towards the solid.

gas, and which has but few nearest neighbours, atom j has a great many neighbours, most of which act to pull the gas atom towards the solid. To find this force we must, of course, know the nature of the interaction between a gas atom and an atom of the solid. One would normally proceed by finding the total potential energy $V_j(x)$ of atom j when at a distance x from the surface by writing down the separate potential energies between j and k, between j and l, etc., and then summing these expressions. The net force $F(x)$ on j due to the atoms of the solid is then found from $F(x) = -dV_j(x)/dx$. It turns out that $F(x) \propto -1/x^6$ when the force between a gas atom and an atom of the solid is of van der Waals type (in which, you will recall, $V(r) \propto -1/r^6$). The process of binding an atom or molecule to a surface by these van der Waals forces is referred to

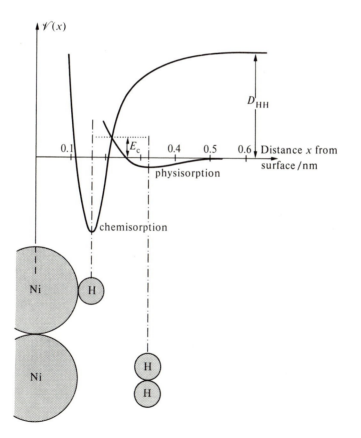

Fig. 5.5. Potential energy curves for the physisorption and chemisorption of hydrogen on nickel, along with a pictorial representation of these adsorbed states.

as *physical adsorption* (often shortened to *physisorption*). In such a process, which can occur with all types of atoms or molecules, the physical integrity of the adsorbed species is preserved. In *chemisorption* the integrity of the absorbed species is lost; there is at least a partial transfer of electrons to or from the solid surface.

The difference between physisorption and chemisorption is neatly exemplified by the adsorption of hydrogen on nickel (Fig. 5.5). When physisorption occurs, the hydrogen *molecule* is loosely bound by van der Waals forces to the nickel surface. When chemisorption occurs, individual hydrogen *atoms* are strongly bound to the nickel. (In Fig. 5.5 the potential energy separation D_{HH} records the energy required to first dissociate the molecule into individual atoms, atoms whose potential energy subsequently drops as each one binds strongly to a nickel atom.) We also see that a physically adsorbed molecule only requires an activation energy E_c for chemisorption to occur. This contrasts with the energy D_{HH} which would have to be provided to first split the H_2 molecule were physisorption not to occur. Thus physisorption allows the hydrogen molecule to get close to the surface and lowers the barrier to chemisorption.

One of the reasons why surfaces frequently act as catalysts is that an adsorbed molecule can often have its shape changed into one that more readily undergoes a reaction. When dissociation occurs, one of the fragments may interact with another incoming molecule. Alternatively, the fragments may migrate over the surface until they meet, and interact with, some other adsorbed species.

5.2. The number of collisions with a solid surface

The number of gas atoms which strike unit area of surface per unit time is an important factor in determining many of the properties of a gas, such as its pressure.

Any surface located in a gas will be continually bombarded by atoms coming in from all directions (over a solid angle 2π) and possessing speeds from near zero to near infinity. Let us therefore consider an area A of such a surface (Fig. 5.6(a)). If we now construct the slanted cylinder shown in Fig. 5.6(a) of length $u\,dt$, and thus of volume $Au\,dt\cos\theta$, then all those atoms moving towards A along directions θ, ϕ with speed u will strike A in time dt. (By way of analogy, the number of cars moving at $40\,\text{m s}^{-1}$ along a one-way street that will pass a fixed point on the road within the next 3 s is equal to the number of cars at present lying up to a distance of 120 m from the point.) However, recalling Chapter 4, we know that the number of atoms with a speed of u *precisely* and moving along *precise* directions θ, ϕ is almost certainly zero. We must therefore

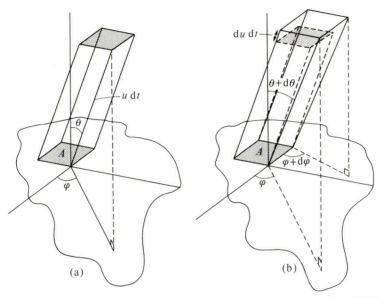

Fig. 5.6. (a) Showing a cylinder of length $u\,dt$ at polar angle θ and azimuth ϕ. (b) Here θ and ϕ range over $d\theta$ and $d\phi$, respectively, and u ranges over du.

evaluate not the number of atoms lying within the cylinder shown in Fig. 5.6(a) but rather the number lying within the cylinder shown in Fig. 5.6(b). Clearly this is equal to the volume of the cylinder (essentially unaltered at $Au\,dt\cos\theta$) multiplied by the number of atoms per unit volume with speeds between u and $u+du$ and moving between θ and $\theta+d\theta$ and between ϕ and $\phi+d\phi$ (this is equal to the total number of atoms per unit volume, n, multiplied by the fraction lying within the desired range of speeds and angles (see eqn (4.63) and problem (5.29)), that is

$$Au\,dt\cos\theta\times\frac{n}{4\pi}f_3(u)\sin\theta\,d\theta\,d\phi\,du, \tag{5.2}$$

where $f_3(u)$ is the speed distribution function in the gas.

To find the total number of collisions per unit area of surface per unit time, I_n, we must integrate eqn (5.2) over θ from $\theta=0$ to $\theta=\pi/2$, over ϕ from $\phi=0$ to $\phi=2\pi$, and over u from $u=0$ to $u=\infty$, and divide by $A\,dt$. This gives

$$I_n=\frac{n}{4\pi}\int_0^{\pi/2}\sin\theta\cos\theta\,d\theta\int_0^{2\pi}d\phi\int_0^{\infty}uf_3(u)\,du$$

$$I_n=\frac{n}{4\pi}[\tfrac{1}{2}\sin^2\theta]_0^{\pi/2}[\phi]_0^{2\pi}\int_0^{\infty}uf_3(u)\,du \tag{5.3}$$

or, since the integral in eqn (5.3) is the mean speed $\bar{u}$ of a gas atom (see eqn (4.48)),

$$I_n = \tfrac{1}{4}n\bar{u}$$. (5.4)

By way of example we will use eqn (5.4) to calculate the time it takes for a monolayer of adsorbed oxygen to form on a solid surface. In so doing we will assume that every molecule striking the surface is adsorbed. Since an oxygen molecule has a cross-sectional area of $1\cdot2\times10^{-19}\,\mathrm{m^2}$ it follows that at least $8\cdot3\times10^{18}$ impacts are required to cover $1\,\mathrm{m^2}$ of surface. Now, at s.t.p., 1 mol of a gas (containing 6×10^{23} molecules) occupies a volume of around $2\cdot24\times10^{-2}\,\mathrm{m^3}$; hence $n = 2\cdot7\times10^{25}\,\mathrm{m^{-3}}$. Since (see eqn (4.50)) $\bar{u} = (8kT/\pi m)^{1/2}$ it follows that $\bar{u} = 424\,\mathrm{m\,s^{-1}}$ for oxygen at 273 K. Thus $I_n = 2\cdot8\times10^{27}\,\mathrm{m^{-2}\,s^{-1}}$ and so the time taken to form a monolayer of oxygen is $(8\cdot3\times10^{18})/(2\cdot8\times10^{27})\,\mathrm{s} = 3\times10^{-9}\,\mathrm{s}$ at s.t.p. At a pressure of 10^{-12} times that of normal atmospheric pressure (regarded as an ultrahigh vacuum) n, and thus I_n, is 10^{-12} times that at s.t.p. Hence it takes $3\times10^3\,\mathrm{s}$ (50 minutes) to form the monolayer at such a pressure. This pinpoints the necessity of having ultrahigh vacuums when studying the surface properties of solids.

We will find it convenient in later sections to have an expression for the number of atoms $I_{n\theta}$ whose speeds between u and $u + du$ and which collide at angles between θ and $d\theta$ per unit area of surface per unit time. This is obtained by integrating eqn (5.2) from $\phi = 0$ to $\phi = 2\pi$ and dividing the result by $A\,dt$. Hence

$$I_{n\theta} = \tfrac{1}{2}nuf_3(u)\,du \sin\theta\cos\theta\,d\theta$$. (5.5)

The one-sixth model

A frequently employed, though rather simplistic, model states that a gas behaves *as if*, at any instant, one-sixth of the atoms are moving along the $+x$-direction, one-sixth are moving along the $-x$-direction, and likewise for the other cartesian directions. To calculate I_n assuming this *one-sixth model* of a gas, we replace the slanting cylinder of Fig. 5.6(b) by one lying perpendicular to the surface, as shown in Fig. 5.7. Since the number of atoms per unit volume with speeds between u and $u + du$ and moving up towards the surface is $\tfrac{1}{6}nf_3(u)\,du$ it follows that the number of atoms with speeds between u and $u + du$ which strike A in time dt is equal to $\tfrac{1}{6}nf_3(u)\,du$ multiplied by the volume of the cylinder (approximately $Au\,dt$), that is $Au\,dt\,(\tfrac{1}{6}n)f_3(u)\,du$. Integrating this expression from $u = 0$ to $u = \infty$ to include atoms of all speeds and dividing by $A\,dt$ gives the

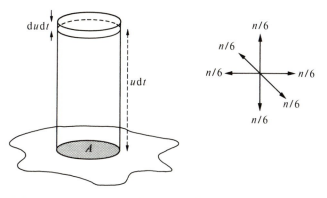

Fig. 5.7. All those atoms within the cylinder of volume $Au\,dt$ and moving towards the surface with speeds of between u and $u+du$ will strike A in time dt.

collision frequency as

$$I_n = \tfrac{1}{6}n\bar{u}\ .\tag{5.6}$$

The one-sixth model will prove useful in considering transport processes (Chapter 7).

Effusion

Although our discussion has centred around the number of atoms about to strike the surface of a solid the same arguments should apply if the surface is absent. In other words if, as shown in Fig. 5.8(a), we cut a small hole of area A in a thin plate the number of atoms $I_n(\theta)$ emerging per unit time within a solid angle $d\omega$ at an angle θ to the normal should equal the number moving up towards the hole per unit time within the solid angle $d\omega$ at θ to the normal. Thus $I_n(\theta)$ is given by eqn (5.2) integrated from $u=0$ to $u=\infty$ and divided by dt, that is

$$I_n(\theta) = \frac{nA}{4\pi} \cos\theta \sin\theta\,d\theta\,d\phi \int_0^\infty uf_3(u)\,du.$$

But $d\omega = \sin\theta\,d\theta\,d\phi$ (see eqn (4.64)) and the integral is equal to $\bar{u}$ (eqn (4.48)). Thus

$$I_n(\theta) = \frac{A\,d\omega}{4\pi}\,n\bar{u}\cos\theta\ .\tag{5.7}$$

For this to apply the diameter of the hole must be very much less than the

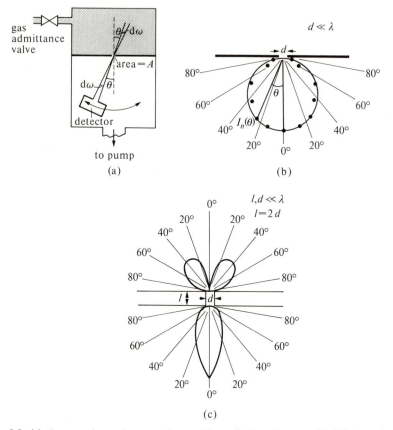

Fig. 5.8. (a) An experimental system for studying effusion of a gas. (b) Effusion of air through a circular aperture of diameter $d \ll \lambda$ (Holland, L. and Priestland, C. (1966). *Nature*, **209**, 274). (c) Predicted effusion through a cylindrical aperture of length $l = 2d$. Both $l \ll \lambda$ and $d \ll \lambda$ (Dayton, B. (1956). *Trans. AVS Vac. Symp.*, **3**, 5).

mean distance travelled by a gas atom between collisions with other gas atoms (the so-called gaseous *mean free path* λ). This ensures that there is a negligible probability that an incoming gas atom will collide with another gas atom, and so be deflected, as it passes through the hole. Under these conditions eqn (5.7) is obeyed in practice (see Fig. 5.8(b)). You will notice that, as expected, Fig. 5.8(b) is the mirror image of Fig. 5.3(a), reflected across the plane of the aperture.

When the hole of Fig. 5.8(b) is replaced by a short cylinder as in Fig. 5.8(c) (here the cylinder's length l is twice its diameter d, and l, $d \ll \lambda$) we see that the emerging beam is strongly directed about $\theta = 0$. Although a long narrow cylinder has been used to produce a near-parallel beam of gas atoms it is more usual to employ a series of circular apertures, as in

Fig. 2.3 (this makes it harder for off-axis atoms to get back in the beam). Fig. 5.8(c) shows that not all the incoming gas atoms succeed in emerging from the output end of the tube. This arises because gas atoms leaving the walls of the tube at the end of their dwell time may do so in any direction. However, summing the magnitudes of the upward and downward-moving beams in Fig. 5.8(c) does lead to a cosine distribution.

So far we have only considered the effusion of a single type of gas through the aperture of Fig. 5.8(a). If there are, say, two gases present on the high-pressure side it follows from eqn (5.4) that the ratio of the rates at which the two species effuse (measured in terms of number of atoms) is given by

$$\frac{I_{n_1}}{I_{n_2}} = \frac{n_1 \bar{u}_1}{n_2 \bar{u}_2} = \frac{n_1}{n_2} \left(\frac{m_2}{m_1}\right)^{1/2}, \tag{5.8}$$

where m_1 and m_2 are the atomic (or molecular) masses of the two gases. Eqn (5.8) has also used the relation $\bar{u} = (8kT/\pi m)^{1/2}$ (eqn (4.50)). Expressed in terms of the masses of the two gases which effuse through the aperture, we have

$$\frac{I_{m_1}}{I_{m_2}} = \frac{m_1 I_{n_1}}{m_2 I_{n_2}} = \frac{n_1}{n_2} \left(\frac{m_1}{m_2}\right)^{1/2}. \tag{5.9}$$

As an example of the application of eqn (5.9) we may mention the first separation of argon from nitrogen by Rayleigh and Ramsay in 1895. They took the stems of twelve clay pipes and evacuated the space outside the stems, the inside of the stems being open to the atmosphere. Some sixteen hours later they collected the gas which has 'transpired' through to the outside of the stems and, after removing O_2, CO_2, and H_2O, compared its density with that of normal air which had been similarly treated to remove O_2, CO_2, and H_2O. Both densities were measured at the same temperature and pressure. They invariably found that the gas that had transpired had the greater density. Rayleigh and Ramsay thus concluded that atmospheric nitrogen is actually a mixture of nitrogen and a 'heavier component' (which they christened argon). The fact that this component is heavier than nitrogen follows immediately from eqn (5.9).

It is customary to reserve the name effusion for transport through holes of negligible length (as in Fig. 5.8(a)) and to call the process *transpiration* if, as in Rayleigh and Ramsays' experiments, the length of the channel is significantly greater than its diameter.

Exercise 5.1

Show that the mean energy of the atoms effusing through an aperture in an oven is $2kT$, where T is the temperature of the oven.

Calculation. The mean energy $\bar{E}$ is found by calculating the total energy E

transported from the oven in a time t and then dividing this energy by the number of atoms N effusing from the oven in time t.

Since each atom of speed u has kinetic energy $\frac{1}{2}mu^2$ it follows that the energy dE transported from the oven in time t by atoms with speeds of between u and $u+du$ and moving at angles between θ and $\theta+d\theta$ to an axis passing through the aperture (of area A) is given by

$$dE = I_{n\theta}(\tfrac{1}{2}mu^2)At.$$

Substituting for $I_{n\theta}$ from eqn (5.5) and integrating from $u=0$ to $u=\infty$ and from $\theta=0$ to $\theta=\theta_1$ (the beam angle passed by the system of apertures—if any—lying outside the oven) gives

$$E = \tfrac{1}{4}Atmn \int_0^\infty u^3 f_3(u)\,du \int_0^{\theta_1} \sin\theta \cos\theta \,d\theta.$$

The total number of atoms N effusing in a time t is $I_{n\theta}At$ integrated over the same limits:

$$N = \tfrac{1}{2}Atn \int_0^\infty u f_3(u)\,du \int_0^{\theta_1} \sin\theta \cos\theta \,d\theta.$$

Dividing E by N and substituting for $f_3(u)$ from eqn (4.46) gives

$$\bar{E} = \frac{E}{N} = \frac{\tfrac{1}{2}m \int_0^\infty u^5\, e^{-mu^2/2kT}\,du}{\int_0^\infty u^3\, e^{-mu^2/2kT}\,du}.$$

To find the numerator and the denominator we use the standard integrals I_5 and I_3 of Table 4.1 with $\lambda = m/2kT$. This leads to

$$\bar{E} = \tfrac{1}{2}m\frac{(1/\lambda^3)}{(1/2\lambda^2)} = \frac{m}{\lambda} = 2kT. \tag{5.10}$$

Comment. The value of $\bar{E}$ is independent of the collimation angle θ_1 of the beam. Irrespective of the precise form of $I_{n\theta}$ in the collimated beam the integrals over θ in the expressions for E and N cancel in $\bar{E}(=E/N)$.

5.3. The pressure required to contain a perfect gas

In discussing the pressure required to contain a perfect gas we will not only assume that $\Delta E = 0$ and that the gas atoms have negligible size (these being the defining features of a perfect gas) but we will also assume that when the gas atoms strike the piston (Fig. 5.9(a)) they stick for sufficient time for there to be no correlation between the angle of incidence θ_i of a particular atom and the angle θ_o at which this atom will leave the piston later on. Of course, on average, the form of $I_n(\theta)$ must be the same for atoms striking and leaving the piston; were the two forms different there would be a net movement of gas within the cylinder.

In essence our explanation as to why we must keep pushing on a

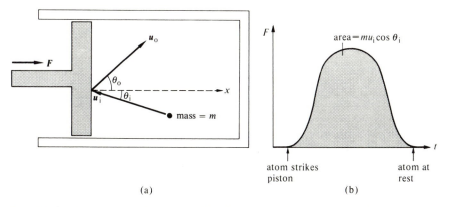

Fig. 5.9. (a) A single atom of mass m strikes the piston with a velocity u_i and sticks. Some time later it leaves the piston with a velocity u_o. (b) Showing how the force F required to bring the atom to rest on the piston might vary with time.

piston in order to contain a gas is that the gas atoms are continually colliding with the piston and we are continually having to send these atoms back into the body of the gas. We will start, however, by considering what happens when a single atom of mass m and moving with a speed u_i at a direction θ_i strikes the piston. To bring this atom to rest we must, according to Newton's second law, provide an impulse $\int F \, dt$ equal to $m(u_i \cos \theta_i)$. Here F is the force directed perpendicular to the face of the piston. The exact shape of the force–time curve will depend on the 'softness', or otherwise, of the gas atom; all we can say is that the area, $\int F \, dt$, under the curve must equal $mu_i \cos \theta_i$ (Fig. 5.9(b)).

There are, of course, a great many atoms striking the piston. According to eqn (5.5) the total number of atoms currently moving towards the piston with speeds between u_i and $u_i + du_i$ lying along directions θ_i to $\theta_i + d\theta_i$ that will strike an area A of the piston within a time t is

$$I_{n\theta} At = \tfrac{1}{2} nAt u_i f_3(u_i) \, du_i \sin \theta_i \cos \theta_i \, d\theta_i. \tag{5.11}$$

Since each of these atoms is brought to rest at the piston it follows that their total change in momentum resolved along a direction perpendicular to the piston face is given by eqn (5.11) multiplied by $mu_i \cos \theta_i$, that is

$$\tfrac{1}{2} mnAt u_i^2 f_3(u_i) \, du_i \sin \theta_i \cos^2 \theta_i \, d\theta_i. \tag{5.12}$$

To take account of the various classes of atoms present eqn (5.12) must now be integrated over θ_i from $\theta_i = 0$ to $\theta_i = \pi/2$ and over u_i from $u_i = 0$ to $u_i = \infty$. (The integration over ϕ from $\phi = 0$ to $\phi = 2\pi$ has already been performed in eqn (5.11); see p. 136). This gives the total change in the perpendicular component of momentum of all the atoms that strike an area A of the piston within a time t as

$$\tfrac{1}{2}mnAt \int_0^\infty u_i^2 f_3(u_i)\, du_i [-\tfrac{1}{3}\cos^3\theta_i]_0^{\pi/2} \tag{5.13}$$

or, recalling that the integral represents $\overline{u^2}$, as

$$\tfrac{1}{6} mn\overline{u^2} At. \tag{5.14}$$

So far we have only, so to speak, brought the gas atoms to a halt. At some later time (equal to the dwell time) each atom will leave the surface and move back into the body of the gas. Now both the ingoing and the outgoing atoms are characterized by the same θ-dependence (recall Figs 5.3(a) and 5.8(b)). Eqn (5.13) with u_i and θ_i replaced by u_o and θ_o therefore also gives the change in the perpendicular momenta of the atoms leaving the surface. Hence the overall change in the perpendicular momenta of all the atoms as they strike, stick, and leave an area A of the piston is twice that given by eqn (5.14), namely

$$\tfrac{1}{3}mn\overline{u^2} At.$$

The person pushing on the piston must thus provide the series of impulses shown schematically in Fig. 5.10. The apparently constant force $\bar{F}$ which is provided over the time t is of such a magnitude that the area $\bar{F}t$ is equal to the total area, $\tfrac{1}{3}mn\overline{u^2} At$, under all the 'blips'. That is, the observed impulse $\bar{F}t$ is equal to the sum of all the individual impulses. Thus,

$$\bar{F} = \tfrac{1}{3}mn\overline{u^2} A$$

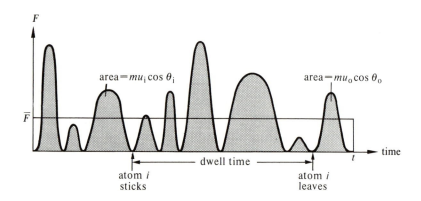

Fig. 5.10. Showing the sequence of impulses necessary to change the momentum of a stream of atoms. Each 'blip' represents the impulse required to either bring an incoming atom to rest on the piston or to expel an atom from the piston (which it leaves at the end of its dwell time).

or, in terms of the gas pressure $p(=\bar{F}/A)$,

$$\boxed{p = \tfrac{1}{3}mn\overline{u^2}}.$$ (5.15)

But $mn = \rho$, the (mass) density of the gas, allowing us to write

$$p = \tfrac{1}{3}\rho\overline{u^2}.$$ (5.16)

Since $n = N/V$, where N is the total number of atoms present and V is the volume of the gas, eqn (5.15) may also be written as

$$pV = \tfrac{1}{3}mN\overline{u^2}.$$ (5.17)

If an amount n of gas is present then $N = nN_A$ and so

$$pV = \tfrac{1}{3}mnN_A\overline{u^2}$$

$$pV_m = \tfrac{2}{3}N_A(\tfrac{1}{2}m\overline{u^2}),$$ (5.18)

where $V_m = V/n$ is the molar volume of the gas.

Attraction of the walls

A potentially serious criticism of our discussion is that it neglected the attractive force F_{pa} (read 'piston on atom') which the piston will exert on an incoming gas atom (Fig. 5.11(a)). Surely (one might argue) the perpendicular component of the momentum of the gas atom as it strikes the piston will have a value of $mu_x + \int |F_{pa}|\,dt$, where the integral is over the period during which the atom is accelerated, rather than the value mu_x which we have been assuming up to now? Since this means that a larger impulse has to be provided to bring the gas atom to rest there must therefore be an increase in the predicted value of the gas pressure. While these arguments are correct, so far as they go, they have failed to consider that as the gas atoms get pulled into the piston by the force $|F_{pa}|$ so the piston will get pulled out by the force $|F_{ap}|$ (Fig. 5.11(a)). By Newton's third law, $|F_{ap}| = |F_{pa}|$ but is oppositely directed. To prevent the piston from moving into the gas the person holding it must pull the other way, that is he must provide an impulse $\int |F_{ap}|\,dt$ directed to the *left*. When the gas atom eventually leaves the piston to return to the body of the gas, it will try to drag the piston with it. To prevent this happening the person holding the piston must provide an impulse $\int |F_{ap}|\,dt$, again directed to the *left*. The overall time dependence of the force provided by the person holding the piston during the collision sequence is shown in Fig. 5.11(b). The net impulse provided by this person is therefore $2mu_x + 2\int |F_{pa}|\,dt - 2\int |F_{ap}|\,dt = 2mu_x$ since $|F_{pa}| = |F_{ap}|$. This is the same as the impulse

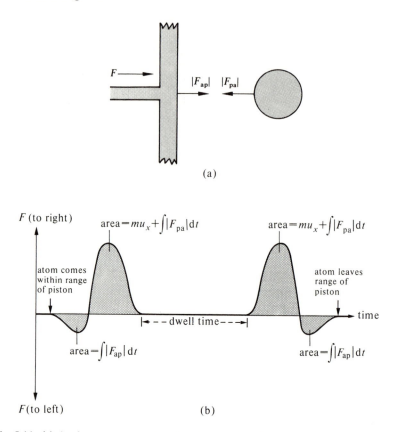

Fig. 5.11. (a) As the atom approaches the piston each pulls on the other partner. (b) The impulses required to reverse the motion of an atom colliding with the piston when the force of attraction between the gas atom and the piston is considered. For simplicity it is here supposed that the gas atom leaves the piston with a perpendicular momentum equal in magnitude to the perpendicular momentum with which it struck the piston.

required when the attraction between the piston and the gas atom is neglected.

Experimental checks

The only direct check on the correctness, or otherwise, of our expressions for the pressure required to contain a gas is to measure $\overline{u^2}$ within the gas and to see whether this agrees with $3p/\rho$, as eqn (5.16) predicts. The measurements of pressure and density are easily accomplished. There are at least two different methods of measuring $\overline{u^2}$ within the gas. The first is to measure $\overline{u^2}$ in a beam of gas effusing through a hole in an oven (using the, by now familiar, time-of-flight techniques) and hence to calculate $\overline{u^2}$ in the oven (exercise 3.1). On comparing the calculated $\overline{u^2}$ with the measured $3p/\rho$ the two generally agree to within the experimental errors; typically some few per cent. This experimental check is not particularly revealing as the gas pressure in the oven must be set at a small fraction of atmospheric pressure. Another method of estimating the r.m.s. atomic speed in a gas is to measure the speed of sound in that gas. Without going into the details of how sound is transmitted in a gas we might hazard the guess that the speed at which it travels along a particular direction, say the x-direction, will be approximately equal to $(\overline{u_x^2})^{1/2}$, that is to $(\frac{1}{3}\overline{u^2})^{1/2}$ (eqn (3.14)). Since the speed of sound in air at sea-level is around 330 m s^{-1} this would suggest that $(\overline{u^2})^{1/2} = 572\text{ m s}^{-1}$. Now air at sea-level has a density of about $1\cdot 2\text{ kg m}^{-3}$ and is at a pressure of $1\cdot 0 \times 10^5\text{ Pa}$. Eqn (5.16) would thus predict $(\overline{u^2})^{1/2} = 500\text{ m s}^{-1}$. The fact that the values of $(\overline{u^2})^{1/2}$ as deduced from the speed of sound and from $p = \frac{1}{3}\rho\overline{u^2}$ agree to within about fifteen per cent strongly suggests that sound waves are indeed propagated through the movement of gas atoms.

Exercise 5.2

There are many different derivations of $p = \frac{1}{3}mn\overline{u^2}$, besides that given above. One such derivation proceeds by first resolving the velocities u_i of atoms moving up towards the wall into their components u_x, u_y, and u_z. As the u_y and u_z-components cannot change when an atom collides with the piston (rather, they cannot change on average) we need only consider the u_x components (Fig. 5.12(a)). The number of collisions with an area A of the piston in a time t by atoms having velocity components of between u_x and $u_x + \mathrm{d}u_x$ is given by $(\frac{1}{2}nf_3(u_x)\,\mathrm{d}u_x)Au_xt$; the term in brackets is the number of atoms per unit volume with the desired range of velocities (we divide by two because only half the atoms with component speeds between u_x and $u_x + \mathrm{d}u_x$ are moving up towards the surface) and Au_xt is the volume of the cylinder shown in Fig. 5.12(b). On being

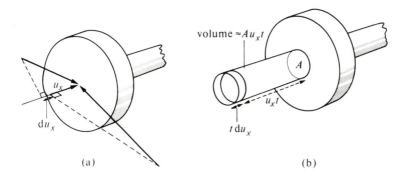

(a)   (b)

Fig. 5.12. (a) Showing the velocity vectors of two incoming gas atoms resolved into component velocities of magnitudes u_x and $u_x + \mathrm{d}u_x$ perpendicular to the piston face. (b) All those atoms within the cylinder shown and having (component) speeds of between u_x and $u_x + \mathrm{d}u_x$ will strike A in time t.

brought to rest at the piston the x-component of the momentum of each of these atoms changes by an amount mu_x. Using these clues show that $p = \frac{1}{3}mnu^2$. You will not find it necessary to use the explicit form of $f_3(u_x)$ (given by eqn (4.69)).

  Solution. The total momenta brought up to an area A of the piston in a time t by atoms whose velocity components perpendicular to the piston lie between u_x and $u_x + du_x$ is

$$\tfrac{1}{2}nf_3(u_x)u_x\, du_x At \times mu_x = \tfrac{1}{2}mnAtu_x^2 f_3(u_x)\, du_x.$$

On integrating this expression from $u_x = 0$ to $u_x = \infty$ we find that the total change in momentum perpendicular to the piston face as these atoms are brought to rest is given by

$$\tfrac{1}{2}mnAt \int_0^\infty u_x^2 f_3(u_x)\, du_x = \tfrac{1}{2}mnAt\overline{u_x^2}.$$

Of course, these atoms eventually leave the piston and return to the body of the gas. Hence the overall change in the perpendicular momenta of all the atoms as they strike, stick, and leave an area A of the piston is given by

$$mnAt\overline{u_x^2} = \tfrac{1}{3}mnAt\overline{u^2},$$

where we have used the fact that $\overline{u_x^2} = \tfrac{1}{3}\overline{u^2}$ (eqn (3.14)). This change in momentum is brought about by the person pushing on the piston with an apparently steady force $\bar{F}$ such that

$$\bar{F}t = \tfrac{1}{3}mnAt\overline{u^2}.$$

Hence the gas pressure $p(=\bar{F}/A) = \tfrac{1}{3}mn\overline{u^2}$, as before.

  Comment. Always remember when applying the speed distribution function $f_3(u_x)$ that half the atoms with a speed component of u_x are moving in the $+x$-direction and half are moving in the $-x$-direction.

5.4. The equation of state of a perfect gas

  Our discussion of gas pressure (eqn (5.15)) has so far made no mention of the temperature T of the gas. This can be remedied by introducing $\tfrac{1}{2}m\overline{u^2} = \tfrac{3}{2}kT$ (eqn (3.26)) into eqn (5.15) to give

$$\boxed{p = nkT} \tag{5.19}$$

or, since $n = N/V$, where N is the total number of molecules present in the volume V,

$$\boxed{pV = NkT}. \tag{5.20}$$

If there is an amount n of gas present $N = nN_A$, and so

$$pV = nN_A kT \tag{5.21}$$

or, since $N_A k = R$ (eqn (3.25)),

$$pV = nRT \qquad (5.22)$$

$$pV_m = RT \quad, \qquad (5.23)$$

where $V_m = V/n$ is the molar volume of the gas. This last equation, which interrelates p, V_m, and T, is known as the *equation of state* of a perfect gas. Eqns (5.19)–(5.23) are, of course, just variations on a common theme! It is usually immediately clear which is the most appropriate one to use in a given situation. For example, if we wish to calculate the number of atoms present per unit volume of gas at a specified pressure and temperature we would use eqn (5.19). Thus at s.t.p., where $p = 1 \cdot 013 \times 10^5$ Pa and $T = 273 \cdot 15$ K, we have $n = p/kT = 2 \cdot 69 \times 10^{25}$ m^{-3}.

Plotted out as a three-dimensional surface eqn (5.23) has the form shown in Fig. 5.20. Along an isotherm pV_m is constant (Boyle's law); p plotted against V_m is a rectangular hyperbola. Along an isochore p is proportional to T, while along an isobar V_m is proportional to T. These last two features are only to be expected since they are implicit in the definition of gas-scale temperatures.

Avogadro's law (equal volumes of all gases measured at the same temperature and pressure contain the same number of molecules) follows immediately from eqn (5.20). As a final corollary, Dalton's law of partial pressures (the pressure exerted by a mixture of gases on the walls of a container is equal to the sum of the pressures that would be exerted by the gases if they were present separately) follows directly from eqn (5.19) written as $p = (n_1 + n_2 + \cdots)kT = n_1 kT + n_2 kT + \cdots = p_1 + p_2 + \cdots$. Any deductions made from eqns (5.19) to (5.23) will, of course, only hold true in so far as eqn (5.15) holds true.

Experimental checks

According to eqn (5.23) pV_m should have a common value for all gases maintained at a common temperature, and that value should be independent of pressure. Fig. 5.13(a) shows that this is not true in general. However each pressure division in Fig. 5.13(a) is 2×10^7 Pa, which is some two hundred times normal atmospheric pressure. At lower pressures pV_m does indeed remain much more nearly constant. Fig. 5.13(b) shows that oxygen, nitrogen, helium, and hydrogen all obey Boyle's law to within one per cent at pressures up to ten atmospheres at $T = 273 \cdot 16$ K. The fact that pV_m is essentially independent of p at one fixed temperature is, of course, no guarantee that this will apply at a

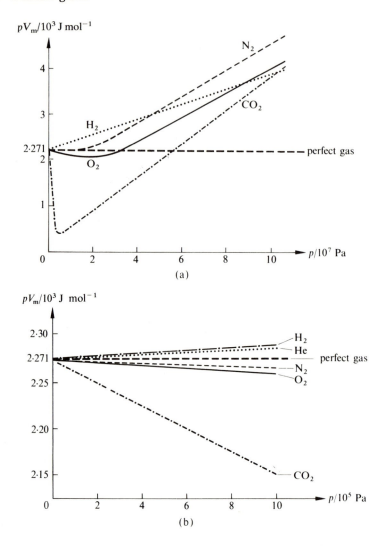

Fig. 5.13. (a) Showing how pV_m varies with p for a number of different gases, all maintained at the triple point of water. (b) As (a) but with a hundredfold smaller range of pressures.

different fixed temperature. One way of exploring this question is to plot graphs of p versus V_m at various temperatures and to see whether the isotherms are rectangular hyperbolas. Such a plot is shown in Fig. 5.14 for carbon dioxide. A more revealing test is to plot (pV_m/RT)—often called the *compression factor* Z—as measured at different temperatures against either $1/V_m$ or p; this is a revealing test because, were the gas

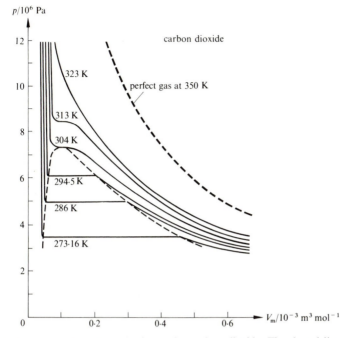

Fig. 5.14. Pressure–molar volume isotherms for carbon dioxide. The dotted line labelled 350 K shows how carbon dioxide would behave were it a perfect gas at this temperature.

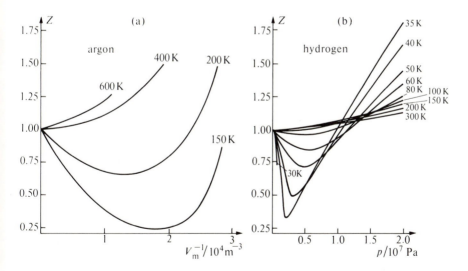

Fig. 5.15. The compression factor $Z(=pV_m/RT)$ plotted (a) against $1/V_m$ for argon at different temperatures, and (b) against p for hydrogen at different temperatures.

perfect, $Z = 1$, independent of p, V_m, and T. The reason for plotting Z against $1/V_m$, rather than V_m, is that the high volume data (where Z approaches unity) can be readily displayed. Figs. 5.15(a) shows Z plotted against V_m^{-1} for argon, and Fig. 5.15(b) shows Z plotted against p for hydrogen. These plots, like Fig. 5.13 and 5.14, demonstrate clear departures from the perfect-gas equation of state. The reasons for the departures are considered in the next chapter.

Exercise 5.3

Show that the number of gas atoms I that escape per unit time from a small hole of area A in an oven containing a vapour at a pressure p and at a temperature T is given by

$$I = \frac{pA}{(2\pi mkT)^{1/2}}, \tag{5.24}$$

where m is the atomic (or molecular) mass of the substance as it occurs in the vapour phase at (p, T). Hence calculate I for caesium atoms $(A_r(Cs) = 133)$ in an oven at 770 K where the vapour pressure of caesium is $1 \cdot 1 \times 10^4$ Pa. Take $A = 2 \cdot 5 \times 10^{-13}$ m^2.

Solution. Since $I = I_n A$ it follows from eqns (5.4), (4.50), and (5.19) that

$$I = \tfrac{1}{4} n\bar{u} A = \frac{1}{4} \left(\frac{p}{kT}\right)\left(\frac{8kT}{\pi m}\right)^{1/2} A = \frac{pA}{(2\pi mkT)^{1/2}}.$$

Substituting the values given for p, A, T, and m (calculated, of course, from $A_r(Cs)$) gives $I = 2 \cdot 3 \times 10^{13}$ s^{-1}.

Comments. Eqn (5.24) assumes that the atoms effuse over 2π sr. When a system of aperture limits the beam we may use eqn (5.5) in place of eqn (5.4) and integrate θ from zero up to the maximum angle θ' passed by the apertures. This procedure will only give a rough-and-ready estimate of I since it assumes that $I_n(\theta)$ has the form shown in Fig. 5.8(b) at $\theta < \theta'$.

5.5. Thermomolecular pressures

It has been known since the 1870s that if two containers at different temperatures T_1 and T_2, but containing the same type of gas, are connected together by an aperture of diameter much less than the gaseous mean free path, then the pressures p_1 and p_1 in the two containers are unequal. To see the reason for this we will write down the condition that, once equilibrium is reached, the number of gas molecules passing through the hole of area A per unit time from side 1 to side 2, namely $\tfrac{1}{4} A n_1 \bar{u}_1$ (eqn (5.4)), must equal the number passing through the hole in the opposite direction, namely $\tfrac{1}{4} A n_2 \bar{u}_2$. Thus

$$n_1 \bar{u}_1 = n_2 \bar{u}_2.$$

Substituting $n = p/kT$ (eqn (5.19)) and $\bar{u} = (8kT/\pi m)^{1/2}$ (eqn (4.50)) leads to

$$\boxed{\frac{p_1}{p_2} = \left(\frac{T_1}{T_2}\right)^{1/2}}.$$ (5.25)

Fig. 5.16(a) shows an arrangement which has been used to check out this relation. The lower chamber was surrounded by liquid refrigerant whose temperature T_1 could be either 77·4 or 232 K. The temperature T_2 of the upper chamber was maintained constant at 295 K with the aid of a heater and a sensing thermocouple. Vacuum gauges were used to measure p_1

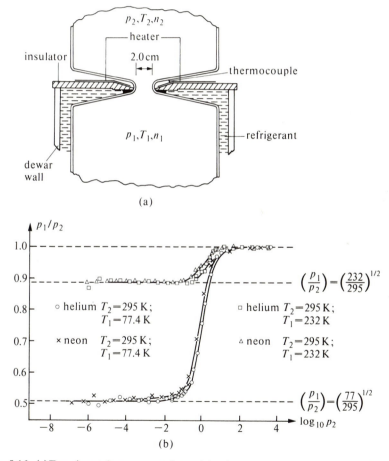

(a)

(b)

Fig. 5.16. (a) Experimental arrangement for studying thermomolecular pressures. (b) Measured values of p_1/p_2 for helium and for neon as a function of $\log_{10} p_2$, where p_2 is in units of the pascal. (From Edmonds, T. and Hobson, J. P. (1965). *J. vac. Sci. Technol.*, **2**, 182.)

and p_2. According to eqn (5.25) p_1/p_2 should depend only on T_1 and T_2 and should thus be independent of p_2. As can be seen from Fig. 5.16(b), p_1/p_2 is indeed independent of p_2 from $p_2 = 10^{-6}$ Pa to $p_2 = 10^{-1}$ Pa and has the predicted value of $(T_1/T_2)^{1/2}$. At higher pressures the gaseous mean free path becomes equal to, and then less than, the diameter of the aperture. At such pressures our analysis breaks down and the condition for equilibrium becomes $p_1 = p_2$. It is also worth pointing out that eqn (5.25) only applies if the containers are connected by a short-length channel. When the channel is much longer than the diameter (as happens in a piece of narrow piping), p_1/p_2 is no longer simply equal to $(T_1/T_2)^{1/2}$. The real importance of the thermomolecular pressures in everyday laboratory practice is that if a pressure gauge at, say, room temperature is connected via narrow piping to an experimental chamber at a different temperature, the gauge will not give a correct reading of the pressure in that chamber.

5.6. The isothermal atmosphere

Up to now we have ignored the pull of gravity on the atoms of a gas. In examining the role of gravity we will consider a vertical column of gas in Earth's atmosphere, assuming it to be isolated from all external disturbances (notably solar heating and convection currents) and to be in a state of equilibrium (meaning there is no net transfer of mass, momentum, or kinetic energy along the column). What we shall do is to express the facts that the total numbers of atoms moving upwards and downwards per unit time through a plane P_1 of unit area (Fig. 5.17) must be equal, as must the total kinetic energy passing upwards and downwards through P_1.

If we consider a region of the column at a distance h below P_1 (for the moment we shall assume $h \leq \lambda$, the gaseous mean free path), only those atoms whose vertical component of velocity u_z at P_0 is such that $\frac{1}{2}mu_z^2 > mgh$ will succeed in reaching P_1. The number of atoms passing through P_0 per unit time with velocity components between u_z and $u_z + du_z$ is given by $(\frac{1}{2}n_0 f_3(u_z) \, du_z)u_z$ (see exercise 5.2). Hence the total

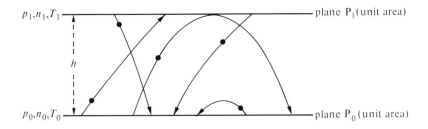

Fig. 5.17. Showing the conditions existing at two planes in Earth's atmosphere.

number of atoms passing upwards through P_1 per unit time is given by

$$\frac{1}{2} \int_{(2gh)^{1/2}}^{\infty} n_0 u_z f_3(u_z)\, du_z$$

or, substituting for $f_3(u_z)$ from eqn (4.69), by

$$n_0 \left(\frac{m}{2\pi kT_0}\right)^{1/2} \int_{(2gh)^{1/2}}^{\infty} u_z\, e^{-mu_z^2/2kT_0}\, du_z = n_0 \left(\frac{kT_0}{2\pi m}\right)^{1/2} e^{-mgh/kT_0}, \quad (5.26)$$

where n_0 and T_0 are, respectively, the local number density of the atoms and the temperature at plane P_0. The total number of atoms moving downwards through P_1 can be written down directly as

$$n_1 \left(\frac{m}{2\pi kT_1}\right)^{1/2} \int_0^{\infty} u_z\, e^{-mu_z^2/2kT_1}\, du_z = n_1 \left(\frac{kT_1}{2\pi m}\right)^{1/2}. \quad (5.27)$$

Equating eqns (5.26) and (5.27) gives

$$n_1 T_1^{1/2} = n_0 T_0^{1/2}\, e^{-mgh/kT_0}. \quad (5.28)$$

The transfer of translational kinetic energy across P_1 will now be considered. An atom leaving P_0 with an upward component of velocity u_z will have an energy $\frac{1}{2}mu_z^2 - mgh$ on passing through P_1. Hence the total energy transferred upwards through P_1 per unit time is, from eqn (5.26),

$$n_0 \left(\frac{m}{2\pi kT_0}\right)^{1/2} \int_{(2gh)^{1/2}}^{\infty} (\tfrac{1}{2}mu_z^2 - mgh)u_z\, e^{-mu_z^2/2kT_0}\, du_z = \frac{n_0(kT_0)^{3/2}}{(2\pi m)^{1/2}} e^{-mgh/kT_0}. \quad (5.29)$$

Likewise, the total energy transferred downwards through P_1 per unit time is

$$n_1 \left(\frac{m}{2\pi kT_1}\right)^{1/2} \int_0^{\infty} \tfrac{1}{2}mu_z^3\, e^{-mu_z^2/2kT_1}\, du_z = \frac{n_1(kT_1)^{3/2}}{(2\pi m)^{1/2}}. \quad (5.30)$$

Equating eqns (5.29) and (5.30) gives

$$n_1 T_1^{3/2} = n_0 T_0^{3/2}\, e^{-mgh/kT_0}. \quad (5.31)$$

Dividing eqn (5.31) by eqn (5.28) gives $T_1 = T_0$; that is the atmosphere is isothermal (at temperature T, say). Hence from eqn (5.31)

$$\boxed{n_1 = n_0\, e^{-mgh/kT}} \quad (5.32)$$

or, multiplying through by kT and appealing to eqn (5.19),

$$\boxed{p_1 = p_0\, e^{-mgh/kT}}. \quad (5.33)$$

Although we originally assumed that $h \le \lambda$, thereby ensuring that a gas atom would not suffer a collision in travelling from P_0 to P_1, this was unnecessarily restrictive. In general, if an equilibrium distribution exists at each point, collisions must throw as many atoms into a particular group of velocities as are removed from that group.

This isothermal model of Earth's atmosphere only really applies at heights of between about 200 and 500 km, over which range the temperature will remain constant to within about ten per cent. (The actual temperature depends on the level of sunspot activity and may vary from 900 to 2500 K.) The temperature of the atmosphere also remains sensibly constant, at around 215 K, at heights between 10 and 20 km. Assuming air to be composed solely of nitrogen ($m = 4 \cdot 65 \times 10^{-26}$ kg), eqn (5.32) predicts that the ratio of the density of air at 20 km to its density at 10 km should be $0 \cdot 215$. The measured ratio is $0 \cdot 222$. The close agreement is specious since eqn (5.32) also predicts that n_1/n_0 should vary with h in a different fashion for different molecular species. In fact the composition of the atmosphere remains reasonably constant at heights up to 100 km, implying a degree of mixing which is inconsistent with the equilibrium model assumed in deriving eqn (5.32).

5.7. The heat capacities of perfect gases

Since $V(r) = 0$ in a perfect gas it follows that the internal energy should be independent of the interatomic separation, and hence the volume, of the gas. That is,

$$\left(\frac{\partial U_m}{\partial V_m}\right)_T = 0. \tag{5.34}$$

As a consequence, the difference between the molar heat capacities at constant pressure and at constant volume (eqn (1.15)) becomes

$$C_{p,m} - C_{V,m} = p\left(\frac{\partial V_m}{\partial T}\right)_p \tag{5.35}$$

or, since $pV_m = RT$, and thus $p(\partial V_m/\partial T)_p = R$,

$$\boxed{C_{p,m} - C_{V,m} = R} . \tag{5.36}$$

Monatomic gases

Since the interatomic potential energy is zero the internal energy of the gas must reside in the kinetic energy of its atoms. Therefore $U_m =$

$N_A(\frac{1}{2}m\overline{u^2})$ or, as $\frac{1}{2}m\overline{u^2}=\frac{3}{2}kT$ and $N_A k = R$,

$$\boxed{U_m=\tfrac{3}{2}RT}. \qquad (5.37)$$

Now $C_{V,m}=(\partial U_m/\partial T)_{V_m}$ (eqn (1.12)), and so

$$\boxed{C_{V,m}=\tfrac{3}{2}R}. \qquad (5.38)$$

The fact that the measured values of $C_{V,m}$ of (low-density) monatomic gases are indeed $\frac{3}{2}R(=12\cdot5$ J K^{-1} mol^{-1}), independent of temperature, is actually rather embarrassing! It is embarrassing because we could well have agreed that the gas atoms also possess rotational kinetic energy and that the atoms' electrons have both kinetic and potential energies. On this basis we should have written

$$U_m = N_A(\tfrac{1}{2}m\overline{u^2}+\text{mean rotational K.E.}+\text{mean electronic K.E.} \\ +\text{mean electronic P.E.}). \qquad (5.39)$$

Because $C_{V,m}$ is constant at $3R/2$, independent of T (that is of $(2/3k)\times(\frac{1}{2}m\overline{u^2})$), we must conclude that

$$\frac{\mathrm{d}(\text{mean rotational K.E.})}{\mathrm{d}(\frac{1}{2}m\overline{u^2})}=0$$

and similarly for all the following terms in eqn (5.39). Thus to get agreement with experiment we must be able to show why, for example, when two gas atoms collide more and more violently (meaning an increasing $\frac{1}{2}m\overline{u^2}$) they do not spin more and more rapidly. Classically, this would indeed be true if the atoms were perfectly smooth. What we cannot explain from the laws of classical mechanics is why the electrons do not acquire more energy as the atoms collide more violently.

Diatomic gases

Fig. 5.18(a) shows a gas composed of rigid diatomic molecules, each of mass m say. As we have already argued in §4.11 each molecule of such a gas will have an average energy given by

$$\bar{E} =\tfrac{1}{2}m\overline{u_x^2}+\tfrac{1}{2}m\overline{u_y^2}+\tfrac{1}{2}m\overline{u_z^2}+\tfrac{1}{2}I_1\overline{\omega_1^2}+\tfrac{1}{2}I_2\overline{\omega_2^2}+\tfrac{1}{2}I_3\overline{\omega_3^2}= 6\times\tfrac{1}{2}kT = 3kT. \quad (5.40)$$

On the assumption that the atoms are smooth we might guess that they cannot be set into rotation about an axis joining their centres. This allows us to set $\frac{1}{2}I_3\overline{\omega_3^2}=0$, so that eqn (5.40) becomes $\bar{E} = 5\times\frac{1}{2}kT =\frac{5}{2}kT$. Assuming that the potential energy *between* molecules is zero we may therefore

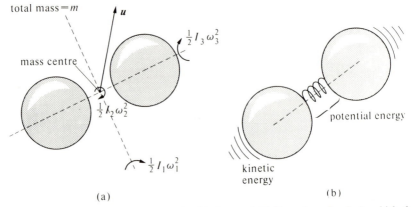

Fig. 5.18. (a) A rigid diatomic molecule. (b) A non-rigid diatomic molecule in which the interatomic forces are schematically shown as arising from a coupling spring.

write $U_m = \frac{5}{2}RT$, so that

$$\boxed{C_{V,m} = \tfrac{5}{2}R}\,. \tag{5.41}$$

If, as shown in Fig. 5.18(b), the two atoms can vibrate back and forth along the line joining their centres they will possess additional kinetic energy and additional potential energy. Each of these energies is proportional to a quadratic term (§3.9) and so, from eqn (4.83), each has an average value of $\frac{1}{2}kT$. Thus for a gas composed of non-rigid diatomic molecules we must add $N_A(\frac{1}{2}kT + \frac{1}{2}kT)$ to eqn (5.40), giving

$$\boxed{C_{V,m} = \tfrac{7}{2}R}\,. \tag{5.42}$$

Fig. 5.19 shows the experimentally determined values of $C_{V,m}$ for Cl_2 and H_2 at different temperatures. Looking at the hydrogen data and

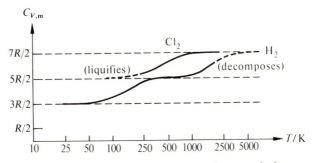

Fig. 5.19. The molar heat capacity at constant volume of gaseous hydrogen and chlorine plotted against temperature.

drawing on eqns (5.38), (5.41), and (5.42) we might conclude that at temperatures below about 50 K hydrogen behaves like a monoatomic gas (devoid of rotation and vibration), that between about 250 and 500 K it behaves like a rigid diatomic gas (devoid of vibration), and that only at the highest temperatures does it behave as a non-rigid diatomic gas. While these conclusions are indeed correct, classical physics provides no grounds for allowing a diatomic molecule to 'switch off' vibrations below a certain temperature, and to 'switch off' rotations below a still lower temperature. (In the case of Cl_2 liquefaction occurs before the expected drop to $C_{V,m} = \frac{3}{2}R$ can be reached.)

When wave mechanics is applied to a diatomic molecule it turns out that only discrete energies of vibration and rotation are possible (see p. 319). A certain minimum energy is required to start rotation; in hydrogen this occurs at around 50 K (when $\frac{1}{2}m\overline{u^2}$ is of the order of this minimum energy). Likewise, the minimum permitted vibrational energy (above the 'ground-state energy') corresponds roughly to the value of $\frac{1}{2}m\overline{u^2}$ at 500 K.

Finally it is worth pointing out that it is usually $C_{p,m}$ rather than $C_{V,m}$ which is measured experimentally. However, since most low-pressure diatomic gases closely satisfy $pV_m = RT$ (Fig. 5.13(b)), we know that $C_{V,m} = C_{p,m} - R$ (eqn (5.36)).

Degrees of freedom

It is customary to speak of the number of *degrees of freedom* of a system, by which is meant the number of independent squared terms which enter into the expression for the energy of the system. Thus one mole of a (classical) monatomic gas has $N_A \times 3$ degrees of freedom (each atom having three squared terms—namely, $\frac{1}{2}mu_x^2$, $\frac{1}{2}mu_y^2$, and $\frac{1}{2}mu_z^2$—in the expression for its energy). Likewise one mole of a gas composed of rigid diatomic molecules has $5N_A$ degrees of freedom, while one mole of a gas composed of non-rigid diatomic molecules has $7N_A$ degrees of freedom.

In general, if a molecule has f degrees of freedom then a mole of the gas will have fN_A degrees of freedom and will have an internal energy of

$$U_m = \frac{1}{2}fRT \tag{5.43}$$

and a molar heat capacity at constant volume of

$$C_{V,m} = \frac{1}{2}fR. \tag{5.44}$$

Assuming that the intermolecular potential energy is negligible in the gas it will closely satisfy $(\partial U_m / \partial V_m)_T = 0$ and $pV_m = RT$, enabling us to write $C_{p,m} - C_{V,m} = R$ (eqn (5.36)) and thus

$$C_{p,m} = R\left(1 + \frac{f}{2}\right). \tag{5.45}$$

It follows from eqns (5.44) and (5.45) that the ratio of $C_{p,m}$ to $C_{V,m}$, normally written γ, is given by

$$\boxed{\frac{C_{p,m}}{C_{V,m}} = \gamma = 1 + \frac{2}{f}} \,. \tag{5.46}$$

5.8. Adiabatic changes in a perfect gas

As was emphasized in Chapter 1, no matter how one changes the p, V_m, and T values of a substance their equilibrium values will always lie on the appropriate p–V_m–T surface of the substance. The only caveats were that there should be no friction between the piston and the cylinder (otherwise the applied pressure p will not equal the value in the system and will be substantially different during an incremental compression $-\delta V$ followed by an incremental expansion δV) and that the changes should be carried out sufficiently slowly for p and T to have a common value throughout the system (otherwise the state of the system cannot be represented by a single triplet of (p, V_m, T) values). When both these caveats are satisfied we say that the change is 'reversible and quasi-static', although one usually shortens this to the single word *reversible*.

We will now suppose that the piston and walls of the apparatus of Fig. 5.9(a) are adiabatic (heat-insulating) and that the cylinder contains a perfect gas. Qualitative arguments suggest that if the piston is allowed to move out any gas atoms striking the piston will have their kinetic energy decreased (a retreating tennis racquet decreases the speed of a returned ball). But by how much?

The most succinct way of discussing the problem is to apply the first law of thermodynamics as expressed in eqn (1.13). This tells us that during an adiabatic process ($đQ = 0$)

$$C_{V,m}\,dT = -\left[p + \left(\frac{\partial U_m}{\partial V_m}\right)_T\right]dV_m \tag{5.47}$$

or, since $(\partial U_m/\partial V_m)_T = 0$ and $p = RT/V_m$ in a perfect gas,

$$\frac{dT}{T} + \frac{R}{C_{V,m}}\frac{dV_m}{V_m} = 0.$$

Integrating this differential equation gives

$$\ln T + \frac{R}{C_{V,m}} \ln V_m = \text{constant}$$

$$TV_m^{R/C_{V,m}} = \text{constant},$$

or, since $R = C_{p,m} - C_{V,m}$ (eqn (5.36)),

$$\boxed{TV_m^{R/C_{V,m}} = TV_m^{\gamma-1} = \text{constant}} .$$
(5.48)

Because $pV_m = RT$ is true at all times during the reversible adiabatic expansion we can write $T = pV_m/R$, or $V_m = RT/p$ in eqn (5.48). This leads to

$$\boxed{pV_m^{(R+C_{V,m})/C_{V,m}} = pV_m^{\gamma} = \text{constant}} ,$$
(5.49)

$$\boxed{Tp^{-R/(R+C_{V,m})} = Tp^{(1-\gamma)/\gamma} = \text{constant}} .$$
(5.50)

The curves represented by eqns (5.48), (5.49), and (5.50) are shown in Fig. 5.20. By measuring how any two of p, V_m, and T change during a reversible adiabatic expansion we can readily deduce γ. For example, if a gas at V_{m1}, T_1 expands adiabatically to V_{m2}, T_2 then, from eqn (5.48), $T_1 V_{m1}^{\gamma-1} = T_2 V_{m2}^{\gamma-1}$, or $(\gamma - 1)\ln(V_{m1}/V_{m2}) = \ln(T_2/T_1)$. Knowing γ, we can apply eqn (5.46) to deduce the number of degrees of freedom f of a molecule of the gas. This value of f can give clues as to the structure of the molecule. Although important historically, there are more direct ways of deducing molecular structures!

As an example of the application of eqn (5.50) we will consider what happens when a wind strikes a sloping terrain, such as a mountain, and is forced to rise. We can imagine a fixed mass of the air, initially at (p_0, T_0) rising rapidly to a height h where its pressure is p_1 and its temperature is T_1. (It often helps to picture the gas as being contained within a light partially-inflated balloon.) Assuming the process to be reversible and adiabatic, eqn (5.50) tells us that

$$\frac{T_1}{T_0} = \left(\frac{p_1}{p_0}\right)^{(\gamma-1)/\gamma} = \left(\frac{p_1}{p_0}\right)^{0.286} ,$$

where γ has been taken as 1.4 (this being the value deduced from laboratory-based adiabatic expansions). Since $p_1/p_0 = 0.9$ at $h = 1$ km it follows that $T_1/T_0 = 0.970$ or, taking $T_0 = 290$ K, that $T_1 = 281$ K. This drop of 9 K km^{-1} is somewhat greater than the measured value of around 6 K km^{-1}. The reason for the discrepancy is that as the air rises and cools so water vapour condenses out. The intermolecular potential energy lost as water molecules coalesce (recall Fig. 2.6(c)) goes to raise the kinetic energy, and thus the temperature, of the surrounding gas atoms. Above 10 km the atmosphere becomes approximately isothermal up to a height of about 20 km (see p. 154).

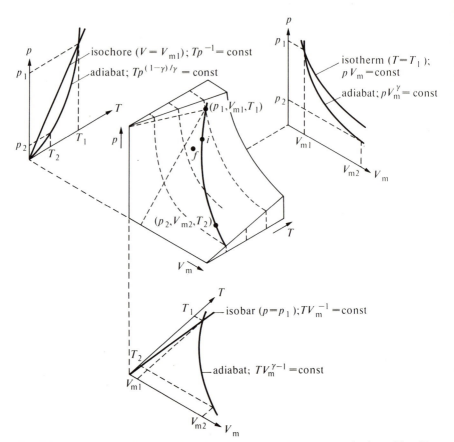

Fig. 5.20. Showing an adiabat on the $p-V_m-T$ surface and as it appears in the $p-T$, $p-V_m$, and $T-V_m$ projections. This surface only applies to a perfect gas.

Exercise 5.4

It can be shown (see virtually any text on the subject of wave motions) that the speed v of propagation of a longitudinal wave (like sound in a gas) is given by

$$v = \left(\frac{\text{bulk modulus}}{\text{density}}\right)^{1/2}. \tag{5.51}$$

The bulk modulus is defined as $-dp/(dV/V)$, where dV/V is the fractional change in volume brought about by a change dp in pressure.

Show that if sound propagation is an isothermal process $v = (p/\rho)^{1/2} = (\overline{u_x^2})^{1/2}$, and that if it is an adiabatic process $v = (\gamma p/\rho)^{1/2}$.

Derivation. It follows from $pV = nRT$ (eqn (5.22)) that during an isothermal change, where $dT = 0$ (by definition), that

$$p\,dV + V\,dp = nR\,dT = 0$$

$$-\frac{dp}{(dV/V)} = p.$$

Substituting this expression for the bulk modulus with p written as $\frac{1}{3}\rho\overline{u^2}$ (eqn (5.16)) and $3\overline{u_x^2} = \overline{u^2}$ (eqn (3.14)) into eqn (5.51) gives $v = (\overline{u_x^2})^{1/2}$.

We know that during an adiabatic process $pV^\gamma = \text{constant}$, so that

$$p\gamma V^{\gamma-1}\,dV + V^\gamma\,dp = 0.$$

Dividing through by $V^{\gamma-1}$ gives $-V\,dp/dV = \gamma p$, so that $v = (\gamma p/\rho)^{1/2}$.

Comment. The fact that the measured speed of sound in a gas agrees with $(\gamma p/\rho)^{1/2}$, rather than with $(p/\rho)^{1/2}$, shows that the compressions and rarefactions which constitute a sound wave are adiabatic in nature.

5.9. Work done in a reversible expansion

We will now calculate the work done by n mol of a gas contained within a piston and cylinder arrangement as the gas expands from an initial volume V_1 to a final volume V_2. As we saw in §1.6, when the volume V of a gas at pressure p changes by dV the gas does work $p\,dV$ on (that is, transfers energy to) the surroundings. Hence the total work W done by the gas as it expands from V_1 to V_2 is given by

$$W = \int_{V_1}^{V_2} p\,dV \tag{5.52}$$

which is, of course, the area under a graph of p plotted against V between $V = V_1$ and $V = V_2$. We will now evaluate this area for a reversible isothermal expansion (Fig. 5.21(a)) and a reversible adiabatic expansion (Fig. 5.21(b)).

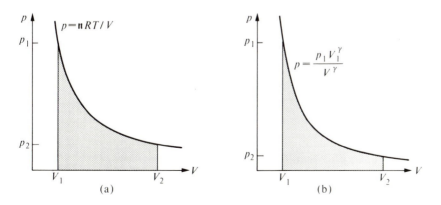

Fig. 5.21. The shaded areas show the work done by a gas as it expands reversibly from V_1 to V_2. In (a) the process is isothermal; in (b) it is adiabatic.

Isothermal expansion

It follows from eqn (5.22) that $p = nRT/V$ and hence, from eqn (5.52), that

$$W = nRT \int_{V_1}^{V_2} \frac{dV}{V}$$

$$\boxed{W = nRT \ln(V_2/V_1)} \ . \tag{5.53}$$

Adiabatic expansion

Writing $V_m = V/n$ (eqn (2.8)) in eqn (5.49) gives $pV^\gamma = $ constant. But since $p_1 V_1^\gamma$ and $p_2 V_2^\gamma$ are also equal to this same constant we have

$$p = \frac{p_1 V_1^\gamma}{V^\gamma} \ .$$

Hence, from eqn (5.52),

$$W = p_1 V_1^\gamma \int_{V_1}^{V_2} \frac{dV}{V^\gamma}$$

$$\boxed{W = \frac{p_1 V_1}{\gamma - 1}\left[1 - \left(\frac{V_1}{V_2}\right)^{\gamma - 1}\right] = \frac{p_1 V_1}{\gamma - 1}\left[1 - \left(\frac{p_2}{p_1}\right)^{(\gamma-1)/\gamma}\right]} \ . \tag{5.54}$$

An alternative way of arriving at W follows on appealing to the first law of thermodynamics (eqn (1.7)). This tells us that the work done by a gas during an adiabatic process $(Q = 0)$ is equal to the decrease in its internal energy. Since, from eqns (5.43) and (5.46),

$$U = \tfrac{1}{2}nfRT = \frac{nRT}{\gamma - 1},$$

it follows that

$$W = \frac{nR}{\gamma - 1}(T_1 - T_2) = \frac{nRT_1}{\gamma - 1}\left(1 - \frac{T_2}{T_1}\right).$$

Substituting $nRT_1 = p_1 V_1$ and obtaining T_2/T_1 from either eqn (5.48) or from eqn (5.50) leads immediately to eqn (5.54).

5.10. The mean free path

Although we already touched on the idea of the mean free path λ of a gas atom we have yet to calculate its value. We will start by using our imagination to follow the motion of any one atom i in the gas over a time

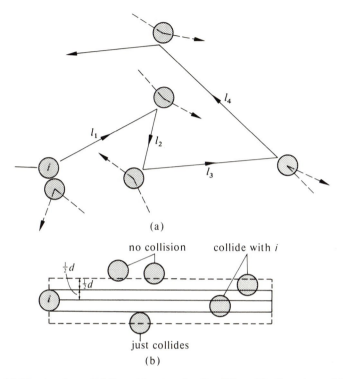

Fig. 5.22. (a) Showing the path followed by atom i as it collides with other atoms of the gas. (b) Atom i will collide with all other atoms whose centres lie within a cylinder of radius d and of length $\bar{u}t$. The cylinder is shown here as a section through a diameter.

t. Clearly, we can expect the atom to follow a zigzag path like that shown in Fig. 5.22(a). Between successive collisions with other atoms the atom i will travel freely through distances l_1, l_2, l_3, etc.; distances known as *free paths*. To find the mean distance λ travelled between collisions we must divide $l_1 + l_2 + l_3 + \cdots$, etc. ($=\bar{u}t$) by the number of collisions occurring in time t, that is

$$\lambda = \frac{\bar{u}t}{\text{number of collisions made by } i \text{ in time } t}. \tag{5.55}$$

Had there been no other atoms present, atom i would have followed the straight-line path shown in Fig. 5.22(b). But other atoms are, of course, present. If one of these other atoms is to collide with atom i, then its *centre* must lie within a distance $\frac{1}{2}d + \frac{1}{2}d = d$ from the path taken by the centre of atom i. Since the number of collisions made by i in a time t will equal the number of atoms whose centres lie within a cylinder of radius d

and length $\bar{u}t$, we have

number of collisions made by i in time $t = (\pi d^2 \bar{u} t) \times n.$ (5.56)

Substituting eqn (5.56) into eqn (5.55) gives

$$\lambda = \frac{1}{\pi d^2 n}$$ (5.57)

or, since $n = p/kT$ (eqn (5.19)),

$$\lambda = \frac{kT}{\pi d^2 p}$$ (5.58)

As an example, air at sea-level is typically at $p = 1 \cdot 0 \times 10^5$ Pa, $T = 300$ K. Assuming $d = 3 \times 10^{-10}$ m, eqn (5.58) tells us that $\lambda = 1 \cdot 46 \times 10^{-7}$ m. A not unknown mistake is to confuse λ with the mean intermolecular spacing of $(1/n)^{1/3} = (kT/p)^{1/3}$ (eqn (5.19)). In air at sea-level $(1/n)^{1/3} = 3 \cdot 5 \times 10^{-9}$ m, which is some ten atomic diameters and some forty times less than λ. At a height of 200 km, $\lambda = 2 \cdot 1 \times 10^2$ m (and the mean intermolecular spacing is $5 \cdot 0 \times 10^{-6}$ m), while at a height of 600 km, $\lambda = 1 \times 10^5$ m (62 miles)!

An improved calculation

Fig. 5.22(a) correctly illustrated the fact that collisions occur between *moving* partners. However, eqn (5.56) effectively assumed that only one atom was moving (atom i in Fig. 5.22(b)) and that all the other atoms were 'sitting ducks'. An actual collision between two atoms is shown in Fig. 5.23(a). To bring atom j to rest—and thus to allow us to apply our previous analysis—we must cancel its velocity, as shown in Fig. 5.23(b), by

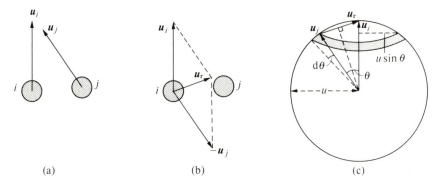

(a)   (b)   (c)

Fig. 5.23. (a) Two gas atoms with velocity $\boldsymbol{u}_i$ and $\boldsymbol{u}_j$. (b) Calculating the velocity $\boldsymbol{u}_r$ of i relative to j. (c) Showing how $\bar{u}_r$ is obtained.

adding a common velocity to both i and j. The resultant velocity of i, namely $\boldsymbol{u}_r$, gives the velocity of i relative to j. What we must now do is to find $\bar{u}_r$ and then use this value in eqn (5.56), and thus in the denominator of eqn (5.55). In this operation we are, so to speak, travelling along with i. The numerator of eqn (5.55) remains unchanged as it represents the total distance i travels, as seen by an observer fixed relative to the vessel's walls.

In finding $\bar{u}_r$ we will make the simplifying assumption that all the atoms have a common speed u. This means that when the tails of their velocity vectors are brought to a common point their tips will lie on a circle of radius u and that $u_r = 2u \sin(\theta/2)$ (Fig. 5.23(c)). To find $\bar{u}_r$ we must thus find the average value of $2u \sin(\theta/2)$. Now because the surface of the sphere shown in Fig. (5.23(c)) will have a constant density of points denoting the vector tips it follows that the fraction of the atoms moving at angles between θ and $\theta + d\theta$ will be the area of thin band shown in Fig. 5.23(c) ($2\pi u \sin\theta \times u \, d\theta$) divided by the area of the sphere ($4\pi u^2$), that is $\frac{1}{2}\sin\theta \, d\theta$. Hence, from eqn (4.6),

$$\bar{u}_r = \int_0^\pi u_r(\tfrac{1}{2}\sin\theta \, d\theta)$$

$$= u \int_0^\pi \sin(\theta/2)\sin\theta \, d\theta$$

$$= u \int_0^\pi \sin(\theta/2)2\sin(\theta/2)\cos(\theta/2) \, d\theta$$

$$= 4u \int_0^{\pi/2} \sin^2\theta' \cos\theta' \, d\theta'$$

$$\bar{u}_r = \tfrac{4}{3}u. \tag{5.59}$$

Hence the denominator of eqn (5.55) becomes $\pi d^2(4u/3)tn$ while the numerator becomes ut (since we are assuming that the speed of i relative to the walls of the container is fixed at u). Thus

$$\lambda = \frac{3}{4}\left(\frac{1}{\pi d^2 n}\right). \tag{5.60}$$

The assumption that all the gas atoms have a common speed u is quite unrealistic since we know that collisions between atoms will soon produce a Maxwell–Boltzmann speed distribution. Replacing Fig. 5.23(c) with the correct vector diagram (hinted at in Fig. 4.7) yields $\bar{u}_r = 2^{1/2}\bar{u}$ and hence

$$\boxed{\lambda = \frac{1}{2^{1/2}}\left(\frac{1}{\pi d^2 n}\right).} \tag{5.61}$$

It is interesting that a value of λ calculated via eqn (5.61) only differs by six per cent from the value calculated via eqn (5.60). As we pointed out earlier, many of the properties of a gas are remarkably immune to the precise form of the speed distribution function.

For many purposes it is quite sufficient to calculate λ via eqn (5.57). Indeed, in view of the supposed billard-ball nature of the collision (hard atoms which never approach closer than d between centres, atoms devoid of any attractive interaction) it is questionable whether there is much to be gained by introducing a factor of 3/4 or $1/2^{1/2}$ into eqn (5.57).

5.11. The collision frequency

We will now return to the simple model with which we began §5.10 and use it to calculate the number of collisions occurring within unit volume of the gas per unit time; the so-called *collision frequency* Z_{AA} (where the subscripts indicate that we are only concerned with collision between like species).

Eqn (5.56) tells us that the number of collisions z made per unit time by a *single* atom i is given by

$$z = \pi d^2 \bar{u} n \qquad (5.62)$$

or, as $n = 1/\pi d^2 \lambda$ (eqn 5.57), by

$$z = \bar{u}/\lambda. \qquad (5.63)$$

Because there are n atoms present per unit volume and because each of them makes z collisions per unit time it follows from eqn (5.62) that

$$Z_{AA} = \tfrac{1}{2}\pi d^2 \bar{u} n^2 \qquad (5.64)$$

and from eqn (5.63) that

$$Z_{AA} = n\bar{u}/2\lambda. \qquad (5.65)$$

The factor of $\tfrac{1}{2}$ has been introduced into eqns (5.64) and (5.65) to prevent each collision being counted twice. Without it atom A_1 striking atom A_2 is regarded as a different event from atom A_2 striking atom A_1; in reality they are, of course, a common event.

Since $\bar{u} = (8kT/\pi m)^{1/2}$ (eqn (4.50)) we can write eqn (5.64) as

$$\boxed{Z_{AA} = \pi d^2 \left(\frac{2kT}{\pi m}\right)^{1/2} n^2} \ . \qquad (5.66)$$

In a more exact analysis $\bar{u}$ in eqn (5.62) would be replaced by $\bar{u}_r = 2^{1/2}\bar{u}$, giving

$$Z_{AA} = \pi d^2 \left(\frac{4kT}{\pi m}\right)^{1/2} n^2. \qquad (5.67)$$

Finally, if there are two different species present with atomic diameters of d_1 and d_2 but with the same atomic mass m we see from Fig. 5.22(b) that d in eqn (5.62) must be replaced by $\frac{1}{2}(d_A + d_B)$. In addition, n^2 in eqn (5.64) must be replaced by $n_A n_B$. Thus eqn (5.67) becomes

$$\boxed{Z_{AB} = \frac{1}{4}\pi(d_A + d_B)^2 \left(\frac{4kT}{\pi m}\right)^{1/2} n_A n_B} \qquad (5.68)$$

This result will presently be used in discussing how chemical reactions can occur in the gas phase.

5.12. The probability of a specified free path

The free paths of a gas atom (l_1, l_2, l_3, etc. in Fig. 5.22(a)) are not, of course, all the same. We will now try to calculate the probability $f(l)$ that an atom, setting off on its travels after suffering a collision, will travel a distance of between l and infinity before it suffers another collision. In other words, we will find the probability that an atom will execute a free path *at least as great* as some specified value l.

After freely travelling a distance l since it last suffered a collision the chance that an atom will suffer a new collision in covering a further dl will be proportional to dl. (If dl is, say, doubled we would expect the chance of a new collision to be doubled.) That is, the chance of a collision in distance dl is $\alpha\, dl$, where α is the constant of proportionality. The chance that the atom will describe the distance dl *without* suffering a collision is therefore $(1 - \alpha\, dl)$. The chance that an atom will describe a distance l and then a further distance dl without suffering a collision is therefore $f(l)(1 - \alpha\, dl)$. This is equal to $f(l + dl)$, that is

$$f(l + dl) = (1 - \alpha\, dl)f(l). \qquad (5.69)$$

The left-hand side of eqn (5.69) may be expanded using Taylor's theorem to give

$$f(l) + \frac{df(l)}{dl}\, dl = f(l) - \alpha\, dl\, f(l)$$

$$\frac{df(l)}{f(l)} = -\alpha\, dl. \qquad (5.70)$$

Integrating eqn (5.70) gives

$$f(l) = \beta\, e^{-\alpha l},$$

where the constant of integration, β, may be evaluated by noting that if $l = 0$ then $f(l) = 1$. (The atom has a complete certainty of colliding with another atom somewhere between 0 and ∞). This gives $\beta = 1$, and so

$$f(l) = e^{-\alpha l}. \tag{5.71}$$

The value of the constant α follows on introducing the gaseous mean free path λ. By definition,

$$\lambda = \frac{\sum (\text{number of atoms with free path } l) \times l}{\text{total number of atoms}}$$

or, using the integral notation of eqn (4.6),

$$\lambda = \int_0^{\infty} l \times (\text{probability of free path between } l \text{ and } l + dl)$$

$$= \int_0^{\infty} l \times (\text{probability of freely travelling } l$$

$$\times \text{probability of a collision in } dl)$$

$$\lambda = \int_0^{\infty} l\, e^{-\alpha l} \alpha\, dl.$$

This may be readily integrated by parts to give $\lambda = 1/\alpha$, that is $\alpha = 1/\lambda$. Thus eqn (5.71) becomes

$$\boxed{f(l) = e^{-l/\lambda}} \tag{5.72}$$

as the required probability that an atom will have a free path at least equal to λ. This equation is shown graphically in Fig. 5.24. It is worth noting that the probability that an atom will travel a distance of at least $l = \lambda$ before suffering a collision is $e^{-1} = 0.37$. Thus there is a probability of 0.63 that an atom will suffer a collision in travelling a distance equal to λ.

Having found $\alpha = 1/\lambda$ we may spell out afresh that the probability that an atom has a free path between l and $l + dl$, namely $f(l)\alpha\, dl$, becomes $e^{-l/\lambda} dl/\lambda$. Hence the fraction of gas atoms with free paths lying between l and $l + dl$ is

$$\boxed{\frac{dN}{N} = \frac{1}{\lambda} e^{-l/\lambda}\, dl}\, . \tag{5.73}$$

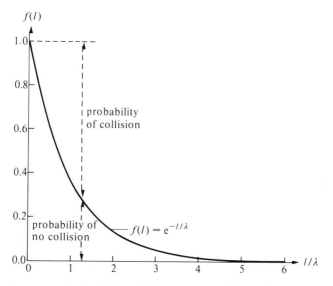

Fig. 5.24. Shows how the probability $f(l)$ that an atom will have a free path at least equal to l varies with l/λ, where λ is the mean free path.

Although we have assumed that l is measured from the site of a previous collision, eqns (5.72) and (5.73) also apply when l (and thus λ) is measured from where a free path crosses an arbitrary plane drawn in the gas. This is so because the basic assumption underlying all our discussions is that the probability of an atom suffering a collision within a distance dl is simply proportional to dl and not to its past history. Thus an atom which has travelled a distance of, say, 5 m without suffering a collision has exactly the same chance of making a collision within a further distance $dl = 10^{-4}$ m, say, as has an atom that has only freely travelled a distance of, say, 0·2 m. (You may know the analogous statement that a radioactive atom which has survived for, say, 10^7 s without decaying has exactly the same chance of decay within the next second as does an atom of the same substance which has only been in existence for, say, 10^2 s.)

Collisions with a solid surface

When we originally derived the expression $I_n = \frac{1}{4}n\bar{u}$ for the number of gas atoms striking unit area of surface per unit time we ignored the possibility that atoms setting off towards the surface from the top end of the slanted cylinder of length $u\,dt$ (Fig. 5.6(b)) might suffer collisions during the time dt, and thereby be scattered out of the cylinder. We will now redo the calculation, allowing interatomic collisions to take their toll.

We will start by considering the number of interatomic collisions

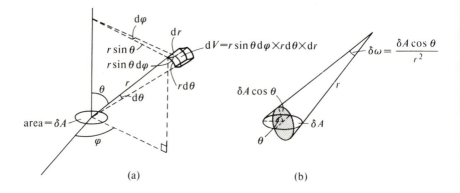

Fig. 5.25. (a) Showing a volume dV of the gas at (r, θ, ϕ). (b) The area δA subtends a solid angle $\delta\omega = \delta A \cos \theta / r^2$ at dV.

occurring within a volume $dV = r^2 \sin \theta \, dr \, d\theta \, d\phi$ at (r, θ, ϕ) from an area δA of the solid surface (Fig. 5.25(a)). Since there are $Z_{AA} = n\bar{u}/2\lambda$ (eqn (5.65)) collisions per unit volume per unit time in the gas it follows that there are $(n\bar{u}/2\lambda) \, dV = (n\bar{u}/2\lambda) r^2 \sin \theta \, dr \, d\theta \, d\phi$ collision per unit time within the volume dV of the gas. Now two atoms are involved in any collision so that the number of atoms per unit time coming straight from a collision within this volume is given by $(n\bar{u}/\lambda) r^2 \sin \theta \, dr \, d\theta \, d\phi$. These atoms will have their velocity vectors distributed uniformly in space. Thus if we now imagine ourselves to be located within the small volume dV, we can see that the fraction of these $(n\bar{u}/\lambda) r^2 \sin \theta \, dr \, d\theta \, d\phi$ atoms which move off towards δA per unit time is given by $(\delta A \cos \theta / r^2)/4\pi$, this being the fraction of 4π subtended at dV by δA (Fig. 5.25(b)). However, out of these

$$\left(\frac{\delta A \cos \theta}{4\pi r^2}\right)\left(\frac{n\bar{u}}{\lambda}\right) r^2 \sin \theta \, dr \, d\theta \, d\phi \tag{5.74}$$

atoms which are moving in the right direction to strike δA per unit time only those with a free path in excess of r can do so. Since the fraction of atoms with a free path in excess of r is $e^{-r/\lambda}$, it follows that the number of atoms originating from dV which strike δA per unit time is

$$\left(\frac{\delta A \cos \theta}{4\pi r^2}\right)\left(\frac{n\bar{u}}{\lambda}\right) r^2 \sin \theta \, dr \, d\theta \, d\phi \times e^{-r/\lambda}$$

$$= \frac{\delta A \, n\bar{u}}{4\pi\lambda} e^{-r/\lambda} \, dr \sin \theta \cos \theta \, d\theta \, d\phi. \tag{5.75}$$

To obtain the total number of collisions I_n per unit area of surface per unit time we must divide eqn (5.75) through by δA and integrate from $r = 0$ to $r = \infty$, from $\theta = 0$ to $\theta = \pi/2$, and from $\varphi = 0$ to $\varphi = 2\pi$. This gives

$$I_n = \frac{n\bar{u}}{4\pi} \int_0^\infty \frac{1}{\lambda} e^{-r/\lambda} \, dr \int_0^{\pi/2} \sin\theta \cos\theta \, d\theta \int_0^{2\pi} d\phi$$

$$I_n = \tfrac{1}{4} n\bar{u} \,,$$

which is the same result as we obtained in §5.2. The reason why our earlier discussion worked is that it implicitly assumed that if an atom was lost by collision from the sloping cylinder of Fig. 5.6 its place was taken by another one possessing the same velocity. Since this actually occurs the simple kinetic theory arguments which ignore intermolecular collisions usually give the same results as do the arguments which explicitly consider these collisions.

Exercise 5.5

Caesium (Cs) vapour at 770 K (where the vapour pressure is $1 \cdot 1 \times 10^4$ Pa) is contained in an otherwise evacuated chamber of size 5 cm × 5 cm × 5 cm. (a) What is the probability that a Cs atom (of diameter $5 \cdot 4 \times 10^{-10}$ m) will travel the length of the chamber without suffering a collision? (b) How many collisions will a single Cs atom make per second in the chamber? (c) How many collisions will there be per second between all the Cs atoms in the chamber. Take $A_r(\text{Cs}) = 133$.

Solutions. Since the answers to (a), (b), and (c) all depend on the number n of Cs atoms present per unit volume of the chamber, it makes good sense to calculate this first. Applying $n = p/kT$ gives $n = 1 \cdot 03 \times 10^{24}$ m^{-3}. (a) As $\lambda = 1/\pi d^2 n$ (you may use eqn (5.61) if you prefer) we have $\lambda = 1/(\pi \times (5 \cdot 4 \times 10^{-10} \text{ m})^2 \times 1 \cdot 03 \times 10^{24} \text{ m}^{-3}) = 1 \cdot 06 \times 10^{-6}$ m. Hence the probability that a Cs atom travels 5×10^{-2} m without making a collision is, from eqn (5.72), $\exp(-5 \times 10^{-2}/1 \cdot 06 \times 10^{-6}) = \exp(-4 \cdot 7 \times 10^4) \approx 10^{-10\,000}$. (b) The number of collisions z made per unit time by a single Cs atom is given by eqn (5.62) with $\bar{u} = (8kT/\pi m)^{1/2}$; that is $z = \pi d^2 n (8kT/\pi m)^{1/2} = 3 \cdot 30 \times 10^8$ s^{-1}. (c) It follows immediately from eqn (5.66) that the number of collisions Z_{CsCs} between Cs atoms is $1 \cdot 70 \times 10^{32}$ m^{-3} s^{-1}. Hence the total number of such collisions within the $1 \cdot 25 \times 10^{-4}$ m^3 chamber is $2 \cdot 13 \times 10^{28}$ s^{-1}.

5.13. Gas-phase chemical reactions

You will have noticed how most of the properties of a gas are remarkably insensitive to the precise form of the speed distribution within the gas. This state of affairs comes about because one is usually performing integrals (as, for example, in eqns (5.3) and (5.13)) where the limits of integration are $u = 0$ and $u = \infty$. If, however, the integration is from a finite speed to infinity the form of the answer does depend on the form of the distribution. Such integrals occur in chemical reaction theory.

As an example of a gas-phase reaction we will consider the reaction between molecular hydrogen (H_2) and atomic deuterium (D, but more properly written ^{2}H). This leads to the formation of DH molecules and atomic hydrogen:

$$H_2 + D \rightleftharpoons H + DH. \qquad (5.76)$$

Clearly, the minimum condition that must be satisfied by an H_2 molecule and a D atom which are to react chemically is that they collide. If this is the only condition, we have from eqn (5.68) (which is applicable since the interacting species have essentially the same diameter) that

$$\text{reaction rate} = Z_{DH_2} = \tfrac{1}{4}\pi(d_{H_2} + d_D)^2\left(\frac{4kT}{\pi m}\right)^{1/2} n_{H_2} n_D, \qquad (5.77)$$

where reaction rate means the rate of change of the number density of DH or H. Eqn (5.77) predicts that the reaction rate should have a $T^{1/2}$ dependence. Experiments, however, show that the rate varies as $\exp(-A/kT)$, where A is a constant. In addition, the observed reaction rates are many orders of magnitude less than those given by eqn (5.77). These discrepancies suggest that mere collision of a D atom and an H_2 molecule is no guarantee that the H_2 molecule will split, with one of its atoms pairing with the D atom (eqn (5.76)). Indeed we know from §2.5 that to increase the interatomic separation within an H_2 molecule (to the value required for the reaction to proceed) calls for energy. Although there is no easy way of calculating this *activation energy* E_a we would expect it to be substantially less than the pair dissociation energy ΔE of a H_2 molecule; the reason being that it is unnecessary to separate completely the two H atoms before one of them can interact with the D atom (eqn (5.76)).

It is thus clear that the reaction rate will be given not by Z_{DH_2}, as in eqn (5.77), but by Z_{DH_2} multiplied by the fraction of the collisions which possess an energy of E_a or greater. To find this fraction we appeal to eqn (4.57) which tells us that the fraction of the atoms in a two-dimensional gas which possess an energy within the range dE at E is given by

$$\frac{dN}{N} = \frac{1}{kT}e^{-E/kT}\,dE. \qquad (5.78)$$

We adopt a two-dimensional gas model on purely pragmatic grounds; eqn (5.78) can be readily integrated. (There are actually more compelling reasons for adopting a two-dimensional, rather than a three-dimensional, model here. However these reasons only become clear on considering the detailed nature of a collision between a H_2 molecule and a D atom.) Since we are interested in the fraction of the atoms with an energy in excess of E_a we integrate eqn (5.78) to obtain

$$\text{fraction of atoms with } E \geq E_a = \int_{E_a}^{\infty} \frac{1}{kT}e^{-E/kT}\,dE = e^{-E_a/kT}. \qquad (5.79)$$

Finally, multiplying the right-hand side of eqn (5.77) by this fraction—the fraction of atoms with the necessary energy to react—gives

$$\text{reaction rate} \left(\frac{dn_H}{dt}\right) = \tfrac{1}{4}\pi(d_D + d_{H_2})^2 n_D n_{H_2}(4kT/m)^{1/2}\, e^{-E_a/kT}. \quad (5.80)$$

This predicts that a graph of $\ln((dn_H/dt)/n_D n_{H_2})$ plotted against $1/T$ should be linear. This is found to be (nearly) true in practice. The measured gradient, when set equal to $-E_a/k$, gives $E_a = 5\cdot8\times10^{-23}\,\text{J}$, which is approximately eight per cent of the pair dissociation energy of a H_2 molecule (thereby justifying our guess that the D atom might interact with the H_2 molecule before the two hydrogen atoms had completely separated). In addition, the reaction rates calculated on the basis of a slightly refined version of eqn (5.80) agree satisfactorily with the measured rates.

PROBLEMS

5.1. Calculate the number of molecules striking the surface of a postage stamp of size $2\,\text{cm}\times2\,\text{cm}$, per second in air at $300\,\text{K}$ and at a pressure $1\cdot0\times10^5\,\text{Pa}$. Remember that the main constituents of air are nitrogen ($M_r(N_2) = 28$) and oxygen ($M_r(O_2) = 32$). By volume, air is roughly 80 per cent nitrogen and 20 per cent oxygen.

5.2. A vessel is partially filled with liquid mercury and sealed, except for a hole of area $10^{-7}\,\text{m}^2$, above the liquid level. This vessel is then placed in a highly evacuated (and continuously pumped) enclosure. After 30 days the vessel is reweighed and is found to be lighter by $2\cdot4\times10^{-5}\,\text{kg}$. If the experiment was performed at a constant temperature of $273\,\text{K}$ what is the vapour pressure of mercury at this temperature? ($A_r(\text{Hg}) = 200$.)

5.3. An oven maintained at $380\,\text{K}$ contains only cadmium vapour at a pressure of $0\cdot133\,\text{Pa}$ (its vapour pressure at $380\,\text{K}$). If the oven contains a hole of radius $r = 1\times10^{-5}\,\text{m}$ (punctured in a piece of metal foil of thickness much less than r), how many atoms will pass through a circular hole of radius $5\times10^{-5}\,\text{m}$, coaxial with the hole in the oven but $3\times10^{-2}\,\text{m}$ from it, over a period of $20\,\text{s}$?

5.4. Fig. 5.26 shows an accurate series of measurements made using a time-of-flight apparatus similar to that described on p. 65. Here the flight time is the time taken for the beam of potassium atoms to travel a fixed distance l. Show that the data given in Fig. 5.26 is consistent with the atoms having a Maxwell–Boltzmann speed distribution within the oven. What is the value of l?

5.5. A stream of balls, each of mass $2\,\text{kg}$ and moving at a speed of $12\,\text{m s}^{-1}$, strike a wall and stop dead. In the course of $10\,\text{s}$ fifty such balls hit the wall at right angles:
(a) What is the total change in momentum of the balls in $10\,\text{s}$? (b) Show graphically how you would expect the force which the wall must exert to stop the balls to vary with time. (c) Can you tell anything about the area

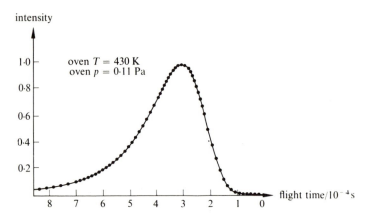

Fig. 5.26. The intensity (shown by dots) as recorded in a time-of-flight study of potassium vapour at 430 K and 0·11 Pa. (Data; P. M. Marcus and J. H. McFee.)

under the graph of part (b)? Anything else? (d) What is the average force which the wall exerts over the 10 s period? (e) If, instead of stopping dead, the balls rebound with unaltered speed back along the direction in which they came in, will this change the mean force which the wall exerts? If so, by how much?

5.6. Show that in a two-dimensional perfect gas the number of atoms striking unit length of the boundary per unit time is $n\bar{u}/\pi$, where n is the number of atoms per unit area.

5.7. Show that the force F required per unit length of boundary wall to contain a two-dimensional perfect gas is given by $F = RT/A$, where A is the molar area of the gas and R is the gas constant.

5.8. You might try to simulate how atoms are scattered at the surface of a solid by throwing ping-pong balls at the surface of an egg carton (or other dimpled supermarket pack). Try to plot a graph showing the number coming off in different directions. It will help to get someone to look 'sideways on' at what is happening. It also helps to carve up 180° into, say, six 30° sector! What type of surface gives specular reflection?

5.9. How many molecules are present in a $10^3 \, m^3$ gas holder which is at a pressure of $1·2 \times 10^5$ Pa and a temperature of 300 K? Assume the gas is perfect?

5.10. Investigate the consequences of introducing the explicit form of $f_3(u)$, as given by eqn (4.46), into eqn (5.13). Likewise, introduce eqn (4.69) into exercise 5.2.

5.11. A gas is at a pressure of 10^{-3} Pa and the r.m.s. speed of its constituent molecules is 200 m s^{-1}. What is the density of the gas? Assume the gas is perfect.

5.12. A container of volume $0·2 \, m^3$ is evacuated. Then 3×10^{-3} kg of water vapour is introduced. The pressure of the water vapour is found to be 5×10^5 Pa. What is the m.s. and r.m.s. speed of the water molecules?

5.13. Consider a perfectly elastic ball of mass m bouncing up and down above a horizontal surface under the influence of gravity. Prove that the downward force exerted by the ball upon the surface (time-averaged over an integral number of bounces) is equal to the weight of the ball. Does the result depend on the smoothness of the surface? (You might consider the effect of the ball striking a 'dimple' inclined at an angle of, say, 30° to the horizontal.) A simple extension of these arguments shows that the pressure exerted by Earth's atmosphere on the ground is *equal* to the weight of the overhead column of air (of unit cross-sectional area): It does *not* show that the pressure is *due* to the weight of air overhead. If Earth's atmosphere consisted of an incompressible fluid whose density was constant at 1.25 kg m^{-3} (approximately the value at sea-level), how high would it have to extend to account for the observed sea-level pressure of 1.01×10^5 Pa?

5.14. Assuming an isothermal atmosphere at a temperature of 240 K and with a sea-level pressure of 1×10^5 Pa, estimate the heights above sea-level at which the air pressure is 1×10^4, 1×10^3, and 1×10^2 Pa.

5.15. Using microscopic-level arguments to explain why (a) a hydrogen balloon, and (b) a hot-air balloon rises in Earth's atmosphere. Hence show, again using arguments at the atomic level, that the upthrust on the hydrogen balloon is equal to the weight of the air displaced by the balloon (an example, at the macroscopic level, of the application of Archimedes' principle).

5.16. In 1908 Perrin determined the value of the Avogadro constant N_A by observing how the number density $n(h)$ of colloidal particles suspended in a liquid varied with the height h (above some level where the number density is n_0). Show (by suitably extending eqn (5.32)) that

$$\ln(n(h)/n_0) = -\frac{(m \times N_A)g}{RT}\left(1 - \frac{\rho'}{\rho}\right)h,$$

where m is the mass of each colloidal particle, ρ' is the density of the material of the particle, and ρ is the density of the liquid. A graph of $\ln(n(h)/n_0)$ plotted against h allows $m \times N_A$ to be deduced (how?). Knowing m from the size and density of the particle, N_A follows immediately.

5.17. Using the values of γ given in Table 2.1 (p. 32) deduce the number of effective degrees of freedom of (a) an argon atom, (b) a nitrogen molecule, and (c) a carbon dioxide molecule. Try to describe the nature of these degrees of freedom.

5.18. A loudspeaker connected to a sine-wave generator is located in a sealed chamber containing only iodine (I) vapour at 400 K. When the generator is set at 1000 Hz it is found that standing waves are produced with nodes 6.77×10^{-2} m apart. Arguing via γ, deduce whether iodine vapour consists of I_2 molecules or I atoms. Take $A_r(I) = 127$.

5.19. Shortly after detonation the 'ball of fire' (sphere of hot gases) produced in a certain nuclear explosion was found to have a radius of 15 m and to be at a temperature of 3×10^5 K. Estimate the radius to which the ball will have to grow before its temperature drops to 3×10^3 K.

5.20. A perfect monatomic gas at a temperature of 400 K and a pressure of 8.0×10^5 Pa occupies a volume of 4 m^3. The gas is now allowed to expand to a final pressure of 1.0×10^5 Pa. Calculate the final temperature and volume of the gas, the work done by the gas, the heat absorbed by the gas, and the change in its internal energy assuming (a) that the expansion is reversible and isothermal, and (b) that the expansion is reversible and adiabatic.

5.21. Two mole of air ($\gamma = 1.40$), initially at s.t.p., is compressed isothermally and reversibly until it occupies one-third of its original volume. It is then expanded adiabatically and reversibly to its original pressure. What is (a) the work done on the gas during the initial compression, (b) the work done by the gas during its adiabatic expansion, (c) the change in the internal energy of the gas during its initial compression, and (d) the change in the internal energy of the gas during its adiabatic expansion?

5.22. Two vessels are separated from one another by a sheet of foil which is perforated with many holes each with a diameter and a length which is very much less than the gaseous mean free path. One chamber contains a gas of relative molecular mass M_1; the other contains a gas of relative molecular mass M_2. If both gases are at the same pressure, show that the initial rate at which the gas of relative molecular mass M_1 effuses into the gas of relative molecular mass M_2 is $(M_2/M_1)^{1/2}$ times the rate at which effusion occurs in the reverse direction. Assume the vessels to be at the same temperature.

5.23. One way to simulate a two-dimensional gas is to vibrate randomly a tray of marbles back and forth over the surface of a table. If a tray has dimensions of $0.3 \text{ m} \times 0.3 \text{ m}$ and it contains 50 marbles, each of diameter 10^{-2} m, what will be the mean free path of a marble on the tray? What is the probability that a marble will travel the length of the tray without colliding with another marble?

5.24. What is the mean free path of an electron moving freely among gas atoms (of diameter $d = 3 \times 10^{-10} \text{ m}$) when the gas is at a pressure of 10 Pa and a temperature of 300 K?

5.25. Using the data given in Fig. 5.16 evaluate the ratio of λ to the diameter of the aperture connecting the two chambers (2 cm) when eqn (5.25) ceases to be valid.

5.26. It is planned to defend a certain town against attack from low-flying bombers by flying barrage balloons (static balloons supporting a lethal cable). If a would-be bomber has a wingspan of 30 m and if the town has a radius of $2 \times 10^3 \text{ m}$, how many (randomly located) balloons should be flown to ensure that an incoming aircraft has only a ten per cent chance of reaching the town's centre? Assume that the pilot has no knowledge of the balloons' locations.

5.27. Show that in a three-dimensional gas the mean square free path ($\overline{l^2}$) is related to λ by $\overline{l^2} = 2\lambda^2$.

5.28. How many collisions will a single atom of argon make per second when the gas is at a pressure of 1.0×10^5 Pa and a temperature of 298 K? How many argon–argon atom collisions will occur per cubic metre per second in the gas? Take the diameter of an argon atom as $3.5 \times 10^{-10} \text{ m}$.

5.29. An alternative derivation of eqn (4.63) to that given on pp. 119–21 proceeds by considering the distribution of velocity vectors in a three-dimensional gas (Fig. 4.24). Out of all the atoms whose speeds lie between u and $u + du$, namely $Nf_3(u) \, du$ (eqn (4.1)), the number whose speeds lie in this range *and* which move between θ and $\theta + d\theta$ and between φ and $\varphi + d\varphi$ is given by $Nf_3(u) \, du$ multiplied by the ratio of the volume of the segment of velocity space between u and $u + du$, θ and $\theta + d\theta$, φ and $\varphi + d\varphi$ to the volume of the hemispherical shell of radii u and $u + du$. Using these ideas, show how eqn (4.63) follows.

6. Real gases

In the previous chapter we predicted how a perfect gas should behave. You will recall that we defined such a gas as one having atoms whose pair dissociation energy $\Delta E = 0$ and whose diameter is effectively zero. Both these assumptions are embraced by the interatomic potential $\mathscr{V}(r)$ shown in Fig. 6.1(a). Using these assumptions we then deduced the equation of state of a perfect gas and discussed how p, V_m, and T change during a reversible adiabatic process. At other times—notably when we discussed interatomic collisions and molecular heat capacities—we relaxed the assumption of zero atomic diameter while retaining the assumption that $\Delta E = 0$. This was equivalent to assuming that $\mathscr{V}(r)$ had the form shown in Fig. 6.1(b).

In this chapter we are going to assume the more realistic form of $\mathscr{V}(r)$ shown in Fig. 6.1(c). While this 'weakly-attracting hard sphere' model fails to reproduce the 'softness' of the repulsive forces (Fig. 2.6) it does incorporate the key facts that atoms actually have a finite ΔE and a finite diameter d. Using this potential model to describe an *imperfect* gas, we will derive its equation of state (paying particular attention to conditions near the critical point) and discuss how p, V_m, and T change during a reversible adiabatic process. We will also discuss the molar heat capacities of an imperfect gas and will look at some of the methods used to cool gases.

6.1. The van der Waals equation of state

In 1873 the Dutch physicist Johannes van der Waals succeeded in modifying the equation of state of a perfect gas so that it could be applied to real gases. Although he did not state the fact explicitly, van der Waals tacitly adopted the interatomic potential of Fig. 6.1(c). His success lay in assuming that the effect of portion BC of the curve, attributable to hard-sphere repulsion, could be treated quite separately from the effect of portion AB, attributable to the weakly attractive force.

The effect of hard-sphere repulsion

Physical intuition suggests that as the size of the gas atoms within a piston and cylinder arrangement is increased from zero to a finite diameter, so there will be less free space in which the atoms can move. Consequently, the number of collisions with the piston in unit time will increase, as will therefore the gas pressure. In quantifying these hunches we will start by considering a one-dimensional gas. (It often makes good

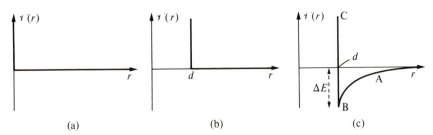

Fig. 6.1. The interatomic potential energy $V(r)$ in (a) a perfect gas, (b) a gas composed of hard atoms of diameter d but between which there are no attractive forces, and (c) an imperfect gas.

sense to start with a one-dimensional representation of a problem rather than with the almost invariably more difficult three-dimensional case. If the one-dimensional discussion yields results we may be able to make an intelligent guess about how a three-dimensional gas behaves.)

Fig. 6.2(a) shows a piston-and-cylinder arrangement in which an atom of effectively *zero size* moves back and forth between the piston and the end of the cylinder. The time between successive collisions with the piston is $2l/u$, where l is the length of the cylinder. Thus the number of collisions per unit time with the piston is given by

$$f_0 = \frac{u}{2l}. \tag{6.1}$$

Fig. 6.2(b) shows that when the single atom of Fig. 6.2(a) is given a finite diameter d the distance travelled by this atom between successive collisions with the piston is reduced from the perfect gas value of $2l$ to $2l - 4(\frac{1}{2}d) = 2(l - d)$. With $N - 1$ additional atoms present (Fig. 6.2(c)), atom i is no longer able to travel freely from one end of the cylinder to the other. Before long atom i will collide with another atom j. At the

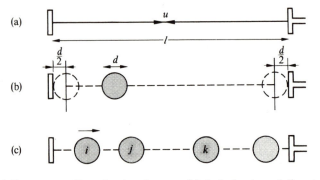

Fig. 6.2. (a) Shows a one-dimensional perfect gas. (b) A single atom of diameter d moves back and forth between the piston and the cylinder end. (c) A one-dimensional gas containing many atoms of finite size.

moment of impact j may be moving to the right, or to the left. Since each alternative is equally likely we will hazard the guess that, on average, the effect is the same as if j were at rest when struck by i. (This is just the same assumption as we made in discussing mean free paths in §5.10.) When i strikes j with speed u the two atoms 'swap places'; atom i stops dead and j moves off to strike atom k. Next j stops dead and k moves off with speed u.... and so on. The same effect is neatly demonstrated in the conversation piece known as Newton's cradle.

When atom i, say, strikes atom j then atom j moves off *immediately* (assuming, as in Fig. 6.1(c), that the spheres are perfectly hard). In other words, the momentum carried by i jumps a distance of $\frac{1}{2}d + \frac{1}{2}d = d$ in such a collision. Therefore in travelling between the piston and the end of the cylinder an atom will make $(N-1)$ jumps, each of length d, with a similar number of jumps on the return journey. Thus the effective distance which an atom has to travel between successive collisions with the piston is reduced from the single-atom value of $2(l-d)$ to $2(l-d)-2(N-1)d = 2(l-Nd)$. As the speed of an atom remains constant at u throughout the sequence of collisions the number of collisions which this one atom makes with the piston per unit time is given by

$$f_d = \frac{u}{2(l-Nd)}.$$

Thus the number of collisions with the piston will be (f_d/f_0) times the value f_0 which applies to a perfect one-dimensional gas (eqn. (6.1)), where

$$\frac{f_d}{f_0} = \frac{l}{l-Nd}. \tag{6.2}$$

We must now try to modify this relation so that it can plausibly be applied to the three-dimensional gas contained within the piston-and-cylinder arrangement of Fig. 5.9(a). If you look back to where the N in the denominator of eqn. (6.2) was first introduced you will discover that (ignoring the difference between N and $N-1$) it actually represents the number of collisions which a single atom makes in travelling the length l of the cylinder. In a three-dimensional gas the corresponding number of collisions will be l/λ or $l\pi d^2 n$, on substituting $\lambda = 1/\pi d^2 n$ (eqn. (5.57)). Hence eqn (6.2) becomes

$$\frac{f_d}{f_0} = \frac{1}{1-\pi d^3 n},$$

$$\frac{f_d}{f_0} = \frac{1}{1-6nv_a}, \tag{6.3}$$

where $v_a (= (4\pi/3)(\frac{1}{2}d)^3)$ is the volume of an individual atom.

We may straight away multiply the right-hand side of our perfect gas relation, $p = \frac{1}{3}m(N/V)\overline{u^2}$ (eqn (5.17)) by this factor. (If the number of collisions with the piston were to be increased by a factor of 1.2 then p would increase by 1.2.) Alternatively, you may prefer to return to eqn (5.11)—which gives the number of collisions with an area A of the piston in a time t—and to multiply it by f_d/f_0. Either way the conclusion is the same:

$$p = \frac{1}{3}mN\overline{u^2}\left(\frac{1}{V - 6nVv_a}\right)$$

or, since $nV = N$,

$$p(V - 6Nv_a) = \frac{1}{3}mN\overline{u^2}.$$

If there is one mole of gas present, this becomes

$$p(V_m - b) = \frac{1}{3}mN_A\overline{u^2} = RT, \tag{6.4}$$

where $b = 6N_Av_a$ is six times the total volume of the (hard-sphere) gas atoms. Other treatments give different values for the *excluded volume* b; values which range from N_Av_a to $8N_Av_a$. The most detailed analyses, based on a proper discussion of collision dynamics, show that $b = 4N_Av_a$. This is the value which we shall adopt from now on. To argue at length over the precise value of the correction factor is to miss the point of the factor.

The effect of ΔE

Consider an atom i which is just about to strike the piston (Fig. 6.3). To have got there it will necessarily have left the vicinity of atoms j, k, l, etc. During this process the potential energy stored in the fields between i and j, k, l, etc, will increase, while the kinetic energy of i will decrease. (We have already shown in §5.3 that the force which the piston atoms exert on i can justifiably be neglected.) In fact, the only gas atoms which affect i on its journey to the piston are those whose centres within a hemisphere of radius D, where D is of the same order as the interatomic separation d in an isolated pair of atoms. To see why the *range D* of the forces on i approximates to d we consider the $-1/r^6$ van der Waals interaction illustrated in Fig. 6.3(b). This shows that when $r = 1 \cdot 2d$ the potential energy has risen by 66 per cent of ΔE. Hence $D \approx 1 \cdot 2d \approx d$, and so the number of atoms which i leaves behind as it moves to the piston is approximately $\frac{2}{3}\pi D^3(N_A/V_m)$. In leaving each of these atoms, atom i loses kinetic energy of order ΔE. Thus if atom i sets off from the bulk of the gas with a speed u_g and arrives at the piston face with a speed u_f then, by the law of conservation of energy,

$$\frac{1}{2}mu_g^2 - \frac{1}{2}mu_f^2 = \frac{2}{3}\pi D^3\left(\frac{N_A}{V_m}\right)\Delta E.$$

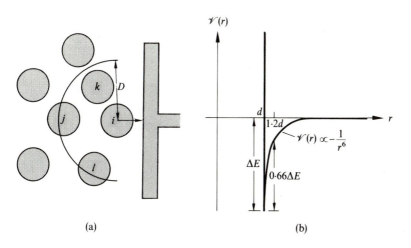

Fig. 6.3. (a) As atom i leaves the vicinity of atoms j, k, l, etc. and moves up towards the piston its kinetic energy falls. (b) The interatomic potential energy $\mathcal{V}(r)$ of a pair of hard-sphere atoms with a $-1/r^6$ van der Waals interaction.

The right-hand side of this equation really only tells us the *average* number of atoms within the hemisphere of radius D; the actual number will, of course, fluctuate about the mean. Hence we should have written

$$\tfrac{1}{2}m\overline{u_g^2} - \tfrac{1}{2}m\overline{u_f^2} = \tfrac{2}{3}\pi D^3\left(\frac{N_A}{V_m}\right)\Delta E. \tag{6.5}$$

Combining both corrections

In deducing $p = \tfrac{1}{3}mn\overline{u^2}$—and, by implication, the partially-corrected form given by eqn (6.4)—we asked what happened as a gas atom struck the piston. This means that the speed u in eqn (6.4) is really the speed u_f, so this equation states

$$p(V_m - b) = \tfrac{2}{3}N_A(\tfrac{1}{2}m\overline{u_f^2}).$$

We may now substitute for $\tfrac{1}{2}m\overline{u_f^2}$ from eqn (6.5). In so doing we shall take it that the range D of the interatomic force is equal to d, the hard-sphere diameter (see Fig. 6.3(b)). This gives

$$\left[p + \frac{N_A^2}{3V_m(V_m - b)}\tfrac{4}{3}\pi d^3\,\Delta E\right] = \frac{2N_A}{3(V_m - b)}(\tfrac{1}{2}m\overline{u_g^2})$$

or, remembering that the temperature of a gas is related to the mean kinetic energy of its atoms by $\tfrac{1}{2}m\overline{u_g^2} = \tfrac{3}{2}kT$,

$$\left[p + \frac{N_A^2}{3V_m(V_m - b)}\tfrac{4}{3}\pi d^2\,\Delta E\right](V_m - b) = N_A kT = RT.$$

TABLE 6.1
Properties of real gases

gas	van der Waals constants[a]		T_B/K	T_c/K	T_i/K
	$a/10^{-1}\,m^6\,Pa\,mol^{-2}$	$b/10^{-5}\,m^3\,mol^{-1}$			
^{4}He	0·035	2·4	16	5·2	25
Ne	0·21	1·71	122	44·5	231
Ar	1·36	3·22	411	151	779
Kr	2·35	3.98	575	209	1089
Xe	4·25	5·10	768	290	1456
O_2	1·38	3·18	406	155	764
N_2	1·41	3·91	327	126	621
H_2	0·25	2·66	114	33	190
CH_4	2·28	4·28	510	191	968

[a] Deduced from critical point data via eqns (6.18) to (6.20).

Because the correction term inside the square brackets is but a correction term to p it introduces little error to replace $(V_m - b)$ in its denominator by V_m, giving

$$\left(p + \frac{a}{V_m^2}\right)(V_m - b) = RT \tag{6.6}$$

as van der Waals' famous equation of state of a real gas, in which a and b are constants. Our discussion has indicated

$$a = \frac{8N_A^2 v_a}{3}\Delta E, \tag{6.7}$$

$$b = 4N_A v_A, \tag{6.8}$$

where v_a is the volume of each hard-sphere atom of the van der Waals gas and ΔE is the binding energy of a pair of atoms. Values of a and b for some representative gases are given in Table 6.1. These values were deduced from the macroscopic behaviour of the gases rather than from eqns (6.7) and (6.8).

Exercise 6.1

Show that when an amount n of gas is present, van der Waals' equation takes the form

$$\left(p + \frac{n^2 a}{V^2}\right)(V - nb) = nRT \tag{6.9}$$

Calculate the pressure exerted by 2×10^{-2} kg of nitrogen gas ($M_r(N_2) = 28$) held at

298 K in a vessel of volume 3×10^{-5} m^3. The van der Waals constants for nitrogen are $a = 1.41 \times 10^{-1}$ m^6 Pa mol^{-2}, $b = 3.9 \times 10^{-5}$ m^3 mol^{-1}.

Solution. The most direct method of arriving at eqn (6.9) is to substitute $V_m = V/n$ into eqn (6.6). Alternatively one can go back to the start of our discussion of van der Waals' equation and consider a volume V of gas containing $N = nN_A$ molecules. Following this route, N_A in eqns (6.7) and (6.8) is replaced by nN_A, implying that a is replaced by n^2a and b by nb. These correction factors are then used to modify $pV = nRT$.

In applying eqn (6.9) we must clearly first evaluate n. Since, by definition, $M_r(N_2)g(= 28 \times 10^{-3}$ kg) of nitrogen contain one mole of nitrogen molecules it follows that 2×10^{-2} kg of nitrogen contains $(2 \times 10^{-2}$ kg$)/(28 \times 10^{-3}$ kg)mol $= 0.714$ mol of molecules. Substituting $n = 0.714$ mol, $T = 298$ K, $a = 1.41 \times 10^{-1}$ m^6 Pa mol^{-2}, $V = 3 \times 10^{-5}$ m^3 and $b = 3.9 \times 10^{-5}$ m^3 mol^{-1} into eqn (6.9) gives

$$\left(p + \frac{(0.714)^2 \times 1.41 \times 10^{-1}}{(3 \times 10^{-5})^2} \text{ Pa}\right)(3 \times 10^{-5} \text{ m}^3 - (0.714 \times 3.9 \times 10^{-5})\text{m}^3)$$

$$= 0.714 \times 8.31 \times 298 \text{ J}$$

$$p = 7.41 \times 10^8 \text{ Pa}.$$

Comment. Had we assumed that the gas was perfect we would have applied $pV = nRT$ and obtained $p = 5.89 \times 10^7$ Pa.

6.2. Dieterici's equation of state

A serious criticism of the van der Waals equation is that it—or, rather, the arguments underlying it—suggest that the mean temperature of a gas in the vicinity of the walls of a container is less than within the bulk of the gas (see eqn (6.5)). Now we know from our discussion of how atoms behave when subjected to an external force (§5.6), albeit a gravitational force, that the temperature of the gas is constant throughout the region where the force acts, and that what varies is the number density of the atoms. Taking over eqn (5.32) and writing $n(h)$ for the number density at the point where the potential energy of an atom is $\mathscr{V} = mgh$, we have

$$n(h) = n_0 e^{-\mathscr{V}/kT},$$

where n_0 is the number density at the plane $h = 0$. Returning to Fig. 6.3(a) we therefore see that the number density n_f of the gas atoms within the immediate vicinity of the piston face is related to the number density n_g at a distance D away by

$$n_f = n_g \exp[-(\tfrac{2}{3}\pi D^3 N_A \Delta E)/kTV_m]$$

or, on multiplying the numerator and the denominator of the exponent by N_A and taking $D = d$, by

$$n_f = n_g e^{-\alpha/RV_m T}, \tag{6.10}$$

where

$$\alpha = 4N_A^2 v_a \, \Delta E = \tfrac{3}{2}a. \tag{6.11}$$

Hence, the actual pressure $p(=n_t kT)$ as measured in a Dieterici gas will be related to the ideal gas pressure $p_i(=n_g kT)$ by

$$p = p_i \, e^{-\alpha/RV_m T}. \tag{6.12}$$

Now the pressure which occurs in the 'volume-corrected' eqn (6.4) is the ideal gas pressure p_i (which we subsequently corrected by the a/V_m^2 term). It can thus be replaced by p_i as given by eqn (6.12), yielding

$$\boxed{(p \, e^{\alpha/RV_m T})(V_m - b) = RT} \tag{6.13}$$

as Dieterici's equation of state. It can easily be shown that when $\alpha \ll RV_m T$ eqn (6.13) reduces to the van der Waals equation, but with a replaced by α $(=3a/2;$ eqn (6.11)).

Many other equations of state have been proposed; one such is the subject of problem 6.13. Out of all the analytic equations involving only two unknowns the one which best fits the experimental values of (p, V_m, T) is Dieterici's. His equation also has, as we have seen, a secure theoretical basis. Despite these undoubted advantages, Dieterici's equation has not enjoyed the popularity of the closely-related van der Waals equation. The reasons are perhaps partly historical (van der Waals' equation appeared twenty-six years before Dieterici's) and partly that van der Waals' equation is easier to handle mathematically.

6.3. The virial equations

As we saw in §5.4 and in Fig. 5.15 it is particularly telling to plot the compression factor Z $(=pV_m/RT)$ of a gas against either $1/V_m$ or against p at different temperatures. Such plots are known as virial plots. Each of the curves of Fig. 5.15(a) can be empirically fitted to an equation of the form

$$\boxed{Z = \frac{pV_m}{RT} = 1 + \frac{B(T)}{V_m} + \frac{C(T)}{V_m^2} + \frac{D(T)}{V_m^3} + \cdots,} \tag{6.14}$$

where the functions of temperature $B(T)$, $C(T)$, $D(T)$, etc, are called the second, third, fourth, etc. *virial coefficients*. The more coefficients that are included the more accurate will the fit become. Likewise, each of the curves of Fig. 5.15(b) can be fitted to an equation of the form

$$Z = \frac{pV_m}{RT} = 1 + B'(T)p + C'(T)p^2 + D'(T)p^3 + \cdots \quad , \qquad (6.15)$$

where $B'(T)$, $C'(T)$, $D'(T)$, etc. are also known as the second, third, fourth, etc., virial coefficients.

It follows from eqn (6.14) that

$$\left(\frac{\partial Z}{\partial(1/V_m)}\right)_T = B(T) + \frac{2C(T)}{V_m} + \frac{3D(T)}{V_m^2} + \cdots$$

and hence that $B(T)$ records the slope of the virial plot at $1/V_m = 0$. We see from Fig. 5.15(a) that $B(T)$ is negative at low temperatures and positive at high temperatures. At one particular temperature (410 K in the case of argon) $B(T) = 0$. This is known as the *Boyle temperature* since it represents the temperature at which the gas most closely obeys Boyle's law (see eqn (6.14)).

It similarly follows from eqn (6.15) that

$$\left(\frac{\partial Z}{\partial p}\right)_T = B'(T) + 2C'(T)p + 3D'(T)p^2 + \cdots$$

and that $B'(T)$ records the slope of Z plotted against p at $p = 0$. Again, $B'(T) = 0$ at one particular temperature (106 K in the case of hydrogen). This temperature has the same value as that given by the isotherm with zero slope on the Z versus $1/V_m$ plot.

Although eqn (6.14) was originally proposed by H. Kamerlingh Onnes in 1901 as an empirical equation, it does arise naturally in certain microscopic theories of fluids. In these theories $B(T)$ arises on considering the interactions between pairs of molecules, $C(T)$ on considering the interactions between triples of molecules, and so on.

6.4. Properties of the van der Waals equation

Isotherms

Rather than examining the shape of the p–V_m–T surface it is easier to look at isotherms in the p–V_m plane. With this end in view we rewrite van der Waals' equation (eqn (6.6)) as

$$p = \frac{RT}{V_m - b} - \frac{a}{V_m^2} . \qquad (6.16)$$

This immediately tells us that at high temperature *and* at high molar

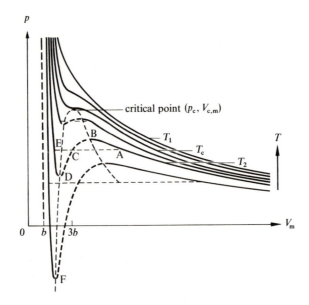

Fig. 6.4. Pressure–volume isotherms as predicted by the van der Waals equation.

volumes $p \approx TR/V_m$, so the isotherms will approximate to rectangular hyperbolas at these temperatures and volumes. However, as V_m approaches b the pressure becomes very large. Such a high-temperature isotherm, labelled T_1, is shown in Fig. 6.4. At a lower temperature T_2 the pressure dips down with decreasing V_m before increasing towards infinity as V_m approaches b. The dipping portion DB of the isotherm T_2 (Fig. 6.4) does not describe the behaviour of any real gas for it implies that an increase in p causes an increase in V_m (have you ever observed this?). It is possible to associate portion AB with a supercooled vapour and portion ED with a superheated liquid (cf. Fig. 1.10). However, we know from Chapter 1 that supercooling and superheating are the exception rather than the rule; phase separation normally occurs along some such line as AE in Fig. 6.4.

The critical point

At some temperature T_c which is intermediate in value between T_1 and T_2 the maxima and minima of Fig. 6.4 coalesce to yield a horizontal tangent at point $(p_c, V_{c,m}, T_c)$ which is also a point of inflexion. We will now apply calculus to determine the values of point $(p_c, V_{c,m}, T_c)$; a point which we can readily associate with the critical point of the gas.

We begin by looking for maxima and minima. Differentiating eqn

(6.16) to produce $(\partial p/\partial V_m)_T$, and equating the resulting expression to zero to give the turning points of the curve, we obtain

$$\frac{RT}{(V_m-b)^2} = \frac{2a}{V_m^3}$$

or, after eliminating T by substituting for RT from eqn (6.6),

$$p = a\left(\frac{1}{V_m^2} - \frac{2b}{V_m^3}\right). \tag{6.17}$$

This equation represents the locus of the maxima and minima of the family of curves produced by giving T different values in van der Waals' equation. It is shown dashed in Fig. 6.4. To find the position of the maximum of this curve, and hence the values of $(p_c, V_{c,m}, T)$, we must differentiate eqn (6.17) and set $dp/dV_m = 0$. This gives

$$V_{c,m} = 3b \tag{6.18}$$

which, when substituted back into eqn (6.17), yields

$$p_c = a/27b^2. \tag{6.19}$$

Substituting eqns (6.18) and (6.19) back into eqn (6.6) gives

$$T_c = 8a/27Rb. \tag{6.20}$$

One can readily confirm that the isotherm at $T = T_c$ does indeed have a horizontal tangent at point $(p_c, V_{c,m})$ and that this is also a point of inflexion by showing that $(\partial p/\partial V_m)_T$ and $(\partial^2 p/\partial V_m^2)_T$ are both zero at the coordinates given by eqns (6.18) to (6.20).

There are several ways of checking out the correctness, or otherwise, of eqns (6.18) to (6.20). The first follows on combining all three equations to give the compression factor at the critical point as

$$Z_c = \frac{p_c V_{c,m}}{RT_c} = \tfrac{3}{8}(=0\cdot375). \tag{6.21}$$

The experimental values of Z_c generally lie between $0\cdot2$ and $0\cdot3$ (for example, it is $0\cdot29$ for argon and $0\cdot275$ for carbon dioxide). They thus lie closer to the value of $Z_c = 2/e^2 = 0\cdot271$ predicted from Dieterici's equation (see problem 6.12) than to the value predicted from the van der Waals equation. Another test is to feed the measured values of $(p_c, V_{c,m}, T_c)$ into eqns (6.18) to (6.20) and thereby deduce the values of a and b. These are next fed into van der Waals' equation, which is now used to predict (p, V_m) values for isotherms other than $T = T_c$. Such (p, V_m) values are usually twenty to thirty per cent away from the measured values. Alternatively, the value deduced for b may be equated to $4N_A v_a$ (eqn (6.8)) thereby giving v_a and hence the diameter of the gas atoms. As can be seen from Table 2.1, diameters calculated by this method are in broad agreement with those obtained by other techniques. But perhaps the most telling test is to put the values of a and b deduced from microscopic-level arguments (eqns (6.7) and (6.8)) into eqns (6.18) to (6.20). This leads, for example, to

$$T_c = \frac{16}{81} \frac{N_A \, \Delta E}{R}.$$

To obtain ΔE we recall §2.6 and problem 2.16 where we argued that the molar enthalpy of evaporation of a liquid in which each molecule is surrounded by ten nearest neighbours is given by $H_{m,e} = 5 N_A \, \Delta E$. Hence

$$\frac{RT_c}{H_{m,e}} = \frac{16}{405} \quad (=0 \cdot 039). \tag{6.22}$$

The experimental values of this ratio are usually several times greater than 16/405. (for example, it is $0 \cdot 19$ for argon and $0 \cdot 13$ for water). However, it is worth bearing in mind that our route to eqn (6.22) involved taking a 'slightly corrected perfect gas' equation—which still assumed that atomic movement was largely unhindered—and applying it to a situation where the atoms are in close proximity ($V_{c,m} = 12 N_A v_a$). Under the circumstances, the discrepancy between theory and experiment is not embarrassing.

The Boyle temperature

The van der Waals equation may be rewritten in virial form by rearranging eqn (6.6) as

$$p = \frac{RT}{V_m} \left(1 - \frac{b}{V_m} \right)^{-1} - \frac{a}{V_m^2}$$

and expanding $(1 - b/V_m)^{-1}$ by means of the binomial expansion, to give

$$p = \frac{RT}{V_m} \left(1 + \frac{b}{V_m} + \frac{b^2}{V_m^2} + \cdots \right) - \frac{a}{V_m^2}$$

$$\frac{pV_m}{RT} = 1 + \left(b - \frac{a}{RT} \right) \frac{1}{V_m} + \frac{b^2}{V_m^2} + \cdots \tag{6.23}$$

$$\frac{pV_m}{RT} = 1 + \frac{B(T)}{V_m} + \frac{C(T)}{V_m^2} + \cdots$$

where

$$B(T) = b - \frac{a}{RT} \tag{6.24}$$

$$C(T) = b^2.$$

At large molar volumes, b^2/V_m^2 and all following terms can be ignored in eqn (6.23). Furthermore, we can write $1/V_m = p/RT$ in the second term on the right-hand side to give

$$\frac{pV_m}{RT} = 1 + B'(T)p,$$

where

$$B'(T) = \left(\frac{b}{RT} - \frac{a}{R^2T^2}\right). \tag{6.25}$$

Therefore the Boyle temperature T_B (where $B(T) = 0$, $B'(T) = 0$) is given by

$$\boxed{T_B = a/Rb} \ . \tag{6.26}$$

This may be combined with $T_c = 8a/27Rb$ (eqn (6.20)) to predict that $T_B/T_c = 27/8 = 3 \cdot 375$. By comparison, the experimental values of T_B/T_c are usually lower than 3 (as you can verify using the data given in Table 6.1).

The law of corresponding states

If we define the so-called *reduced pressure* p_r, *reduced volume* V_r, and *reduced temperature* T_r by

$$p_r = p/p_c, \qquad V_r = V_m/V_{c,m}, \qquad T_r = T/T_c, \tag{6.27}$$

then substituting these expressions combined with eqn (6.19) for p_c, eqn (6.18) for $V_{c,m}$, and eqn (6.20) for T_c into the van der Waals equation (eqn (6.6)) gives

$$\boxed{\left(p_r + \frac{3}{V_r^2}\right)(3V_r - 1) = 8T_r} \tag{6.28}$$

as the *reduced equation of state*. We immediately notice that a and b have conveniently disappeared. With no constants depending on the nature of the particular substance under study, eqn (6.28) should apply to all substances. This, in turn, implies that the p–V_m–T surfaces of all substances can be reduced to one common surface by dividing the three sets of coordinates on each surface by linear scaling factors (three factors for each substance). This is known as the *law of corresponding states*. Although deduced for the van der Waals equation it applies to any equation of state which can be expressed in the form

$$F(p, V_m, T, R, c_1, c_2) = 0$$

where c_1 and c_2 are constants (such as a and b) which are characteristic of the substance. As you would expect, the law of corresponding states is only approximately correct.

By way of example of the application of eqn (6.28), if a variety of substances have the same *reduced* volume of 2/3 (the *actual* molar volumes, being $2V_{c,m}/3$, are all different, of course) and the same *reduced* pressure of 3/4 they will all have the same *reduced* temperature of 15/16.

6.5. Molecular aggregates in gases

It was argued in §3.1 that the molecules of a gas should be able to form aggregates, even though each such aggregate may have only a fleeting existence. It was also argued that the average size of these aggregates should decrease as the temperature of the gas is raised. We will now look at the experimental evidence for the existence of these molecular aggregates, starting with a gas whose state lies close to $(p_c, V_{c,m}, T_c)$.

To recall §1.2, if a sealed tube of constant volume containing a liquid and its vapour is heated, one of three things will happen: The volume of the liquid will shrink until it disappears (route $1 \rightarrow 2$ of Fig. 1.7). The volume of the liquid will grow until it fills the tube (route $3 \rightarrow 4$ of Fig. 1.7). The volume of the liquid remains stationary while the density of the vapour increases until, at the critical temperature, it equals that of the 'liquid' beneath. This last alternative is illustrated in Fig. 1.8. One way of studying how conditions change as the critical point is approached is to measure the fluid density as a function of position along the tube at various temperatures close to T_c. A particularly convenient way of measuring the density of a transparent, but inaccessible, material is to measure its refractive index. (Density and refractive index are related by a well-tested equation; the Clausius–Mosotti equation.) This can easily be done by measuring the angle through which a ray of light is deviated in passing through a prism-shaped cell of the liquid. By measuring the angle of deviation as a function of vertical position along such a cell one can readily deduce the density of the fluid as a function of position. Fig. 6.5(a) shows the results of such a study on carbon dioxide at temperatures ranging from $T_c - 0.025$ K (where $T_c = 304.154$ K) to $T_c + 1.93$ K. This sequence shows how the sharp discontinuity in density at the interface between the liquid and its vapour persists right up to within 0.007 K of T_c, even though the dividing meniscus may no longer be clearly seen by eye (Fig. 1.8). The experimental points are omitted in Fig. 6.5(a) as their scatter is less than the line thickness.

It is tempting to suppose that the gas density decreases with increasing height in Fig. 6.5(a) for the same reasons as the gas density decreases

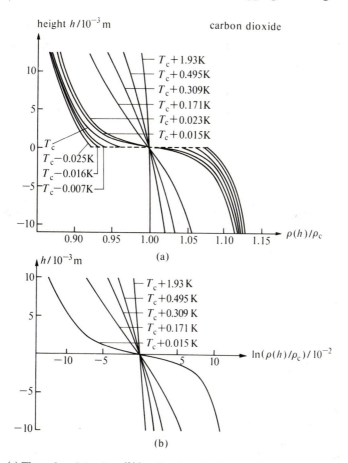

Fig. 6.5. (a) The reduced density $\rho(h)/\rho_c$ of carbon dioxide as a function of height at various temperatures close to T_c. (Data from Schmidt, E. H. W. and Traube, K. (1962). *Prog. int. Res. Thermodyn. Transport Prop.*, A.S.M.E. Academic Press, New York and London.) (b) Data from (a) replotted as $\ln(\rho(h)/\rho_c)$ against h.

with height in the isothermal region of Earth's atmosphere. If this is so, then according to eqn (5.32),

$$\frac{\rho(h)}{\rho_c} = \frac{mn(h)}{mn_c} = e^{-mgh/kT}, \tag{6.29}$$

where m is now the mass of the molecular aggregates present and h is the height measured from the meniscus. Thus a graph of h plotted against $\ln(\rho(h)/\rho_c)$ should be linear and with a gradient of $-kT/mg$. As Fig. 6.5(b) shows this prediction is largely borne out at temperatures down to $T_c + 0 \cdot 171$ K. Equating the measured gradient at $T_c + 1 \cdot 93$ K to $-kT/mg$

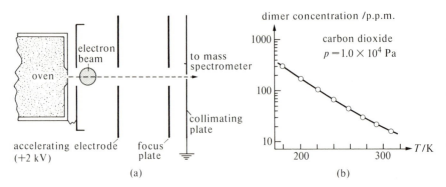

Fig. 6.6. (a) Showing how gas molecules emerging from an oven are ionized by a transverse electron beam prior to analysis in a mass spectrometer. Apart from the oven the entire apparatus is maintained at a high vacuum. (b) The dimer concentration expressed as parts per million in carbon dioxide at $p = 1 \cdot 0 \times 10^4$ Pa as a function of temperature. (Data from Leckenby, R. E. and Robbins, E. J. (1965). *Proc. R. Soc. A* **291**, 389.)

gives $m = 4 \cdot 3 \times 10^{-22}$ kg. Dividing this by the mass of a CO_2 molecule $(7 \cdot 3 \times 10^{-26}$ kg) tells us that the aggregates present at $T_c + 1 \cdot 93$ K contain $5 \cdot 9 \times 10^3$ molecules. At $T_c + 0 \cdot 495$ K the figure is $1 \cdot 7 \times 10^4$ molecules; at $T_c + 0 \cdot 309$ K it is $2 \cdot 4 \times 10^4$ molecules; at $T_c + 0 \cdot 171$ K it is $4 \cdot 4 \times 10^4$ molecules. This trend is much as we might have expected from the qualitative arguments of §3.1.

Different techniques must be used to study aggregation in low-density gases. One such technique involves ionizing a molecular beam as it emerges from an oven and then analyzing the masses of the ionized species in a mass spectrometer (Fig. 6.6(a)). To ensure that there is no adiabatic cooling of the gas as it emerges from the oven (a possible source of molecular aggregates) the length and diameter of the orifice ($<2 \times 10^{-6}$ m) are made less than or equal to the gaseous mean free path in the oven. Fig. 6.6(b) shows how the measured dimer ($(CO_2)_2$) concentration in carbon dioxide varies with temperature at a fixed pressure of $1 \cdot 0 \times 10^4$ Pa. Again, the trend is what we would expect.

6.6. Thermodynamics of a van der Waals gas

Internal energy

Probably the simplest way of arriving at an expression for the internal energy $U_m(V_m, T)$ of a monatomic van der Waals gas is to imagine ourselves setting up 1 mol of such a gas from scratch. Since the internal energy depends only on the final state and not on the route by

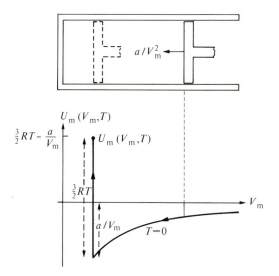

Fig. 6.7. In deducing $U_m(V_m, T)$ for a van der Waals gas we imagine the gas to contract at $T \approx 0$ until it reaches the desired value of V_m and then to be heated to the desired value of T.

which that state is reached we can choose the route to suit ourselves. Let us start therefore with the atoms at a near-infinite distance apart and at rest (meaning, classically, that $T = 0$). We next allow them to come together until they occupy the desired volume V_m (see Fig. 6.7). As they do so they will exert a pressure of $-a/V_m^2$ (eqn (6.6) with $T = 0$); that is, the piston is pulled inwards with a pressure $-a/V_m^2$ (due, of course, to the intermolecular attractions). The energy *removed* from the gas during this phase of the operations is

$$-\int_{\infty}^{V_m} \frac{a}{V_m^2} \, dV_m = \frac{a}{V_m}. \tag{6.30}$$

Finally, we set the atoms moving by heating the gas until each atom has a mean energy of $\frac{3}{2}kT$, where T is the desired temperature of the system. This calls for an energy $\int_0^T C_{V,m} \, dT = (\frac{3}{2}kT) \times N_A = \frac{3}{2}RT$. Hence the *total* energy fed into the gas in setting it up at (V_m, T) is given by

$$U_m(V_m, T) = \int_0^T C_{V,m} \, dT - \frac{a}{V_m} = \frac{3}{2}RT - \frac{a}{V_m}. \tag{6.31}$$

Exercise 6.2

Use eqn (6.31) to predict the internal energy of argon at temperatures of 75, 151, and 300 K on the isochore $V_m = V_{c,m}$. The critical point data for argon is $p_c = 4 \cdot 9 \times 10^6$ Pa, $V_{c,m} = 7 \cdot 3 \times 10^{-5}$ m^3 mol^{-1}, and $T_c = 151$ K.

Solution. Before using eqn (6.31) we must know a. This is found by applying eqns (6.19) and (6.20) to the given critical pressure and temperature. Eliminating b from these equations gives $a = 27R^2T_c^2/64p_c$, or, substituting for T_c and p_c, $a = 1 \cdot 36 \times 10^{-1}$ m^6 Pa mol^{-2}. Using this value of a along with $V_m = V_{c,m} = 7 \cdot 3 \times 10^{-5}$ m^3 mol^{-1} in eqn (6.31) gives $U_m = -9 \cdot 2 \times 10^3$ J mol^{-1} at 75 K, $U_m = 24 \cdot 3$ J mol^{-1} at 151 K, and $U_m = 1 \cdot 9 \times 10^3$ J mol^{-1} at 300 K.

Comment. The negative internal energy of the liquid (how do we know argon is a ~uid at $T = 75$ K, $V_m = V_{c,m}$?) is consistent with the observation that energy is ː d to disperse the liquid into its constituent atoms. The positive internal energy of the gas is consistent with the fact that the gas can do work as its atoms separate adiabatically.

The molar heat capacities

We have just argued that for a monatomic van der Waals gas

$$
\boxed{C_{V,m} = \left(\frac{\partial U_m}{\partial T} \right)_{V_m} = \tfrac{3}{2}R} \ . \tag{6.32}
$$

To find $C_{p,m} - C_{V,m}$ (and hence $C_{p,m}$) we appeal to eqn (1.15). This tells us that we must find the values of $(\partial U_m / \partial V_m)_T$ and $(\partial V_m / \partial T)_p$. The value of $(\partial U_m / \partial V_m)_T$, obtained by differentiating eqn (6.31), is

$$
\left(\frac{\partial U_m}{\partial V_m} \right)_T = \frac{a}{V_m^2} \ . \tag{6.33}
$$

The simplest method of finding $(\partial V_m / \partial T)_p$ is first to multiply out the van der Waals equation (eqn (6.6)) to give

$$
pV_m - pb + \frac{a}{V_m} = RT , \tag{6.34}
$$

where the second-order term ab/V_m^2 has been ignored. Hence

$$
p \, dV_m + V_m \, dp - b \, dp - \frac{a}{V_m^2} dV_m = R \, dT
$$

and so

$$
\left(\frac{\partial V_m}{\partial T} \right)_p = \frac{R}{p - (a/V_m^2)} \ . \tag{6.35}
$$

Substituting eqns (6.33) and (6.35) into eqn (1.15) gives

$$
C_{p,m} - C_{V,m} = R \left(1 + \frac{a}{pV_m^2} \right) \left(1 - \frac{a}{pV_m^2} \right)^{-1}
$$

which becomes, on applying the binomial theorem and ignoring second-order terms,

$$C_{p,m} - C_{V,m} = R\left(1 + \frac{2a}{pV_m^2}\right),$$

or, writing $V_m = RT/p$ in the correction term $2a/pV_m^2$,

$$\boxed{C_{p,m} - C_{V,m} = R + \frac{2ap}{RT^2}}. \tag{6.36}$$

The extra term $2ap/RT^2$ (over the perfect gas value of $C_{p,m} - C_{V,m} = R$) arises because additional energy is required to increase the mean interatomic separation in a van der Waals gas as it expands while being heated at constant pressure.

Adiabatic changes in an imperfect gas

To find the change in state of a van der Waals gas during a reversible adiabatic process we follow the same procedures as we used when dealing with a perfect gas, except that we employ different relations for the equation of state and the internal energy of the gas. Thus eqn (5.47) becomes, on substituting for $(\partial U_m/\partial V_m)_T$ from eqn (6.33), and then for $p + (a/V_m^2)$ from the equation of state (eqn (6.6)),

$$\frac{\mathrm{d}T}{T} = -\frac{R\,\mathrm{d}V_m}{C_{V,m}(V_m - b)},$$

which readily integrates to yield

$$\boxed{T(V_m - b)^{R/C_{V,m}} = \text{constant}}. \tag{6.37}$$

This may be expressed in terms of p and V by substituting for T from eqn (6.6) to give

$$\boxed{\left(p + \frac{a}{V_m^2}\right)(V_m - b)^{(R + C_{V,m})/C_{V,m}} = \text{constant}}. \tag{6.38}$$

Alternatively, eqn (6.37) may be rewritten in terms of p and T by substituting for $(V_m - b)$ from eqn (6.6), to give

$$T\left(\frac{RT}{p + (a/V_m^2)}\right)^{R/C_{V,m}} = \text{constant}$$

which becomes, upon expanding it by the binomial theorem and using

$V_m = RT/p$ in correction terms,

$$T\left(\frac{RT}{p} - \frac{a}{RT} - \frac{b^2 p}{RT} + \cdots\right)^{R/C_{V,m}} = \text{constant} \quad . \qquad (6.39)$$

While eqns (6.37) and (6.38) are equivalent to the first statements in eqns (5.48) and (5.49) but with p replaced by $p + (a/V_m^2)$ and with V_m replaced by $(V_m - b)$ we cannot take the additional step of using the perfect gas relation $R = C_{p,m} - C_{V,m}$ in eqns (6.37) to (6.39). However, this relation is approximately true at high temperatures and low pressures (see eqn (6.36)), allowing us to write eqn (6.38), for example, as

$$\left(p + \frac{a}{V_m^2}\right)(V_m - b)^\gamma = \text{constant}, \qquad (6.40)$$

a result first derived by van der Waals in 1899.

Work done in a reversible expansion

To find the work W done by an amount n of a van der Waals gas contained within a piston and cylinder arrangement as the gas expands reversibly and isothermally at temperature T from an initial volume V_1 to a final volume V_2 we apply eqn (5.52) with p given by eqn (6.9). Thus

$$W = \int_{V_1}^{V_2} \frac{nRT}{(V - nb)} \, dV - \int_{V_1}^{V_2} \frac{n^2 a}{V^2} \, dV$$

$$W = nRT \ln\left(\frac{V_2 - nb}{V_1 - nb}\right) + n^2 a\left(\frac{1}{V_2} - \frac{1}{V_1}\right) \qquad (6.41)$$

which, of course, reduces to eqn (5.53) when $a = 0$, $b = 0$.

To find the work done during an adiabatic expansion we can directly appeal to the first law of thermodynamics. This law tells us that during an adiabatic process the work done by a gas is equal to the decrease in its internal energy. It therefore follows from eqn (6.31), as applied to an amount n of the gas, that

$$W = nC_{V,m}(T_1 - T_2) + n^2 a\left(\frac{1}{V_2} - \frac{1}{V_1}\right) \quad . \qquad (6.42)$$

If the expansion is reversible eqn (6.37) can now be used to relate T_2 to T_1.

6.7. Cooling by adiabatic processes

To cool any gas—be it perfect or imperfect—we must lower the translational kinetic energy of its component molecules. Thus if we change the kinetic energy by $\delta(\frac{1}{2}m\overline{u^2})$ the temperature will change by $\delta T = (2/3k)\, \delta(\frac{1}{2}m\overline{u^2})$. In Chapter 3 we discussed the mechanism whereby the gas molecules' kinetic energy can be lowered by placing the gas in contact with a solid whose atoms have kinetic energy $\frac{1}{2}M\overline{U^2} < \frac{1}{2}m\overline{u^2}$. We will now look at the other methods available for cooling a gas. In so doing we shall assume that the gas is either perfect or that it satisfies van der Waals' equation.

Cooling by internal work alone

Fig. 6.8(a) shows an adiabatic enclosure with one mole of a gas confined to the left-hand chamber and occupying a volume V_{m1}; the right-hand chamber is evacuated. If the dividing partition is removed by pulling on it sideways, no work will be done on (or by) the gas as it expands irreversibly to a volume V_{m2}. The gas is then said to have undergone a *free* (or *Joule*) *expansion*, characterized by $Q = 0$, $W = 0$. It

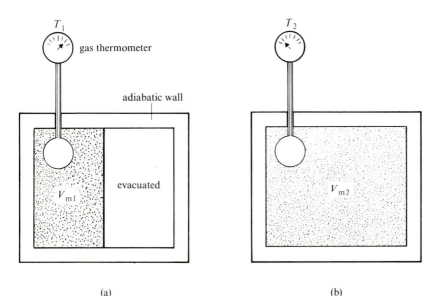

Fig. 6.8. On removing the dividing partition shown in (a) the gas undergoes a free expansion into the right-hand chamber, as shown in (b).

follows from eqn (6.42) (which was derived assuming $Q = 0$) that

$$T_2 - T_1 = \frac{a}{C_{V,m}} \left(\frac{1}{V_{m2}} - \frac{1}{V_{m1}} \right) . \tag{6.43}$$

As expected, there is no temperature change when a perfect gas ($a = 0$) undergoes a free expansion. Viewed microscopically, this arises because the atoms of a perfect gas simply move further apart and, unlike the atoms of a real gas, do not have to use part of their translational kinetic energy to increase the intermolecular potential energy.

Cooling by external work

If a gas undergoes an adiabatic expansion (whether reversible or not) from an initial volume V_{m1} to a final volume V_{m2} during which it does work W on its surroundings, then according to eqn (6.42),

$$T_2 - T_1 = \frac{1}{C_{V,m}} \left[\frac{-W}{n} + a \left(\frac{1}{V_{m2}} - \frac{1}{V_{m1}} \right) \right] . \tag{6.44}$$

To evaluate W we imagine ourselves standing outside the piston and cylinder arrangement. A suitable Newton balance attached to the piston allows a graph of the pressure p exerted by the gas on the surroundings to be plotted against V_m; the area under this indicator diagram giving the work done by the gas. Only if the change is reversible can we evaluate W via eqns (5.52) and (6.38). This step is unnecessary because eqn (6.37) holds true under reversible adiabatic conditions, predicting

$$T_2 = T_1 \left(\frac{V_{m1} - b}{V_{m2} - b} \right)^{R/C_{V,m}} \tag{6.45}$$

which reduces, on setting $b = 0$, to the perfect gas value of

$$T_2 = T_1 (V_{m1}/V_{m2})^{R/C_{V,m}} . \tag{6.46}$$

This is the same relation as that obtained directly from eqn (5.48). It may seem curious that eqn (6.45) involves b but not a. Surely, one might argue, as the gas atoms move further apart they will lose kinetic energy and the gas must cool more than would a perfect gas. Therefore a should feature in eqn (6.45)! The reason why it fails to appear is that while it is true that the atoms of a real gas do slow down as the gas expands this also means that they exert less pressure on the piston, thereby doing less external work, than they would in a perfect gas undergoing an expansion between the same initial and final volumes.

In practice the external work is often done by allowing a high-pressure gas to drive a turbine rather than a piston engine. This technique is commonly used in liquifying air. Unlike the Joule and Joule–Kelvin expansions, it can cool a perfect gas.

Joule–Kelvin cooling

This technique, the product of a collaboration between Joule and Kelvin in 1853, makes use of both internal and external work. Gas, at pressure p_1 and at a molar volume V_{m1} (Fig. 6.9(a)), is forced through a porous plug (or fine nozzle) to a region where the pressure is p_2 and where the molar volume is V_{m2} (Fig. 6.9(b)). What we will do is to follow the fate of one mole of the gas as it passes through the equipment. To simplify the discussion we will, at least initially, suppose that there is a heater (or refrigerator) on the right-hand side of the plug and that this is adjusted to ensure that the gas has the same temperature on both sides of the plug.

The external work done *by* the one mole of gas in passing through the equipment is $p_2 V_{m2} - p_1 V_{m1}$. The corresponding internal work done *by* the gas against the attractive intermolecular forces is given by

$$\int_{V_{m1}}^{V_{m2}} \frac{a}{V_m^2} \, dV_m = -\frac{a}{V_{m2}} + \frac{a}{V_{m1}}.$$

Therefore the total work W done by the gas is given by

$$W = (p_2 V_{m2} - p_1 V_{m1}) + \left(\frac{a}{V_{m1}} - \frac{a}{V_{m2}} \right). \tag{6.47}$$

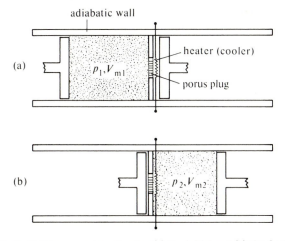

Fig. 6.9. The Joule–Kelvin expansion, showing (a) the initial and (b) the final situation after one mole of gas has passed through the equipment.

Now, to the first-order of small quantities, we can write the van der Waals equation as in eqn (6.34). Substituting this relation for pV_m, along with the approximately correct equation $1/V_m = p/RT$, into eqn (6.47) gives

$$W = \left(\frac{2a}{RT} - b\right)(p_1 - p_2). \tag{6.48}$$

Since the temperature is kept at the same value on both sides of the plug this total work cannot come from the translational kinetic energy of the gas atoms. It must therefore be provided by the heater (or removed by the refrigerator) located on the right-hand side of the plug (Fig. 6.9). Eqn (6.48) in fact tells us that, since $p_1 - p_2$ is always positive, W is positive if $2a/RT > b$; thus if no heat is provided the gas will cool. Conversely, if $2a/RT < b$ then W will be negative; thus if no heat is provided, the gas will heat up. At one particular temperature, known as the *inversion temperature* T_i and given by

$$T_i = \frac{2a}{Rb} \tag{6.49}$$

there will be no change in the gas temperature as a result of the Joule–Kelvin expansion. If the gas is subjected to a Joule–Kelvin expansion starting from a temperature $T < T_i$ (meaning $2a/Rb > T$, or $2a/RT > b$), the gas will cool. If $T > T_i$, the gas will warm up. Looking at Table 6.1 we see that, apart from helium, neon, and hydrogen, all these gases will cool if subjected to a Joule–Kelvin expansion in which the gas on the high-pressure side of the plug is at room temperature; helium, neon, and hydrogen must be precooled to below their inversion temperature before they are—to use the jargon—throttled. This precooling is usually done by making the gas undergo an adiabatic expansion (as in a turbine engine) before it meets the porous plug.

   If the gas cools from T_1 to T_2 during the real throttling process, then the heat which must have been provided initially, when we supposed the heater to have been present, will be given by $C_{p,m}(T_1 - T_2)$. Equating this to eqn (6.48) gives

$$T_2 - T_1 = \frac{1}{C_{p,m}}\left(b - \frac{2a}{RT}\right)(p_1 - p_2). \tag{6.50}$$

Finally, if we compare eqns (6.49) and (6.26) we see that $T_i = 2T_B$ in a van der Waals gas. The data given in Table 6.1 shows that this prediction approximates to the truth.

Exercise 6.3

The purpose of this exercise is to demonstrate the relative effectiveness of each of the three cooling processes we have just considered. By how much will, say, one mole of argon cool if it (a) undergoes a free expansion from an initial temperature of 300 K and an initial volume of 10^{-3} m^3 to a final volume of 2×10^{-3} m^3, (b) undergoes a reversible adiabatic expansion over the same range as in (a), and (c) undergoes a Joule–Kelvin expansion from the same initial conditions as in (a), expanding to a final pressure of $1 \cdot 2 \times 10^6$ Pa (chosen so that the final volume is again approximately 2×10^{-3} m^3)? You can assume that $C_{V,\mathrm{m}} = 12 \cdot 6$ J K^{-1} mol^{-1}, $C_{p,\mathrm{m}} = 20 \cdot 9$ J K^{-1} mol^{-1}, $a = 0 \cdot 136$ m^6 Pa mol^{-2}, and $b = 3 \cdot 22 \times 10^{-5}$ m^3 mol^{-1}.

Solutions. This is largely an exercise in the correct use of eqns (6.43), (6.45), and (6.50). As always when substituting data into formulae it is good practice to include the units. This guards against using erroneous formulae and making erroneous substitutions!

(a) The temperature change produced in the free expansion is, from eqn (6.43),

$$T_2 - T_1 = \frac{0 \cdot 136 \text{ m}^6 \text{ Pa mol}^{-2}}{12 \cdot 6 \text{ J K}^{-1} \text{ mol}^{-1}} \left(\frac{1}{2 \times 10^{-3} \text{ m}^3 \text{ mol}^{-1}} - \frac{1}{1 \times 10^{-3} \text{ m}^3 \text{ mol}^{-1}} \right)$$

$$T_2 - T_1 = -5 \cdot 4 \text{ K}.$$

In checking out the units recall that $1 \text{ Pa} = 1 \text{ N m}^{-2}$ and that $1 \text{ J} = 1 \text{ N m}$.

(b) The temperature change produced in the reversible adiabatic expansion is, from eqn (6.45),

$$T_2 = 300 \text{ K} \left(\frac{10^{-3} - 3 \cdot 22 \times 10^{-5}}{2 \times 10^{-3} - 3 \cdot 22 \times 10^{-5}} \right)^{(8 \cdot 31/12 \cdot 6)}$$

$$T_2 = 188 \text{ K}$$

$$T_2 - T_1 = -112 \text{ K}.$$

(c) Before applying eqn (6.50) we will need to know the initial pressure p_1. This is found by substituting $T = 300$ K and $V_{\mathrm{m}} = 10^{-3}$ m^3 mol^{-1} into van der Waals equation, giving $p_1 = 2 \cdot 44 \times 10^6$ Pa. Eqn (6.50) now becomes

$$T_2 - T_1 = \frac{1}{20 \cdot 9 \text{ J K}^{-1} \text{ mol}^{-1}} \left(3 \cdot 22 \times 10^{-5} \text{ m}^3 \text{ mol}^{-1} - \frac{2 \times 0 \cdot 136 \text{ m}^6 \text{ Pa mol}^{-2}}{8 \cdot 31 \times 300 \text{ J mol}^{-1}} \right)$$

$$\times (2 \cdot 44 \times 10^6 - 1 \cdot 2 \times 10^6) \text{Pa}$$

$$T_2 - T_1 = -4 \cdot 5 \text{ K}.$$

Comments. The only safe generalization which can be made is that a free expansion is an inefficient way to cool a gas. Using the particular initial and final conditions specified in this problem the cooling produced by the Joule–Kelvin expansion is considerably lower than the cooling produced by the reversible adiabatic expansion.

PROBLEMS

6.1. Thirty moles of xenon is introduced into a $2 \times 10^{-3} \, m^3$ pressure vessel at a temperature of 300 K. Calculate the pressure of the zenon assuming (a) that it obeys the perfect gas law, and (b) that it obeys van der Waals' equation. Use the data given in Table 6.1.

6.2. Calculate the volume occupied by 1 mol of water at 770 K and at a pressure of $2 \times 10^7 \, Pa$. Assume that water obeys van der Waals' equation with $a = 0 \cdot 546 \, m^6 \, Pa \, mol^{-2}$ and $b = 3 \cdot 04 \times 10^{-5} \, m^3 \, mol^{-1}$. *Clues.* Start off by finding a very approximate value of V_m using the perfect gas law. Then use this value of V_m in a/V_m^2 and solve for V_m. Use this new value of V_m in a/V_m^2 and solve again for V_m. Repeat the procedures until successive answers only differ by, say, one per cent.

6.3. What is the temperature of $14 \times 10^{-3} \, kg$ of N_2 $(A_r(N) = 14)$ if it occupies a volume of $5 \times 10^{-5} \, m^3$ and is at a pressure of $4 \times 10^6 \, Pa$? Use the data given in Table 6.1.

6.4. Using the van der Waals equation with a and b as given in Table 6.1, plot $p-V_m$ isotherms for methane (CH_4) at $T = 150$, 190, and 230 K. Replot the data as Z against p at these three temperatures.

6.5. Calculate the second virial coefficient, $B(T)$, for hydrogen at 273 K given that, at this temperature, its molar volume is $4 \cdot 634 \times 10^{-4} \, m^3 \, mol^{-1}$ at $5 \times 10^6 \, Pa$, $2 \cdot 386 \times 10^{-4} \, m^3 \, mol^{-1}$ at $1 \times 10^7 \, Pa$, $1 \cdot 271 \times 10^{-4} \, m^3 \, mol^{-1}$ at $2 \times 10^7 \, Pa$, and $0 \cdot 9004 \times 10^{-4} \, m^3 \, mol^{-1}$ at $3 \times 10^7 \, Pa$. Proceed by plotting a suitable graph.

6.6. There are several different methods of obtaining the virial coefficients, $B(T)$, $C(T)$, etc. from measured (p, V_m, T) data. One method starts by rearranging eqn (6.14) as

$$A = \left(\frac{pV_m}{RT} - 1 \right) V_m = B(T) + \frac{C(T)}{V_m} + \cdots .$$

Show that

$$B(T) = \lim_{p \to 0} A$$

$$C(T) = \lim_{p \to 0} (A - B(T)) V_m.$$

6.7. Using the data given in Fig. 5.15(a) plot $B(T)$ against T for argon. Superimpose on this plot the theoretical value of $B(T)$ as given by eqn (6.24) with $a = 0 \cdot 136 \, m^6 \, Pa \, mol^{-2}$ and $b = 3 \cdot 22 \times 10^{-5} \, m^3 \, mol^{-1}$.

6.8. Estimate the critical pressure, temperature, and volume for CO_2 given that it has the van der Waals constants $a = 0 \cdot 359 \, m^6 \, Pa \, mol^{-2}$ and $b = 4 \cdot 27 \times 10^{-5} \, m^3 \, mol^{-1}$.

6.9. What is the mean intermolecular separation and the mean free path in a van der Waals gas at its critical point? Express your answers in terms of the molecular diameter d. Use $\lambda = 1/\pi d^2 n$ when calculating the mean free path.

6.10. Show that at the Boyle temperature the volumes calculated assuming the perfect gas equation of state do not differ by more than $0 \cdot 01$ per cent from the volumes calculated assuming van der Waals' equation of state at pressures p between $p = 0$ and $p = 0 \cdot 27 p_c$. Also show that over the range $p = 0$ to $p = p_c$ the differences do not exceed $0 \cdot 14$ per cent.

6.11. Show that Dieterici's equation of state can be expanded in virial form (eqn (6.14)) with

$$B(T) = b - \frac{\alpha}{RT}$$

$$C(T) = b^2 - \frac{\alpha b}{RT} + \frac{\alpha^2}{2R^2T^2}.$$

6.12. Show that Dieterici's equation of state predicts that at the critical point $p_c = \alpha/4e^2b^2$, $V_{c,m} = 2b$, and $T_c = \alpha/4bR$. What is the compression factor at the critical point? *Clues.* The locus of the maxima and minima of the isotherms is obtained in the usual way by setting $(\partial p/\partial V_m)_T = 0$. This gives a quadratic in V_m. This curve will have a maximum at the critical point, at which the two roots of the quadratic are equal.

6.13. In 1897 Berthelot proposed the equation of state

$$\left(p + \frac{A}{TV_m^2}\right)(V_m - b) = RT,$$

where A and b are constants. Show that this can be expressed in reduced form as

$$\left(p_r + \frac{3}{T_r V_r^2}\right)(3V_r - 1) = 8T_r.$$

6.14. If three different substances are all at the same *reduced* volume of 0·6 and at the same reduced pressure of 0·8 what is their common *reduced* temperature? Assume the reduced form of the van der Waals equation (eqn (6.28)). If the substances are argon, carbon dioxide, and hydrogen, what will be their actual temperatures? Use the data given in Table 1.1.

6.15. Using the data given in Table 6.1, calculate $C_{p,m} - C_{V,m}$ for argon at $T = 160$ K, $V_m = 6 \times 10^{-4}$ m³ mol⁻¹.

6.16. How much work can be done by allowing 4×10^{-3} kg of H_2 gas to expand isothermally at 40 K from 1×10^{-4} to 2×10^{-4} m³? How much work can be done if the expansion is adiabatic, between the same volumes, and starting at 40 K? Use Fig. 5.19 and the data given in Table 6.1.

6.17. Investigate the constancy of T_i/T_B and T_B/T_c for the inert gases listed in Table 6.1. What, if anything, does this suggest about the form of the interatomic potential $V(r)$ in this sequence of gases?

6.18. Using the data given in Table 6.1 calculate the molar internal energy of hydrogen at 5 K intervals from 10 to 50 K and at $V_m = 6·5 \times 10^{-5}$ m³ mol⁻¹. Using the same temperature intervals repeat the calculations at 4×10^{-5} and at 9×10^{-5} m³ mol⁻¹. Plot your results graphically. Note that $V_{c,m} = 6·5 \times 10^{-5}$ m³ mol⁻¹ for hydrogen.

6.19. The molar enthalpy H_m of a system is defined as $H_m = U_m + pV_m$. Show that the enthalpy of a gas does not change as it undergoes a Joule–Kelvin expansion.

6.20. Cooling by adiabatic expansion was first observed by Clement and Désormes in 1819. Olszewski applied this method in 1895 to liquify hydrogen. He compressed the gas to 190 atmospheres ($1·9 \times 10^7$ Pa) and precooled it to 62 K (with liquid oxygen boiling under reduced pressure). On releasing the pressure he observed a fog of liquid hydrogen drops. Calculate the expected temperature drop if the process is regarded as a reversible adiabatic expansion. Since the process is actually an irreversible adiabatic expansion, will the observed temperature drop be greater or less than the calculated temperature drop? Give reasons for your answer. *Note*: This is a tricky problem. At very least you should calculate the temperature drop assuming hydrogen is a perfect gas.

6.21. In an early helium liquifier designed by Kapitza, helium at $2 \cdot 5 \times 10^6$ Pa (25 atmospheres) and at a temperature of 14 K was expanded through a Joule–Kelvin valve to a pressure of $1 \cdot 5 \times 10^5$ Pa. Calculate the expected temperature drop. You may take $C_{p,m}$ to be constant with a value appropriate to the mean of the input and output pressures. Take $C_{V,m} = \frac{3}{2}R$ and use the data given in Table 6.1. Comment on your answer if it seems unreasonable.

7. Transport processes in gases

So far we have been mainly concerned with equilibrium situations. We will now look at situations which are deliberately contrived to be non-equilibrium in character. We might have two different gases, as shown in Fig. 7.1(a), separated by a partition. On removing the partition the two gases interdiffuse; a process characterized, we shall shortly discover, by the transport of mass. Alternatively, we might have one gas sandwiched between two plates, as shown in Fig. 7.1(b). If the plates are at a different temperature, heat is conducted across the gas; a process characterized by the transport of energy. If, say, the upper plate is moved parallel to the fixed lower plate the gas is found to be 'viscous' (meaning that a shearing stress is required to maintain the relative movement of the plates); a process characterized by the transport of momentum. A final possibility is to have the two plates at a different electrostatic potential. If the gas contains ions, an electric current will flow; a process characterized by the transport of charge. These different transport processes—the focus of our attention in this chapter—are summarized in Table 7.1. Throughout this chapter we will assume that the gases behave in perfect-gas fashion. If you study the subject further, particularly in more advanced texts, you may well find that the numerical factors which occur in the transport coefficients (for example the factor of $\frac{1}{3}$ in D; eqn (7.4)) differ from those derived here. We have already seen an example of how different numerical factors can arise (in the discussion of the gaseous mean free path in §5.9). Although these different factors may signal a more—or even less—thorough treatment to the one given here they do not change the nature of the underlying mechanism responsible for the transport process in question.

7.1. Diffusion

The diffusion coefficient

For simplicity, we will only consider the diffusion of a gas into a radioactive isotope of the same gas. In this case both gases have atoms of effectively the same size and mass. One method of studying the inter-diffusion is to start with the two gases separated by a dividing partition. Once the partition is removed (Fig. 7.2(a)) the diffusion proceeds until at time $t = \infty$ the concentration of both species is constant throughout the chamber (Fig. 7.2(c)). At some intermediate time t (Fig. 7.2(b)) the radioactive gas will have partially penetrated the non-radioactive gas, and

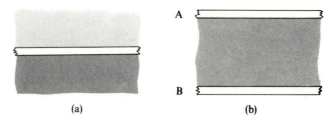

Fig. 7.1. (a) On removing the partition shown, the two gases will interdiffuse. (b) If the plate A is moved to the right and B is held stationary, momentum is fed into the gas at plate A and removed at plate B. If A is at a higher temperature than B, energy is fed into the gas at A and removed at B. If A is at a higher electrical potential than B and if the gas is at least partially ionized, the movement of positive ions toward B and of negative ions towards A will induce a current in the external circuit.

vice versa. The concentration of the radioactive gas can be found by moving a suitable detector along the axis of the cylinder. Such experiments allow one to measure the concentration gradient dn/dx and the *flux* j, defined as the number of atoms of one species crossing unit area per unit time, at, say, $x = x_P$ in Fig. 7.2(b). Measurements show that $|j| \propto |dn/dx|$, with diffusion occurring down the concentration gradient. These observations are summed up in the relation

$$j = -D\frac{dn}{dx}, \qquad (7.1)$$

a result known as Fick's law of diffusion. The constant of proportionality D is called the *diffusion coefficient* (here the *self-diffusion* coefficient). Since dn/dx is positive at $x = x_P$ in Fig. 7.2(b), and since D is always supposed to have a positive value, it follows that the minus sign in eqn (7.1) indicates diffusion in the $-x$-direction. Conversely, if dn/dx is negative (as it is for the non-radioactive atoms of Fig. 7.2(b)), eqn (7.2) gives a positive value to j, indicating diffusion in the $+x$-direction. As a numerical example, if a concentration gradient of $1 \cdot 6 \times 10^{22}$ m^{-4} produces

TABLE 7.1
Characteristics of the main transport processes

Process	Requirement at macroscopic level	Characteristic
Diffusion	Density gradient	Transport of mass
Viscous flow	Shearing stress	Transport of momentum
Thermal conduction	Temperature gradient	Transport of energy
Electrical conduction	Potential gradient	Transport of charge

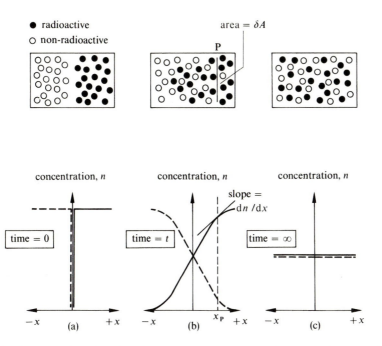

Fig. 7.2. Showing the interdiffusion of a gas and its radioactive equivalent. (a) The separating barrier is removed at time $t = 0$. (b) At some later time t the interdiffusion is evident. (c) Eventually both atomic species attain the same uniform concentration.

a flux of $-2 \cdot 5 \times 10^{15} \, \mathrm{m}^{-2} \, \mathrm{s}^{-1}$ then $D = (2 \cdot 5 \times 10^{15} \, \mathrm{m}^{-2} \, \mathrm{s}^{-1})/(1 \cdot 6 \times 10^{22} \, \mathrm{m}^{-4})$ $= 1 \cdot 67 \times 10^{-7} \, \mathrm{m}^2 \, \mathrm{s}^{-1}$.

Our goal is to predict the value of D in terms of atomic parameters. Now there are, as we saw in Chapter 5, many different ways of discussing gas kinetic processes. What we shall do here is to look at three different ways of calculating D. We shall find that all three ways yield the same expression for D. This will then allow us to adopt the simplest of these approaches in discussing the remaining transport processes listed in Table 7.1, being fairly confident that more sophisticated approaches would also lead to the same answer.

The mechanism: method 1

In this approach we shall use the 'one-sixth' model of a gas (§5.2), adding to it the assumption that the free paths in the gas are all equal and of value λ. This crude model therefore tells us that all those atoms which cross a plane of area δA drawn in the gas (Fig. 7.3) will have come from a region of the gas at a distance λ away from δA. It is clear that, because of the concentration gradient, more radioactive atoms will pass through δA

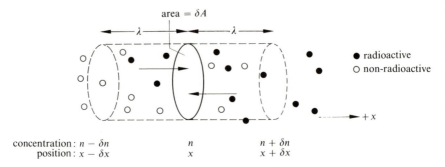

concentration: $n - \delta n$ n $n + \delta n$
position: $x - \delta x$ x $x + \delta x$

Fig. 7.3. Atoms crossing δA will have come from a region of the gas a distance λ away. Because of the concentration gradient more radioactive atoms will cross δA moving to the left per unit time than will cross δA moving to the right.

moving to the left than will pass through δA moving to the right. (The reverse will, of course, be true for the non-radioactive atoms.) If the local concentration of radioactive atoms at point $x - \delta x = x - \lambda$ (Fig. 7.3) is $(n - \delta n)$ then, according to the 'one-sixth' model, the number of radioactive atoms moving to the right through δA per unit time is $\frac{1}{6}(n - \delta n)\bar{u}\,\delta A$. The number passing to the left through δA per unit time is $\frac{1}{6}(n + \delta n)\bar{u}\,\delta A$, where $(n + \delta n)$ is the concentration of radioactive atoms at point $x + \delta x = x + \lambda$. Therefore the net number of radioactive atoms passing through δA per unit time is

$$\tfrac{1}{6}(n + \delta n)\bar{u}\,\delta A - \tfrac{1}{6}(n - \delta n)\bar{u}\,\delta A = \tfrac{1}{3}\delta n\,\bar{u}\,\delta A. \tag{7.2}$$

The number crossing per unit area per unit time—the flux—is thus given by

$$j = -\tfrac{1}{3}\bar{u}\,\delta n, \tag{7.3}$$

where we have inserted a minus sign to make this equation conform to our sign convention which labels diffusion in the $-x$-direction as negative. Multiplying the right-hand side of eqn (7.3) by λ/λ ($= 1$) gives

$$j = -\tfrac{1}{3}\lambda\bar{u}\,\frac{\delta n}{\lambda}.$$

Since $\delta x = \lambda$, this can be rewritten as

$$j = -D\frac{\delta n}{\delta x},$$

where (cf. eqn (7.1)) the coefficient of self-diffusion D is given by

$$\boxed{D = \tfrac{1}{3}\lambda\bar{u}}\ . \tag{7.4}$$

To see whether this is a plausible result we can consider the analogous problem of would-be passengers 'diffusing' towards a waiting train against a crowd of disembarking passengers bent on 'diffusing' in the opposite direction. Our everyday experiences tell us that the fewer people there are about (meaning the larger the 'mean free path') and the faster everyone moves the faster do people get places. This is in accord with the predictions of eqn (7.4).

The mechanism: method 2

This method acknowledges the fact that gas atoms move in all directions while retaining the fiction that all the free paths have the common value λ. It starts by recalling that eqn (5.5) gives the number of atoms striking an area δA per unit time as

$$\tfrac{1}{2}\delta A\, n \sin \theta \cos \theta\, d\theta\, u f_3(u)\, du, \tag{7.5}$$

where θ is defined in Fig. 7.4. According to this model each atom last suffered a collision when at a distance λ from δA. We therefore draw in the hemisphere of radius λ as shown (in section) in Fig. 7.4(a). This tells us that those atoms moving towards δA along directions θ to $\theta + d\theta$ have come from a region in the gas where the concentration of radioactive atoms is

$$n + \lambda \cos \theta \frac{dn}{dx}. \tag{7.6}$$

We must therefore replace n in eqn (7.5) by eqn (7.6). Integrating from

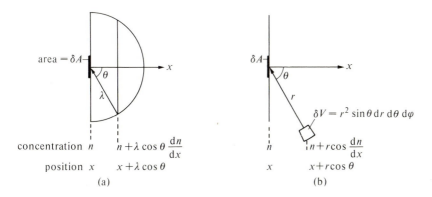

Fig. 7.4. (a) All those atoms passing through δA and moving in the $-x$-direction are assumed to have suffered their last collision when at a distance λ from δA. (b) Atoms passing through δA will consist of those which last suffered a collision within all the elements of volume like δV and which succeed in reaching δA without suffering a further collision.

$\theta = 0$ to $\theta = \pi/2$ and from $u = 0$ to $u = \infty$ gives the total number of atoms passing through δA per unit time and moving in the $-x$-direction as

$$\tfrac{1}{2}\delta A \int_0^{\pi/2} \left(n + \lambda \cos \theta \frac{dn}{dx}\right) \sin \theta \cos \theta \, d\theta \int_0^{\infty} u f_3(u) \, du. \tag{7.7}$$

Likewise, the number of atoms passing through δA per unit time and moving in the $+x$-direction is

$$\tfrac{1}{2}\delta A \int_0^{\pi/2} \left(n - \lambda \cos \theta \frac{dn}{dx}\right) \sin \theta \cos \theta \, d\theta \int_0^{\infty} u f_3(u) \, du. \tag{7.8}$$

Subtracting eqn (7.8) from eqn (7.7) and dividing the result by δA gives the flux j as

$$j = -\frac{dn}{dx} \int_0^{\pi/2} \lambda \sin \theta \cos^2\theta \, d\theta \int_0^{\infty} u f_3(u) \, du$$

$$j = -\frac{dn}{dx} [\tfrac{1}{3} \cos^3 \theta]_0^{\pi/2} \lambda \bar{u} \tag{7.9}$$

where, as before, we have inserted a minus sign to ensure that flow in the $-x$-direction is labelled negative. Hence

$$j = -\tfrac{1}{3}\lambda \bar{u} \frac{dn}{dx}$$

and thus $D = \tfrac{1}{3}\lambda \bar{u}$; exactly the result obtained using the one-sixth model.

Another, but less convincing, line of argument merely replaces the terms $\tfrac{1}{6}n\bar{u}$ and λ, as they occur in method 1, by $\tfrac{1}{4}n\bar{u}$ and $\tfrac{2}{3}\lambda$, respectively. (The term $\tfrac{1}{4}n\bar{u}$ is, you will recall, the correct expression for the number of atoms striking unit area of surface per unit time in the gas. The term $\tfrac{2}{3}\lambda$ represents the mean value of $\lambda \cos \theta$ in Fig. 7.4(a); a result you will establish in problem 7.1.) Both these terms occur as a product in the expression for j. Thus, since $\tfrac{1}{4} \times \tfrac{2}{3} = \tfrac{1}{6} \times 1$, the 'one-quarter' and the 'one-sixth' models yield the same result.

The mechanism: method 3

This method acknowledges the fact that gas atoms move in all directions *and* have a range of free paths, distributed according to eqn (5.72). Its starting point is eqn (5.75). This tells us the number of atoms which, setting off fresh from a collision within a volume $\delta V = r^2 \sin \theta \, dr \, d\theta \, d\varphi$, succeed in reaching an area δA per unit time (Fig. 7.4(b)). For convenience we immediately integrate eqn (5.75) from $\phi = 0$ to $\phi = 2\pi$ and obtain

$$\frac{\delta A \bar{u} n}{2\lambda} \sin \theta \cos \theta \, d\theta \, e^{-r/\lambda} \, dr.$$

To allow for the concentration gradient we must replace n by $n + r\cos\theta(dn/dx)$. Integrating from $\theta = 0$ to $\theta = \pi/2$ and from $r = 0$ to $r = \infty$ gives the total number of atoms passing through δA per unit time and moving in the $-x$-direction as

$$\frac{\delta A\,\bar{u}}{2\lambda}\left[n\int_0^{\pi/2}\sin\theta\cos\theta\,d\theta\int_0^\infty e^{-r/\lambda}\,dr\right.$$

$$\left.+\frac{dn}{dx}\int_0^{\pi/2}\sin\theta\cos^2\theta\,d\theta\int_0^\infty r\,e^{-r/\lambda}\,dr\right]. \quad (7.10)$$

Likewise, the number of atoms passing through δA per unit time and moving in the $+x$-direction is

$$\frac{\delta A\,\bar{u}}{2\lambda}\left[n\int_0^{\pi/2}\sin\theta\cos\theta\,d\theta\int_0^\infty e^{-r/\lambda}\,dr\right.$$

$$\left.-\frac{dn}{dx}\int_0^{\pi/2}\sin\theta\cos^2\theta\,d\theta\int_0^\infty r\,e^{-r/\lambda}\,dr\right]. \quad (7.11)$$

Subtracting eqn (7.11) from eqn (7.10) and dividing the result by δA gives the flux j as

$$j=-\frac{\bar{u}}{\lambda}\frac{dn}{dx}\int_0^{\pi/2}\sin\theta\cos^2\theta\,d\theta\int_0^\infty r\,e^{-r/\lambda}\,dr$$

where, once again, we have inserted a minus sign to ensure that flow in the $-x$-direction is labelled negative. The first integral is as in eqn (7.9): the second integrates by parts to λ^2. Hence

$$j=-\tfrac{1}{3}\lambda\bar{u}\frac{dn}{dx}$$

which, yet again, gives $D=\tfrac{1}{3}\lambda\bar{u}$.

Predictions

Substituting $\lambda = kT/\pi d^2 p$ (eqn (5.58)) and $\bar{u} = (8kT/\pi m)^{1/2}$ (eqn (4.50)) into eqn (7.4) gives

$$D=\frac{1}{p}\frac{1}{3d^2 m^{1/2}}\left(\frac{2kT}{\pi}\right)^{3/2}. \quad (7.12)$$

This predicts that D is proportional to p^{-1} at a fixed temperature and that D is proportional to $T^{3/2}$ at a fixed pressure.

Testing the predictions

Fig. 7.5 shows an apparatus used by T. R. Mifflin and C. O. Bennett in 1958 to study self-diffusion in argon at a fixed temperature but at

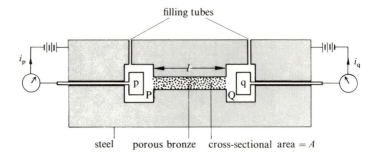

Fig. 7.5. Initially, chamber P contains only radioactive argon and chamber Q contains only non-radioactive argon. Interdiffusion occurs via the porous bronze plug.

pressures up to 3×10^7 Pa (300 times normal atmospheric pressure). Essentially, the apparatus consisted of two chambers P and Q, each of volume $V = 3 \cdot 6 \times 10^{-5}$ m³, separated by a porous bronze plug of length $l = 3 \cdot 8 \times 10^{-2}$ m. To perform the experiment, argon at the same pressure was introduced into the two chambers. Initially chamber Q contained natural non-radioactive argon and chamber P contained natural argon with a trace of ^{37}Ar. As time progressed the radioactive argon diffused from P to Q. The concentration of ^{37}Ar in each chamber was obtained by monitoring the currents i_p and i_q flowing between the electrodes p and q and the surrounding steel case (Fig. 7.5). These currents are proportional to the concentration of ^{37}Ar in chambers P and Q, respectively. Currents of order 10^{-12} A were typical. To see how D is related to i_p and i_q we will consider the situation at some time t when the concentration of ^{37}Ar is n_P in P and is n_Q in Q. According to eqn (7.1) the flux j of ^{37}Ar atoms through the plug is

$$j = -D(n_Q - n_P)/l = D(n_P - n_Q)/l. \tag{7.13}$$

The total number of ^{37}Ar atoms flowing per unit time through the plug of area A is, of course, given by Aj. Clearly, Aj must also equal the gain in Q of ^{37}Ar atoms per unit time and the loss in P per unit time. Thus

$$-\frac{\mathrm{d}}{\mathrm{d}t}(Vn_P) = Aj = AD\frac{(n_P - n_Q)}{l} \tag{7.14}$$

$$\frac{\mathrm{d}}{\mathrm{d}t}(Vn_Q) = Aj = AD\frac{(n_P - n_Q)}{l}, \tag{7.15}$$

where we substituted for j from eqn (7.13). Combining eqns (7.14) and

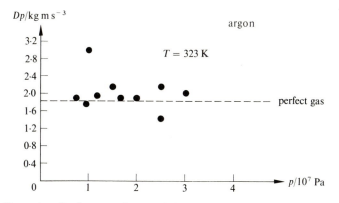

Fig. 7.6. Shows that Dp for argon is essentially independent of gas pressure. (Data from Mifflin, T. R. and Bennett, C. O. (1959). *J. Chem. Phys.* **29**, 975.)

(7.15) gives

$$\frac{d(n_P - n_Q)}{(n_P - n_Q)} = -\frac{2AD}{Vl} \, dt.$$

Integrating this equation, and making use of the fact that the currents i_p and i_q are proportional to n_P and n_Q, respectively, (a fact we shall later justify) leads to

$$\ln(i_p - i_q) = \text{constant} - \left(\frac{2AD}{Vl}\right) t. \tag{7.16}$$

As predicted by eqn (7.16), a graph of $\ln(i_p - i_q)$ plotted against t was found to be linear, allowing D to be obtained from its gradient $(= -2AD/Vl)$. When the experiment was repeated at different pressures p (which must, of course, be the same in both chambers P and Q), the results shown in Fig. 7.6 were obtained. Here Dp is plotted against p. (Why is this a better way of presenting the data than plotting D against $1/p$?) According to eqn (7.12), at any fixed temperature Dp should be constant (with the value indicated by the dashed line in Fig. 7.6), independent of p. The data shows that Dp is indeed constant (to within about 25 per cent) over a fourfold change in pressure. Although Mifflin and Bennett did not change the temperature of the gas in their experiments other measurements show that, at constant pressure, D increases rather more rapidly with increasing T than the $\frac{3}{2}$-power suggested by eqn (7.12). We will shortly discover the reason for this discrepancy.

7.2. Viscous flow

The dynamic viscosity

Fig. 7.7 shows a gas contained between two plates of effectively infinite area. The upper plate is pulled with a force F and the lower plate is kept stationary by applying an equal, but oppositely directed, force to that applied to the top plate. Experiments show that, at low values of F, the velocity gradient $\delta v_x/\delta y$ is the same throughout the gas, that is at all y. Here δv_x is the change in the value of the velocity v_x at which the gas *flows* as we shift a distance δy towards the moving plate. (One way to measure the velocity gradient is to introduce smoke at various levels between the plates and to photograph its position at two different times.) It is found that, at low flow rates,

$$\frac{F}{A} \propto \frac{dv_x}{dy},$$

$$\boxed{\frac{F}{A} = \eta \frac{dv_x}{dy},} \qquad (7.17)$$

where η is called the *dynamic viscosity* of the gas. You should be able to establish for yourself that, in the SI system, η is measured in units of Pa s or, the same thing, in units of $\mathrm{kg\,m^{-1}\,s^{-1}}$. It is important to note that dv_x/dy is the velocity gradient *as measured in the gas*. As we shall shortly see, this is *not*, in general, equal to v_1/l, where v_1 is the speed of the moving plate and l is the separation of the plates.

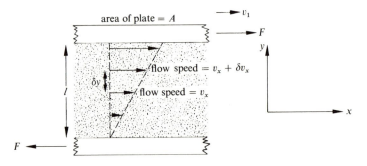

Fig. 7.7. A gas is contained between two parallel plates of (effectively) infinite area. The upper plate is pulled with a force F and the lower plate is kept stationary by applying an equal, but oppositely directed, force to that applied to the top plate. The vectors indicate the velocity of flow of the gas at various positions between the plates.

The mechanism

Fig. 7.8 shows a close-up of conditions near the plates of Fig. 7.7. Any atom which strikes the plate will, on average, have come from a distance of about λ into the gas. Confining our attention to the top plate for the moment, an atom (mass m) which strikes this plate *and sticks* will have its flow (or drift) velocity increased from v', the flow velocity of the layer of gas at distance λ from the top plate, to v_1, the velocity of the top plate. In other words, its momentum to the right will have increased by $m(v_1 - v')$, calling for an impulse of magnitude $m(v_1 - v')$ at each impact.

Adopting a 'one-sixth' model, the total number of atoms impacting on the top plate in time t is $\frac{1}{6} At n \bar{u}$, where n is the number of atoms per unit volume of the gas. Hence the total momentum brought up to the top plate in time t is $\frac{1}{6} At n \bar{u} m (v_1 - v')$. Equating this to $\bar{F}t$, where $\bar{F}$ is the apparently constant force provided by the person pulling the plate, gives

$$\frac{\bar{F}}{A} = \frac{1}{6} n \bar{u} m (v_1 - v'). \tag{7.18}$$

Now the mean flow velocity of the gas atoms immediately adjacent to the top plate will be the average of the flow velocity of an atom just before impact (v') and just after impact (v_1), that is $\frac{1}{2}(v_1 + v')$ (see Fig. 7.8(a)). Since the velocity at a distance λ away is v', it follows that the velocity gradient is given by

$$\frac{\delta v_x}{\delta y} = \frac{\frac{1}{2}(v_1 + v') - v'}{\lambda} = \frac{v_1 - v'}{2\lambda}. \tag{7.19}$$

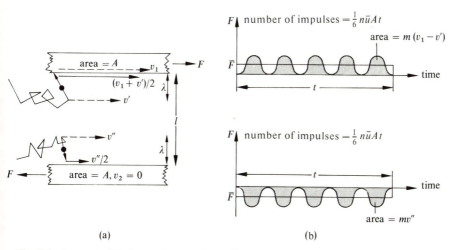

(a)   (b)

Fig. 7.8. An atom diffusing to the top plate will gain tangential momentum $m(v_1 - v')$. (b) Shows the series of impulses required to keep the top plate moving with speed v_1, and to keep the lower plate stationary.

Substituting for $v_1 - v'$ from eqn (7.19) into eqn (7.18) gives

$$\frac{\bar{F}}{A} = \eta \frac{\delta v_x}{\delta y},$$

where η, the dynamic viscosity, is given by

$$\boxed{\eta = \tfrac{1}{3}nm\bar{u}\lambda} . \tag{7.20}$$

or, writing $n = p/kT$ (eqn (5.19)), $\bar{u} = (8kT/\pi m)^{1/2}$ (eqn (4.50)), and $\lambda = kT/\pi d^2 p$ (eqn (5.58)),

$$\eta = \frac{1}{\pi d^2}\left(\frac{8kTm}{9\pi}\right)^{1/2}. \tag{7.21}$$

Although we chose to look at the top plate in Fig. 7.8 the same result would have been obtained by looking at conditions at the lower plate.

What we actually measure

To relate the *true* velocity gradient dv_x/dy present in the gas to the *apparent* velocity gradient v_1/l (as measured experimentally) we looked at the flow velocities of the gas immediately adjacent to each plate and at a distance λ from each plate (Fig. 7.9). Looking in turn at triangles *NPS*, *OPQ*, and *TRS* we see that

$$\frac{dv_x}{dy} = \tan\theta = \frac{1}{l}\left(\frac{v_1 + v'}{2} - \frac{v''}{2}\right) \tag{7.22}$$

$$\frac{dv_x}{dy} = \tan\theta = \frac{1}{\lambda}\left(\frac{v_1 - v'}{2}\right) \tag{7.23}$$

$$\frac{dv_x}{dy} = \tan\theta = \frac{1}{\lambda}\left(\frac{v''}{2}\right). \tag{7.24}$$

Substituting for v' and v'' from eqns (7.23) and (7.24), respectively, into eqn (7.22) gives

$$\frac{dv_x}{dy} = \frac{v_1}{l + 2\lambda}.$$

Putting this result into our operational definition of dynamic viscosity (eqn (7.17)) gives

$$\frac{F}{A} = \eta\frac{v_1}{l}\left[\frac{1}{1 + (2\lambda/l)}\right]. \tag{7.25}$$

Experiments tell us the apparent velocity gradient v_1/l with a known applied shear stress F/A. If we divide our value of F/A by our value of

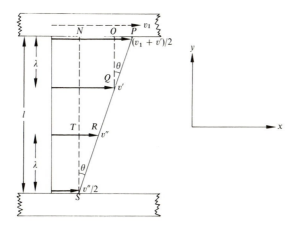

Fig. 7.9. Showing the true velocity gradient present during viscous flow. The apparent velocity gradient is v_1/l.

v_1/l we will be calculating an *apparent viscosity* η_{app}, where

$$\eta_{app} = \frac{(F/A)}{(v_1/l)}.$$ (7.26)

On substituting for F/A from eqn (7.25) this becomes

$$\boxed{\eta_{app} = \eta \frac{1}{1+(2\lambda/l)}}.$$ (7.27)

Pressure dependence of η_{app}

When $\lambda \ll l$, eqns (7.27) and (7.21) predict that η_{app} should be independent of gas pressure, p. To see the underlying reasons for this prediction we will consider the effect of halving p, keeping T fixed. Halving p will halve n and double λ. This means that the number of atoms arriving at the top plate ($\frac{1}{6}(Atn\bar{u})$) is halved. However, because λ is doubled, each atom striking the plate has come from a point which is twice as far into the gas as before. It thus arrives at the plate with double the flow velocity it had at pressure p. Therefore the person pulling the top plate must provide *double* the impulse each time an atom arrives. As there are only *half* the number of arrivals as at pressure p, this means that $\bar{F}$ is unaltered.

Fig. 7.10 shows that the apparent viscosity of air is indeed constant, independent of pressure, from 10^3 to 10^6 Pa. As predicted by eqn (7.27), the viscosity starts to fall when λ becomes comparable with l. However,

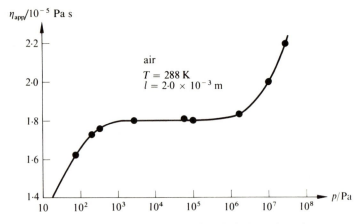

Fig. 7.10. Shows how the apparent viscosity of air depends on its pressure.

when careful measurements are made η_{app} is found to decrease even more rapidly with falling pressure than is allowed for in eqn (7.27). Something has evidently been overlooked in our discussions.

The effect of non-sticking atoms

In discussing viscous flow we have assumed that all the gas atoms striking a surface 'stick' for sufficient time for them to acquire the velocity of the plate. However, we know from §5.1 that, while a large fraction, say σ, of atoms incident on a surface do indeed stick, a certain fraction $(1-\sigma)$ are specularly reflected and thus suffer no change in momentum parallel to the surface. The consequence is that the flow momentum of an atom striking the moving plate will, on average, increase by

$$m[\sigma v_1 + (1-\sigma)v'] - mv' = m\sigma(v_1 - v').$$

You can check that this makes sense by setting $\sigma = 1$ (corresponding to all the atoms sticking) and $\sigma = 0$ (no atoms sticking). If we now replace $m(v_1 - v')$ in eqn (7.18) by $m\sigma(v_1 - v')$ we obtain

$$\frac{\bar{F}}{A} = \tfrac{1}{6} n \bar{u} m \sigma (v_1 - v'). \tag{7.28}$$

Following through exactly the same arguments as before leads (see problem 7.8) to the result

$$\boxed{\eta_{app} = \eta \frac{1}{1 + (2\lambda/l)[(2-\sigma)/\sigma]}}. \tag{7.29}$$

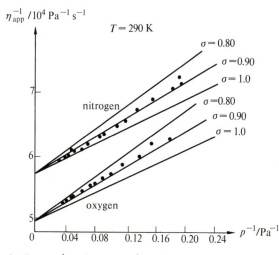

Fig. 7.11. Showing how η_{app}^{-1} varies with p^{-1} for N_2 and O_2 maintained at 290 K. The lines labelled $\sigma = 0.8$, $\sigma = 0.9$, and $\sigma = 1.0$ show the behaviour to be expected at these different values of σ. The walls of the viscometer were made of aluminium. (Data from Hurlbut, F. C. (1960). *Phys. Fluids* **3**, 541.)

Cross-multiplying this equation and substituting for η from eqn (7.21) and for λ from eqn (5.58) gives

$$\frac{1}{\eta_{app}} = \pi d^2 \left(\frac{9\pi}{8kTm}\right)^{1/2} + \frac{1}{p}\left(\frac{9\pi kT}{2ml^2}\right)^{1/2}\left(\frac{2-\sigma}{\sigma}\right). \tag{7.30}$$

This predicts that a graph of η_{app}^{-1} plotted against p^{-1} at a fixed temperature should be linear, with a gradient of $(9\pi kT/2ml^2)^{1/2}(2-\sigma)/\sigma$. Fig. 7.11 shows that this is indeed the case. The best fit between theory and experiment is obtained by assigning $\sigma = 0.90$; that is, by assuming that ten per cent of the nitrogen and oxygen molecules are specularly reflected from the aluminium surfaces of the viscometer used in obtaining the data of Fig. 7.11.

As a practical application of these arguments we may note that the tangential drag force per unit area $(\bar{F}/A)$ on a plate moving through a gas is proportional to σ (eqn (7.28)). Thus the lifetime of a satellite will depend on σ, and this, in turn, can influence the choice of surface material. Earth-bound simulations of conditions likely to be found in space give, for example, $\sigma = 0.5$ when gold is struck by ionized inert gas atoms.

The effect of interatomic forces

Experiments show that η_{app} varies more rapidly with temperature than the $T^{1/2}$-dependence predicted by eqns (7.29) and (7.21) when $\lambda \ll l$.

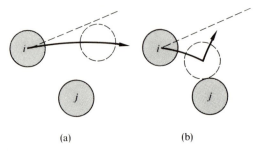

(a)   (b)

Fig. 7.12. With no attractive forces present atom i will follow the dashed path relative to j. With an attractive force present atom i may miss j if it is moving fast, as in (a). If moving more slowly it may hit j, as in (b).

Looking at eqn (7.21) we see that we could account for this discrepancy if πd^2, and hence the atomic diameter d, were to *decrease* as T (that is, $\frac{1}{2}m\overline{u^2}$) increases. This makes sense if we picture the two colliding atoms as behaving like tennis balls; the faster they collide the smaller is their separation between centres. Alternatively, we can argue that the $1/\pi d^2$ term in eqn (7.21) actually arose from $\lambda = 1/\pi d^2 n$, leaving us to explain why λ *increases* as T (that is, $\frac{1}{2}m\overline{u^2}$) increases. Fig. 7.12 should make this plausible. Here the dotted trajectory shows how atom i moves relative to j if there are no attractive forces present. With the attractive forces present atom i may now collide with j (Fig. 7.12(b)); thus the attractive forces have led to a reduction in the free path of i. At higher relative velocities i may miss j (Fig. 7.12(a)). Hence, as required, λ increases as T increases.

We therefore conclude that both the repulsive and the attractive portions of the interatomic force characteristic are involved in the temperature dependence of viscosity (the same statement is true of diffusion and of thermal conduction). In the past the constants occurring in the Lennard-Jones potential (eqn (2.18)) were actually determined by studying how the viscosity of a gas varies with temperature. Nowadays these constants can be obtained more directly by studying how two crossed beams of atoms scatter off one another (cf. Fig. 7.12).

Flow down tubes

Practical laboratory work frequently involves transporting gases through tubes. At high pressures, where the gaseous mean free path λ is very much less than the tube's radius r, there is a velocity gradient across the tube, the gas at the centre flowing more rapidly than the gas nearer the walls (Fig. 7.13(a)). Under these circumstances the rate of flow of gas down the tube is found by suitably integrating eqn (7.17). As is shown in

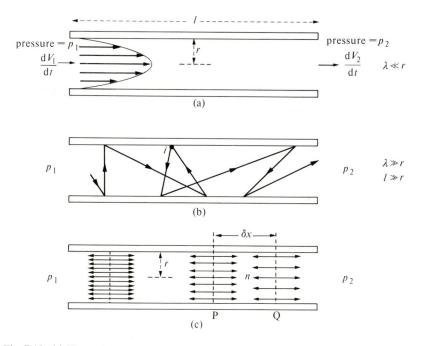

Fig. 7.13. (a) Illustrating viscous flow down a tube. (b) Knudsen flow. (c) The mechanism of Knudsen flow: Because there are more atoms at P than at Q there is a net flow along direction PQ.

practically any book on fluid mechanics, the *throughput* Q, defined by

$$Q = p_1 \frac{dV_1}{dt} = p_2 \frac{dV_2}{dt}, \tag{7.31}$$

is given by *Poiseuille's equation*, namely

$$\boxed{Q = \frac{\pi r^4}{16\eta l} (p_1^2 - p_2^2)}, \tag{7.32}$$

where p_1 and p_2 are the gas pressures at the input and output ends of the tube. To see the significance of eqn (7.31) we return to our familiar gas law relation $pV = NkT$ (eqn (5.20)) and note that at constant pressure and temperature

$$p \frac{dV}{dt} = kT \frac{dN}{dt}. \tag{7.33}$$

Thus Q is kT times the number of atoms entering, and therefore leaving, the pipe per unit time. Hence it has the same value at both ends of the pipe.

It is often convenient to rewrite eqn (7.32) as

$$Q = \frac{\pi r^4}{8\eta l} p_{av}(p_1 - p_2) \tag{7.34}$$

where the average pressure $p_{av} = \frac{1}{2}(p_1 + p_2)$. Going one step further we can introduce the *conductance* U of the tube, defined by

$$U = \frac{Q}{p_1 - p_2}. \tag{7.35}$$

It follows from eqn (7.34) that

$$U = \frac{\pi r^4}{8\eta l} p_{av}. \tag{7.36}$$

The advantage of conductances are that they can be treated in much the same way as electrical conductances. Thus the total conductance, U, of two tubes of conductance U_1 and U_2, respectively, is given by $U = U_1 + U_2$ when these tubes are connected in parallel and by $1/U = (1/U_1) + (1/U_2)$ when these tubes are in series.

At even moderate pressures, such as those reached by rotary pumps, λ may well exceed r. When $\lambda \gg r$ flow down a tube proceeds by quite a different mechanism. What happens in such a, so-called, *Knudsen gas*, is that the atoms can travel without collision from one side of the tube to the other (Fig. 7.13(b)). At every such impact a majority of the gas atoms (we will assume 100 per cent of them; corresponding to $\sigma = 1$) will stick, albeit temporarily. When they do leave they will do so according to the cosine-distribution (Fig. 5.3(a)). This means that an *individual* atom, such as i in Fig. 7.13(b), is equally likely to leave heading up or down the tube. Flow only happens because there are more atoms near the high-pressure end of the tube than there are at the low-pressure end of the tube. This is shown in Fig. 7.13(c) where the number of vectors at points such as P and Q indicates the number density of the gas atoms at these points; the length of each vector being $\bar{u}_x$.

Focusing our attention on the length δx of tube shown in Fig. 7.13(c), we know that the number of gas atoms colliding with the surface area $2\pi r\, \delta x$ of tube per unit time is $\frac{1}{4}n\bar{u}(2\pi r\, \delta x)$, where n is the (mean) number density of the gas atoms at a point midway between P and Q. At the moment of impact each of these atoms has, on average, an x-component of velocity given by $\bar{u}_x$. Hence the total horizontal momentum transferred to the length δx of the tube per unit time is given by $\frac{1}{4}n\bar{u}(2\pi r\, \delta x)m\bar{u}_x$. Integrating this expression from $x = 0$ to $x = l$ tells us that the total momentum transferred to the tube per unit time, which must also equal the net force F which the gas exerts on the tube, is given by

$$F = \tfrac{1}{2}\pi r m n \bar{u}_x \bar{u} \int_0^l dx$$

or, substituting for $\bar{u}$ from eqn (4.50) and for $n\bar{u}_x$ from the relation $dN/dt = (\pi r^2)n\bar{u}_x$,

$$F = \frac{l}{r}\left(\frac{2kTm}{\pi}\right)^{1/2}\frac{dN}{dt}.$$

By Newton's third law, the force which the gas exerts on the tube must have the same magnitude as the force which the tube exerts on the gas, namely $F = \pi r^2(p_1 - p_2)$. It therefore follows that $Q = kT(dN/dt)$ (eqn (7.33)) is given by

$$\boxed{Q = \left(\frac{\pi^3 kT}{2m}\right)^{1/2}\frac{r^3}{l}(p_1 - p_2)} \tag{7.37}$$

from which we conclude that the tube conductance U, as defined by eqn (7.35), is given by

$$U = \left(\frac{\pi^3 kT}{2m}\right)^{1/2}\frac{r^3}{l}. \tag{7.38}$$

This relation, like its high-pressure equivalent (eqn (7.36)), emphasizes the importance of using tubes of short length and of large bore to transport gases. Clearly, the bore of the tube is a more important consideration than its length.

Exercise 7.1

A tube of length 0.2 m and of inner diameter 1×10^{-3} m connects a chamber in which the gas pressure is 1.5 Pa to one in which the gas pressure is 6.0 Pa. (a) Assuming that both chambers contain nitrogen at 300 K what is the throughput of the tube? (b) How many molecules pass down the tube per second? (c) How would the answers differ if the two pressures were 1.5×10^4 Pa and 6×10^4 Pa? Take the diameter d of a nitrogen molecule ($M_r(N_2) = 28$) as 3.5×10^{-10} m. Use the data given in Fig. 7.10.

Calculations. The first step is to determine the mean free path λ at, say, the average of the pressures in the two chambers. When $p = 3.75$ Pa and $T = 300$ K, it follows that $\lambda = kT/\pi d^2 p = 2.87 \times 10^{-3}$ m. Likewise, when $p = 3.75 \times 10^4$ Pa, $\lambda = 2.87 \times 10^{-7}$ m. Hence eqn (7.37), which only applies when $\lambda \gg r$, is to be used in answering parts (a) and (b) while eqn (7.32), which applies when $\lambda \ll r$, is to be used in answering part (c).

(a) Substituting $m = 4.65 \times 10^{-26}$ kg, $r = 5 \times 10^{-4}$ m, $l = 0.2$ m, $p_1 = 6.0$ Pa, $p_2 = 1.5$ Pa, and $T = 300$ K into eqn (7.37) gives $Q = 3.3 \times 10^{-6}$ Pa m^3 s^{-1}. (b) It follows from eqn (7.33) that $dN/dt = Q/kT = 7\cdot98 \times 10^{14}\,\text{s}^{-1}$. (c) Substituting $p_1 = 6 \times 10^4\,\text{Pa}$, $p_2 = 1\cdot5 \times 10^4\,\text{Pa}$, $r = 5 \times 10^{-4}\,\text{m}$, $l = 0\cdot2\,\text{m}$, and $\eta = 1\cdot8 \times 10^{-5} \times (300/288)^{1/2}\,\text{Pa s}$ (that is, the value of η ($= \eta_{\text{app}}$ here) obtained from Fig. 7.10 and modified to $T = 300\,\text{K}$ according to eqn (7.21)) into eqn (7.32) gives $Q = 11\cdot27\,\text{Pa m}^3\,\text{s}^{-1}$. Hence $dN/dt = Q/kT = 2\cdot72 \times 10^{21}\,\text{s}^{-1}$.

Comments. The pressures used in this example were chosen so that either $\lambda \gg r$ or $\lambda \ll r$. When $\lambda \approx r$ an appropriate empirical relation must be used for the throughput. Texts on vacuum techniques should be consulted for such relations.

7.3. Thermal conduction

The thermal conductivity

If one encloses a gas between two parallel plates, each of area A and separated by a distance l (Fig. 7.14), and if the plates are at temperatures T_1 and T_2 with $T_1 > T_2$ then one finds that heat is conducted through the gas at a rate Φ which is proportional to A and to the temperature gradient dT/dx as measured in the gas. Thus we can write

$$\Phi = -\kappa A \frac{dT}{dx}, \tag{7.39}$$

where the constant of proportionality κ is known as the *thermal conductivity* of the gas. The negative sign ensures that the heat flow which occurs to the right in Fig. 7.14 (where dT/dx is negative) has a positive value. Since Φ is measured in J s^{-1}, A in m^2, and x in m, it follows that κ is measured in units of $\text{W m}^{-1}\,\text{K}^{-1}$.

How κ is measured

Eqn (7.39) only applies if there are no convection currents present. These currents can be avoided by having large, parallel, closely-

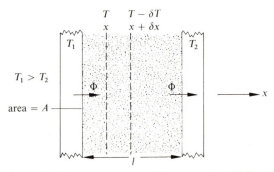

Fig. 7.14. A gas is contained between two parallel plates each of (effectively infinite) area A. The left-hand plate is maintained at temperature T_1; the right-hand plate at a lower temperature T_2. Heat energy flows at a rate Φ from the left to the right-hand plate.

separated, horizontal plates with the upper plate at the higher temperature.

Almost all recent measurements of κ have been made using a *hot-wire cell*. The gas to be studied is contained in a cylindrical tube. A fine wire, lying along the axis of the tube, is heated electrically. The outside of the tube is immersed in a constant-temperature bath. Once a steady state is reached the heat energy Φ produced per unit time in the wire is measured. This is given by $\Phi = VI$, where V is the potential drop along the wire, and I is the current. The temperature of the wire, and hence the temperature drop across the cell, can be deduced from the resistance (V/I) of the wire. The temperature drop and Φ are related to κ by an integrated form of eqn (7.39); see problem 7.15.

The mechanism in a gas

Fig. 7.15(a) show a close-up of conditions near the two plates of Fig. 7.14. Any atom striking these plates will, on average, have come from a distance λ into the gas. On average, each atom striking the hot plate will have its energy increased by $\frac{3}{2}k(T_1 - T')$. Likewise, each atom striking the cold plate will give up energy $\frac{3}{2}k(T'' - T_2)$. There is thus a steady transfer of energy from the hot to the cold plate. In what follows we shall assume that $(T_1 - T_2) \ll T_1$ or T_2. This allows us to take $\bar{u} \propto T^{1/2}$ as constant throughout the gas. We shall also assume that the gas is monatomic.

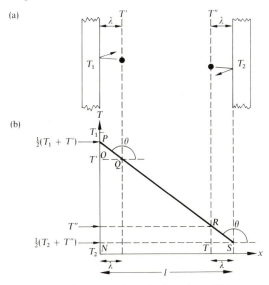

Fig. 7.15. (a) Gas atoms acquiring energy from the hot plate will have come from a distance of about λ into the gas. Atoms giving energy up to the cold plate will likewise have come from a distance of λ into the gas. (b) The temperature gradient present in the gas during thermal conduction.

Adopting, yet again, the 'one-sixth' model, we see that the total energy Φ transferred from the hot plate to the gas per unit time is

$$\Phi = \tfrac{1}{6}(An\bar{u}) \times \tfrac{3}{2}k(T_1 - T') = \tfrac{1}{4}An\bar{u}k(T_1 - T'). \qquad (7.40)$$

Now the mean temperature of the gas atoms immediately adjacent to the hot plate is the average of their temperature (T') just before impact and their temperature (T_1) just after impact, that is $\tfrac{1}{2}(T_1 + T')$. Since the temperature a distance $\delta x = \lambda$ away is T' it follows that

$$\frac{\delta T}{\delta x} = \frac{T' - \tfrac{1}{2}(T_1 + T')}{\lambda} = \frac{T' - T_1}{2\lambda}. \qquad (7.41)$$

Substituting for $T_1 - T'$ from eqn (7.41) into eqn (7.40) gives

$$\Phi = -\kappa A \frac{\delta T}{\delta x},$$

where the thermal conductivity of the monatomic gas is given by

$$\boxed{\kappa = \tfrac{1}{2}n\bar{u}\lambda k} \qquad . \qquad (7.42)$$

Substituting $n = p/kT$ (eqn (5.19)), $\bar{u} = (8kT/\pi m)^{1/2}$ (eqn (4.50)), and $\lambda = kT/\pi d^2 p$ (eqn (5.58)), gives

$$\kappa = \frac{k}{\pi d^2}\left(\frac{2kT}{\pi m}\right)^{1/2}. \qquad (7.43)$$

The discussion may be generalized to embrace polyatomic gases by writing the energy of a gas molecule as $(C_{V,m}/N_A)T$, rather than as $\tfrac{3}{2}kT$, in eqn (7.40). (This follows on integrating eqn (6.32) to give U_m, which is then divided by the N_A molecules present.) This leads to

$$\kappa = \tfrac{1}{3}(C_{V,m}/N_A)n\bar{u}\lambda. \qquad (7.44)$$

What we actually measure

To relate the true temperature gradient dT/dx present in the gas to the apparent temperature gradient $(T_2 - T_1)/l$ as measured experimentally we look in turn at triangles *NPS*, *OPQ*, and *TRS* (Fig. 7.15) and write down three expressions for dT/dx:

$$\frac{dT}{dx} = \tan\theta = \frac{\tfrac{1}{2}(T_1 + T') - \tfrac{1}{2}(T_2 + T'')}{(-l)} \qquad (7.45)$$

$$\frac{dT}{dx} = \tan\theta = \frac{\tfrac{1}{2}(T_1 + T') - T'}{(-\lambda)} \qquad (7.46)$$

$$\frac{dT}{dx} = \tan\theta = \frac{T'' - \tfrac{1}{2}(T_2 + T'')}{(-\lambda)}. \qquad (7.47)$$

Substituting for T' and T'' as given by eqns (7.46) and (7.47), respectively, into eqn (7.45) gives

$$\frac{dT}{dx} = \frac{T_2 - T_1}{l + 2\lambda}.$$

Putting this result into the operational definition of thermal conductivity (eqn (7.39)) gives

$$\frac{\Phi}{A} = -\kappa \frac{(T_2 - T_1)}{l}\left[\frac{1}{1 + (2\lambda/l)}\right]. \tag{7.48}$$

Experiments tell us the heat current flowing per unit area of plate, Φ/A, when the applied temperature gradient is $(T_2 - T_1)/l$. If we divide our measured Φ/A by our measured $(T_2 - T_1)/l$ we will be calculating an *apparent thermal conductivity* κ_{app};

$$\kappa_{app} = -\frac{\Phi/A}{(T_2 - T_1)/l}. \tag{7.49}$$

Substituting Φ/A, as given by eqn (7.48), into eqn (7.49) gives

$$\boxed{\kappa_{app} = \kappa\frac{1}{1 + (2\lambda/l)}}. \tag{7.50}$$

Pressure and temperature dependence of κ_{app}

When $\lambda \ll l$ eqns (7.50) and (7.43) predict that κ_{app} will be independent of pressure. As with η_{app}, this prediction holds true over a wide range

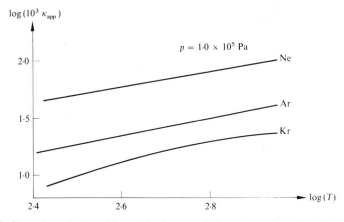

Fig. 7.16. Shows how the logarithm of the (apparent) thermal conductivity of Ne, Ar, and Kr varies with the logarithm of the temperature. The thermal conductivity κ_{app} is measured in units of $W\,m^{-1}\,K^{-1}$ and the temperature T in K. (Data from Kestin, J., Wakeham, W., and Watanabe, K. (1970). *J. Chem. Phys.* **53**, 3773.)

of pressures. At the low-pressure end κ_{app} starts to fall off when λ becomes comparable to the separation l of the plates in the conductivity cell.

Eqns (7.50) and (7.43) also predict that when $\lambda \ll l$ the apparent thermal conductivity κ_{app} (which now equals κ) should vary as $T^{1/2}$. Fig. 7.16 shows that plots of log κ_{app} against log T are indeed nearly linear for neon, argon, and krypton but that their gradients are greater than 1/2. The reasons for this departure are similar to the reasons for the departure of η_{app} from its predicted $T^{1/2}$ dependence.

Energy accommodation

Throughout our discussion we have assumed that a gas atom striking a solid surface comes into complete thermal equilibrium with that surface before it leaves the surface at the end of the dwell time. We will now suppose that a fraction α of the incoming atoms have their energy increased from $\frac{3}{2}kT'$ to $\frac{3}{2}kT_1$ on striking the surface (Fig. 7.15) and that the fraction $(1-\alpha)$ suffer no change in energy. This means that the average change in the energy of a gas atom between striking and leaving the surface is $\frac{3}{2}\alpha k(T_1 - T')$, rather than the value of $\frac{3}{2}k(T_1 - T')$ which we have been using up to now. Introducing this new assumption and following through exactly the same arguments as before leads to

$$\boxed{\kappa_{app} = \kappa \frac{1}{1 + (2\lambda/l)[(2-\alpha)/\alpha]}} \,. \tag{7.50}$$

Cross-multiplying the equation and substituting for κ from eqn (7.43) and for λ from eqn (5.58) gives

$$\frac{1}{\kappa_{app}} = \frac{\pi d^2}{k}\left(\frac{\pi m}{2kT}\right)^{1/2} + \frac{1}{p}\left(\frac{2\pi m T}{kl^2}\right)^{1/2}\left(\frac{2-\alpha}{\alpha}\right).$$

This predicts that a graph of κ_{app}^{-1} plotted against p^{-1} at a fixed temperature should be linear with a gradient of $(2\pi m T/kl^2)^{1/2}(2-\alpha)/\alpha$. Fig. 7.17 shows results obtained using a platinum hot-wire cell. The gradient of this plot tells us that $\alpha = 0 \cdot 8$ for air on platinum at 273 K. As a further example, $\alpha = 0 \cdot 01$ for helium on tungsten at 50 K; meaning that, on average, a helium atom only acquires one per cent of the energy that it would have acquired had it come into thermal equilibrium with the tungsten surface. At the domestic level α influences the time over which coffee will remain hot in a vacuum flask. (Thermal conduction through the residual gas in the evacuated walls is responsible for some of the heat loss.)

In view of the formal similarities between eqns (7.29) and (7.50) it is

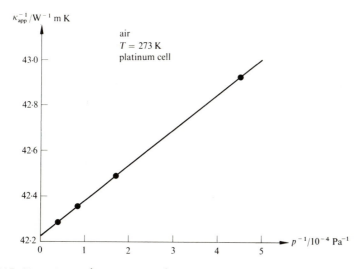

Fig. 7.17. Shows how κ_{app}^{-1} varies with p^{-1} for air. Measurements were made in a platinum hot-wire cell. (Data from Taylor, W. J. and Johnston, H. L. (1946). *J. Chem. Phys.* **14,** 219.)

tempting to equate σ to the energy accommodation coefficient α. These temptations must be resisted since each measures a quite different property. The coefficient σ tells us how momentum—a *vector* quantity—is accommodated at a surface. The coefficient α tells us how energy—a *scalar* quantity—is accommodated at a surface. To change the momentum of an atom it is unnecessary to change its speed; you simply change its direction of motion. To change the kinetic energy of an atom you must change its speed; a more difficult operation.

Relations between D, η, and κ

It follows from eqns (7.4) and (7.20) that

$$\boxed{\frac{\rho D}{\eta} = 1}, \tag{7.51}$$

where $\rho\ (= mn)$ is the density of the gas, and from eqns (7.20) and (7.44) that

$$\boxed{\frac{\kappa}{c_V \eta} = 1}, \tag{7.52}$$

where $c_V = C_{V,m}/N_A m$ (which is thus the heat capacity at constant volume per unit mass of the gas). Experimental values of $\rho D/\eta$ lie between about $1\cdot3$ and $1\cdot5$ (e.g. it is $1\cdot37$ for hydrogen) whereas the values of $\kappa/c_V\eta$ lie close to $2\cdot5$ for the inert gases but decrease with increasing molecular complexity (e.g. it is $1\cdot55$ for ethylene). All these values were obtained at 273 K and apply to the range of pressures over which $D \propto p^{-1}$ and η and κ are constant. The departures of these ratios from unity indicate that the numerical constants occurring in our expressions for D, η, and κ (eqns (7.4), (7.20), and (7.44)) are in error.

The wide range of values noted for $\kappa/c_V\eta$ arises because the relative contribution made by the 3 translational degrees of freedom and by the remaining $(f-3)$ degrees of freedom of a molecule is different in a heat capacity and a thermal conductivity measurement. We already know (§4.11) that, according to classical physics, each degree of freedom contributes $\frac{1}{2}kT$ to the heat capacity at constant volume, that is $C_{V,m} = N_A f(\frac{1}{2}k)$ and hence $c_V = \frac{1}{2}fk/m$. According to Eucken (1913) each of the three translational degrees of freedom transports $\frac{5}{2}$ times the energy transported by each of the remaining $(f-3)$ degrees of freedom. Hence eqn (7.44) should be written not as $\kappa = \frac{1}{3}(3k/2)n\bar{u}\lambda$ but as

$$\kappa = \tfrac{1}{3}n\bar{u}\lambda[\tfrac{5}{2}(\tfrac{3}{2}k)+(f-3)\tfrac{1}{2}k]. \tag{7.53}$$

Thus

$$\frac{\kappa}{c_V\eta} = \frac{1}{f}[\tfrac{15}{2}+(f-3)]$$

or, substituting $f = 2/(\gamma-1)$ from eqn (5.46),

$$\boxed{\frac{\kappa}{c_V\eta} = \tfrac{1}{4}(9\gamma-5)} . \tag{7.54}$$

When the measured values of $\gamma = C_{p,m}/C_{V,m}$ are put into eqn (7.54) the values so predicted for $\kappa/c_V\eta$ usually agree to within a few per cent with the experimental values.

Exercise 7.2

Estimate the diffusion coefficient, dynamic viscosity, and thermal conductivity of air at s.t.p. Assume that air is wholly composed of nitrogen ($M_r(N_2) = 28$) with $C_{V,m} = \frac{5}{2}R$ and with a molecular diameter $d = 3\cdot5\times10^{-10}$ m.

Calculations. Since all three transport coefficients involve λ and $\bar{u}$, it will pay to calculate these straight away. It follows from eqn (5.58) that $\lambda = kT/\pi d^2p = 9\cdot7\times10^{-8}$ m and from eqn (4.50) that $\bar{u} = (8kT/\pi m)^{1/2} = 453$ m s^{-1}. To find n (which occurs in η and κ) we use eqn (5.19) and obtain $n = p/kT = 2\cdot68\times10^{25}$ m^{-3}. Thus $D = \frac{1}{3}\lambda\bar{u} = 1\cdot46\times10^{-5}$ m^2 s^{-1}, $\eta = \frac{1}{3}nm\bar{u}\lambda = 1\cdot83\times10^{-5}$ Pa s. In finding κ we must use eqn (7.44) (rather than eqn (7.42) which assumed a monatomic gas). Thus $\kappa = \frac{1}{3}(C_{V,m}/N_A)n\bar{u}\lambda = 1\cdot35\times10^{-2}$ W m^{-1} K^{-1}.

Comments. The experimentally determined values of the transport coefficients are $D = 1 \cdot 7 \times 10^{-5}\,\mathrm{m^2\,s^{-1}}$, $\eta = 1 \cdot 75 \times 10^{-5}\,\mathrm{Pa\,s}$, and $\kappa = 2 \cdot 4 \times 10^{-2}\,\mathrm{W\,m^{-1}\,K^{-1}}$. We see that the predicted and the measured values agree to within a factor of two. As is often the case, the biggest discrepancy occurs in the thermal conductivity. This discrepancy largely disappears if Eucken's arguments are applied; eqn (7.53) with $f = 5$ predicts $\kappa = 2 \cdot 6 \times 10^{-2}\,\mathrm{W\,m^{-1}\,K^{-1}}$.

7.4. Electrical conduction

The V–I characteristic

All gases are partially ionized. This ionization occurs both as a result of cosmic ray bombardment and as a natural consequence of there being a few atoms in any gas (those at the tail of the Maxwell–Boltzmann distribution) with sufficient energy to ionize other atoms on impact. Ionization involves the removal of one or more electrons from a previously neutral atom (producing a positive ion) and the subsequent deposition of these electrons on to one or more neutral atoms (which thereby become negative ions). At low pressures the electrons may remain unattached.

Fig. 7.18 shows a typical set-up to measure the V–I characteristic of an ionized gas. The gas is enclosed within a glass bulb so that its composition, pressure, and temperature can be easily controlled. A beam of X-rays ionizes the gas uniformly throughout its volume; apertures prevent the beam falling on to the anode (the positive electrode) and the cathode (the negative electrode). Because of the high electrical resistance of the gas we can (usually) equate the potential drop across the chamber to the battery voltage V.

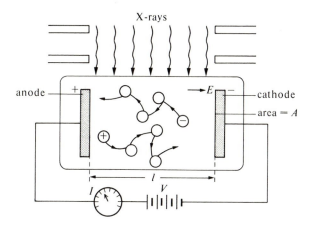

Fig. 7.18. A beam of X-rays uniformly irradiates gas contained between two plates. The current I is measured with different voltages V applied to the tube.

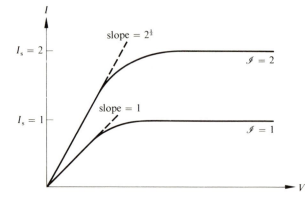

Fig. 7.19. The current–voltage characteristic of a partially-ionized gas. If the intensity $\mathcal{I}$ of the X-ray beam is doubled the saturation current I_s is doubled but the slope of the initial linear portion of the characteristic only increases by a factor of $2^{1/2}$.

If the applied voltage is gradually increased, keeping the intensity $\mathcal{I}$ of the X-ray irradiation constant, the current at first rises linearly, then more slowly, until it reaches a constant (saturation) value I_s (Fig. 7.19). If the intensity of the irradiation is doubled the initial linear rise is $2^{1/2}$ times more steep and I_s is doubled, and so on. We will only look in any detail at the processes occurring in the initial linear region of the V–I characteristic. At saturation all the ions produced by the irradiation are collected at the electrodes. Because the number of ions produced per unit time is proportional to $\mathcal{I}$ we see why $I_s \propto \mathcal{I}$. Incidentally, Mifflin and Bennett measured the saturation currents in chambers P and Q of Fig. 7.5: since the intensity $\mathcal{I}$ of the ionizing irradiation is proportional to the ^{37}Ar concentration, i_p and i_q are proportional to the ^{37}Ar concentrations in chambers P and Q, respectively.

The electrical conductivity

As a prelude to considering the linear region of the V–I characteristic we will look at how an ion in a gas behaves when an electric field is applied to the gas.

In the absence of an electric field, a gas ion will, of course, follow straight-line paths in between collisions with gas atoms. The effect of an electrical field of strength E will be to subject an ion of charge q to a force $F = qE$. This force will bend the paths of the ions into parabolas (Fig. 7.18) with the result that the positive ions tend to move in the field direction and the negative ions in the opposite direction. If an ion of mass m travels freely for a time t_1 before it suffers a collision, it will have acquired a final speed of $(F/m)t_1 = (qE/m)t_1$ (Fig. 7.20) and will have

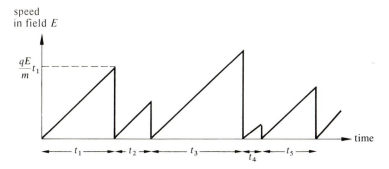

speed
in field E

$\frac{qE}{m}t_1$

time

t_1 t_2 t_3 t_4 t_5

Fig. 7.20. After each collision, which occur at times t_1, t_2, t_3, etc., an ion stops drifting. Between collisions it accelerates at a constant rate.

travelled a distance

$$x_d = \tfrac{1}{2}q\frac{E}{m}t_1^2 \tag{7.55}$$

at the end of this interval. The effect of the collision at $t = t_1$ is to remove the extra speed acquired by q from the field. The ion does *not* stop moving; it merely stops drifting. The ion may now travel freely for a time t_2, say, collide and stop drifting yet again, and so on. This sequence of accelerations and collisions is shown graphically in Fig. 7.20.

We will now define the *drift speed* v_d as the mean distance x_d travelled between collisions divided by the mean time interval τ between collisions;

$$\boxed{v_d = \frac{\bar{x}_d}{\tau}}. \tag{7.56}$$

Substituting for $\bar{x}_d$ from eqn (7.55) gives

$$v_d = \frac{qE}{2m}\frac{\overline{t^2}}{\tau}, \tag{7.57}$$

where (looking at a sequence of n collisions)

$$\overline{t^2} = \frac{t_1^2 + t_2^2 + t_3^2 + \cdots + t_n^2}{n} \tag{7.58}$$

$$\tau = \frac{t_1 + t_2 + t_3 + \cdots + t_n}{n}. \tag{7.59}$$

To evaluate v_d we clearly must know how the various free times, t_1, t_2, etc. are distributed. In fact their distribution is identical to the distribution of free paths in the gas. To satisfy yourself of the truth of this statement you should read §5.12 afresh, substituting t for l (thus reading, for example, 'mean free time τ' in place of 'mean free path λ'). This leads us to the conclusion (cf. eqn (5.73)) that the fraction of the gas atoms with free times between t and $t+dt$ is

$$\frac{dN}{N} = \frac{1}{\tau} e^{-t/\tau} \, dt \qquad (7.60)$$

and hence that

$$\overline{t^2} = \frac{1}{N} \int_0^\infty t^2 \, dN$$

$$= \frac{1}{\tau} \int_0^\infty t^2 e^{-t/\tau} \, dt.$$

On evaluating the integral by parts (applied twice over) we obtain

$$\overline{t^2} = 2\tau^2.$$

Substituting this result into eqn (7.57) gives

$$v_d = q \frac{E\tau}{m}. \qquad (7.61)$$

Assuming that the gas contains n ions per unit volume, each with a charge q, it follows that the current I carried by these particular ions through a plane of area A is given by the number of ions within a cylinder of length v_d and of area A. Thus

$$I = nqv_dA. \qquad (7.62)$$

Note that there is no factor of $\frac{1}{6}$ or $\frac{1}{4}$ as the electric field ensures that all the ions drift along the field direction. Substituting for v_d from eqn (7.61) gives

$$I = \frac{nq^2\tau AE}{m} \qquad (7.63)$$

which can be rewritten as

$$\boxed{J = \sigma E}, \qquad (7.64)$$

where $J = I/A$ denotes the *current density*, and where

$$\boxed{\sigma = \frac{nq^2\tau}{m}} \qquad (7.65)$$

is known as the *electrical conductivity* of the gas. Since, in the SI system, J is measured in units of $A\,m^{-2}$ and E is measured in units of $V\,m^{-1}$, it follows that σ is measured in units of $A\,V^{-1}\,m^{-1}$ or since ohm means $V\,A^{-1}$, in units of $ohm^{-1}\,m^{-1}$ or, using the symbol Ω for ohm, $\Omega^{-1}\,m^{-1}$. It should be noted that E is the electric field *as measured in the gas;* because of boundary effects (a higher than average density of positive ions near the cathode and of negative ions near the anode) E cannot be exactly equated to V/l (Fig. 7.18). Where no such boundary effects exist (as in metals where electrons are the only charge carriers) we can write $E = V/l$. Eqn (7.63) then predicts $I \propto V$; a result well established by experiment and known as *Ohm's law*. However, because $I \propto V$ in metals we cannot say that the proposed mechanism *must* be correct. Indeed we shall discover in §11.5 that electrons in a metal do *not* behave like ions in a classical gas.

The total current

The results which we have just derived (e.g. eqn (7.63)) have implicitly assumed there is only a single type of ion present—those with a charge q. We will now look at the consequences of acknowledging the existence of both positive and negative ions and, in particular, at how these ions determine the current I as measured in the external circuit of Fig. 7.18.

We begin by looking at the consequences of introducing a single positive ion of charge q^+ into the space between the electrodes (Fig. 7.21(a)). If, during a time δt, this positive ion moves a distance δx^+ along the field direction, it will acquire energy $(Eq^+)\,\delta x^+$ from the field E $(= V/l)$. The ion is now closer to the cathode C and further from the anode A than previously. As a result, the electrical potential of C rises

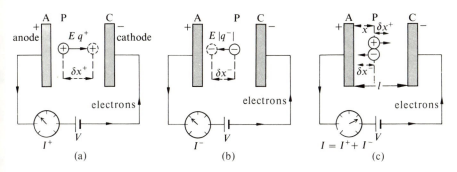

Fig. 7.21. (a) As the positive ion moves through δx^+ from position P, electrons flow in the external circuit. (b) As the negative ion moves through δx^- from position P, electrons flow in the external circuit. (c) When an ion pair is present the total current in the circuit is the sum of that in (a) and (b).

while that of A falls. To restore C and A to their original potentials, electrons must be driven through the external circuit connecting A to C. This is the job of the battery. If the battery e.m.f. is V and the current in the circuit during the time interval δt is I^+ (the $+$ sign is used to indicate that this (electron) current is induced by the movement of a *positive* ion in the gas), the energy provided by the battery is $VI^+ \delta t$. This energy appears as the kinetic energy which the ion acquired in moving through δx^+ in time δt. Conservation of energy demands

$$VI^+ \delta t = Eq^+ \delta x^+$$

or, since $E = V/l$,

$$I^+ = \left(\frac{q^+}{l}\right)\left(\frac{\delta x^+}{\delta t}\right) = \frac{q^+}{l}v^+, \tag{7.66}$$

where v^+ is the speed of the positive ion.

As shown in Fig. 7.21(b), a single negative ion of charge q^- will be subjected to a force of magnitude $E|q^-|$ and will move in the opposite direction to that in which the positive ion moved. This movement will lower the potential of A and raise that of C. To restore A and C to their original potentials electrons must yet again be transferred from A to C. Notice that this electron flow occurs in *the same sense* as it did when a positive ion was present. Denoting the current induced in the external circuit by the movement of the negative ion q^- in the chamber by I^-, we can argue, as before, that

$$I^- = \frac{|q^-|}{l}\frac{\delta x^-}{\delta t} = \frac{|q^-|}{l}v^-, \tag{7.67}$$

where v^- is the speed of the negative ion. With both a positive and a negative ion present (Fig. 7.21(c)) the total current I induced in the external circuit will be

$$I = \frac{q^+}{l}v^+ + \frac{|q^-|}{l}v^-$$

or, substituting from eqns (7.66) and (7.67),

$$\boxed{I = I^+ + I^-} \ . \tag{7.68}$$

Clearly, the same result applies with a great number of ion pairs present. We can now make use of eqn (7.63) to express I^+ and I^- in terms of the charges q^+ and q^- of the ions and of their masses m^+ and m^-. This gives

$$I = nA\left(\frac{(q^+)^2\tau}{m^+} + \frac{(q^-)^2\tau}{m^-}\right)E, \tag{7.69}$$

where n is the common number density of both the positive and negative ions (these numbers must be equal in the case of singly charged ions considered here). As was mentioned earlier, the negative 'ions' may sometimes be free electrons. Under these conditions m^- will be several thousand times less than m^+ (even in hydrogen $m^- = m^+/1836$) and I will be almost wholly due to the motion of the electrons in the gas.

It is often convenient to introduce the, so-called, *mobility* μ of an ion defined as the drift speed of the ion per unit electric field. Hence the mobilities of the positive and negative ions are, from eqn (7.61),

$$\mu^+ = \frac{v_d}{E} = \frac{q^+ \tau}{m^+} \tag{7.70}$$

and

$$\mu^- = \frac{|q^-| \tau}{m^-}. \tag{7.71}$$

Recalling that $|q^-| = |q^+| = q$, eqns (7.70) and (7.71) allow eqn (7.69) to be rewritten as

$$I = nAq(\mu^+ + \mu^-)E. \tag{7.72}$$

Recombination of ions

We will now seek to explain why the linear portion of the V–I characteristic (Fig. 7.19) should be proportional to $\mathscr{I}^{1/2}$, where $\mathscr{I}$ is the intensity of the ionizing radiation. Since the only term in eqn (7.69) which is likely to be a function of $\mathscr{I}$ is n the problem is really to explain why $n \propto \mathscr{I}^{1/2}$.

Our starting point is the notion of dynamic equilibrium. This tells us that in the steady state the rate at which ion pairs are produced per unit volume, dn/dt, in the gas must equal the rate at which the ions are neutralized. Neutralization can occur either at the electrodes (a positive ion will acquire an electron from the cathode; a negative ion will give up an electron at the anode) or within the body of the gas following collisions between ions of opposite sign. The latter is the dominant mechanism in the lower reaches of the V–I characteristic. The rate of such neutralizing collisions will be proportional to the collision frequency in the gas and hence, from eqn (5.66), to n^2. Thus, at equilibrium,

$$\frac{dn}{dt} = \rho n^2, \tag{7.73}$$

where ρ is known as the coefficient of recombination. But since $dn/dt \propto \mathscr{I}$ it follows from eqn (7.73) that $\mathscr{I} \propto n^2$ and thus, as hoped, that $n \propto \mathscr{I}^{1/2}$.

Exercise 7.3

According to Drude (1912) a metal contains electrons which behave like a perfect gas. Thus if the metal is at a temperature T the electrons have a mean speed $\bar{u}$ given by $\bar{u} = (8kT/\pi m)^{1/2}$. Drude supposed that the free paths are terminated by collisions with the positive metal ions. Using this model, make a rough estimate of the mean free path of an electron in metallic sodium. Assume sodium contains 4×10^{28} Na$^+$ ions m^{-3}; formed by each sodium atom contributing one electron to the 'gas'. Take the diameter of a Na$^+$ as 2×10^{-10} m. The electrical conductivity of sodium at 273 K is $2 \cdot 38 \times 10^7 \ \Omega^{-1} \text{m}^{-1}$.

Calculation. Substituting $n = 4 \times 10^{28} \text{ m}^{-3}$, $q = 1 \cdot 6 \times 10^{-19}$ C, $m = 9 \cdot 1 \times 10^{-31}$ kg, and $\sigma = 2 \cdot 38 \times 10^7 \ \Omega^{-1} \text{m}^{-1}$ into eqn (7.65) gives $\tau = m\sigma/nq^2 = 2 \cdot 11 \times 10^{-14}$ s. Guessing that $\lambda = \bar{u}\tau$ (or defining it so) and knowing that $\bar{u} = (8kT/\pi m)^{1/2} = 1 \cdot 03 \times 10^5 \text{ m s}^{-1}$ we obtain $\lambda = 2 \cdot 17 \times 10^{-9}$ m for the mean free path of an electron. Now the gas-kinetic relation $\lambda = 1/\pi d^2 n$, as applied to a very small particle moving amongst ions of radius r, becomes $\lambda = 1/\pi r^2 n = 7 \cdot 9 \times 10^{-10}$ m.

Comments. Despite the order-of-magnitude agreement between λ as predicted from the electrical conductivity and from gas-kinetic arguments, Drude's theory fails to account for the temperature dependence of electrical conductivity. Experiments show that $\sigma \propto T^{-1}$ whereas eqn (7.65), with τ replaced by $\lambda/\bar{u}$, predicts $\sigma \propto T^{-1/2}$.

PROBLEMS

7.1. Show that the average distance at which gas molecules striking a plane surface made their last collision before striking the surface is $2\lambda/3$, measured perpendicular to the surface.

7.2. One way to simulate diffusion, as it occurs in, say, the apparatus of Fig. 7.5 is to line up eight pieces on one side of a chess board. Toss a coin to decide if the first piece advances a square: say 'heads' for advance. Repeat the game for all eight pieces in turn. Now perform a second round of tosses but this time 'tails' withdraw a piece one square (except when on the starting line). A round of tosses corresponds to the (constant) time of $\lambda/\bar{u}$ in the life of the gas as it diffuses. Plot a graph showing how the concentration varies with distance at different times after the commencement of diffusion. Your plot should display the main features sketched for one of the gases in Fig. 7.2. If this is so, the simulation may be taken as vindicating the 'random walk' nature of diffusion, as exhibited by the men on the chess-board.

7.3. Using the data of Mifflin and Bennett shown in Fig. 7.6, calculate the diameter of an argon atom $(A_r(\text{Ar}) = 40)$. Start by drawing the best horizontal line through the data.

7.4. A perfume manufacturer claims that diffusion is the mechanism whereby the scent is transmitted. Drawing on your knowledge of the sort of time that it takes for the perfume worn by someone newly entering a room to become evident, and on the random-walk nature of diffusion, assess the validity of the manufacturer's claims.

7.5. Imagine two planes A and B drawn perpendicular to the axis of the apparatus shown in Fig. 7.5 and used to study interdiffusion of gases. If these planes are separated by a distance δx then if the number density of, say, the radioactive species at plane A is n at time t, then the number density at plane B at the same time t is $n - (\partial n/\partial x) \delta x$. By applying Fick's law at planes A and B, calculate the net gain in the number of radioactive atoms within the volume δV contained between the two planes over a time δt. Setting this gain equal to $\delta V(\partial n/\partial t)\,\delta t$ show that

$$\frac{\partial n}{\partial t} = D\frac{\partial^2 n}{\partial x^2}.$$

This is a quite general equation describing how diffusion proceeds.

7.6. This problem allows you to check up on your understanding of the same processes as those responsible for the viscosity of a gas. Three goods trains carrying coal move in the same direction along parallel tracks; the outer one at $50\ \mathrm{m\,s}^{-1}$, the middle one at $45\ \mathrm{m\,s}^{-1}$, and the inner one at $40\ \mathrm{m\,s}^{-1}$. As they pass, a team of men on each train shovel coal on to their nearest neighbouring train (two trains in the case of the two teams on the middle train). Every 3 s each of the teams shovels $10^3\ \mathrm{kg}$ of coal on to their neighbouring train. The total mass of each train therefore does not change. The driver of the fastest train finds his engine has to pull harder to maintain speed. Why, and by how much? The driver of the middle train finds that his engine does not have to pull any harder. Why? The driver of the slowest train finds he has to apply the brakes. Why, and what is the braking force applied to the train?

7.7. A practical 'viscometer' consists of two coaxial cylinders, both 25 cm long. The outer cylinder rotates once every 55 s. The inner cylinder is fastened to a wire whose angle of twist measures the torque applied to that cylinder. The radius of the outer cylinder is 6 cm and that of the inner cylinder is 5·8 cm. If the torque as measured at the inner cylinder is $4\cdot4\times10^{-7}\ \mathrm{N\,m}$, what is the viscosity of the gas? Since the 0·2-cm gap between the cylinders is small in comparison with the diameter of either cylinder, you can imagine the layer of gas as being 'straightened out' (to conform with Fig. 7.7).

7.8. A layer of gas is sandwiched between two infinite parallel plates a distance l apart. One plate is fixed; the other moves with a velocity V. Assuming that the probability that an atom striking either plate will stick (for long enough for its tangential velocity to be increased to that of the plate) is σ, show that the true velocity gradient in the gas is $V/[l+2\lambda(2-\sigma)\sigma^{-1}]$. *Clue.* Remember to calculate the mean flow velocity in the immediate vicinity of each plate.

7.9. Using the data given in Fig. 7.11 calculate the molecular diameters of nitrogen $(M_r(\mathrm{N}_2)=28)$ and oxygen $(M_r(\mathrm{O}_2)=32)$. *Clue:* eqn (7.30).

7.10. Estimate the drag on a spherical satellite of radius 1 m in Earth orbit at an altitude of 100 km. Assume that the isothermal model of Earth's atmosphere holds up to this altitude. Take the momentum accommodation coefficient σ to be 0·5 on the satellite's skin. *Clue.* First calculate the speed of the satellite. You will find it is much greater than the r.m.s. speed of the air molecules.

7.11. Show that the conductance of a circular hole of radius r in a plate of negligible thickness is $r^2(\pi kT/2m)^{1/2}$ when $\lambda \gg r$. Here m is the molecular mass of the gas.

7.12. A tube of length 0·5 m and inner radius 5×10^{-4} m is joined to another tube of length 1·5 m and inner radius 1×10^{-3} m. (a) If the pressure at one end of the composite tube is 0·3 Pa and is 0·1 Pa at the other end what is the throughput of nitrogen at a temperature of 300 K? (b) What is the throughput if the pressures are 1×10^5 Pa and 4×10^4 Pa? Take $\eta=1\cdot8\times10^{-5}$ Pa s.

7.13. Using the data given in Fig. 7.17 deduce the mean diameter of an air molecule. Take the mean relative molecular mass of an air molecule as 30.

Note that it is unnecessary to know the energy accommodation coefficient α.

7.14. A layer of gas is sandwiched between two infinite parallel plates a distance l apart. One plate is at a temperature T_1; the other is at a lower temperature T_2. Denoting the energy accommodation coefficient by α, show that the true temperature gradient in the gas is $(T_1 - T_2)/[l + 2\lambda(2 - \alpha)\alpha^{-1}]$.

7.15. In the hot-wire cell for measuring the thermal conductivity of a gas two coaxial cylinders of radii a and b are maintained at temperatures T_a and T_b, respectively. The inner cylinder is an electrically-heated wire whose temperature can be deduced from its electrical resistance. The heat Q generated in the wire per unit time is the product of the current flowing in the wire and the potential drop along its length. Show that at pressures such that $\lambda \ll (b - a)$ the thermal conductivity of the gas is given by

$$\kappa = \frac{Q}{2\pi} \frac{\ln(b/a)}{(T_a - T_b)}.$$

Assume the heat flow to be entirely radial.

7.16. Consider two planes A and B within the body of a gas and lying normal to the direction of the heat flow (the x-axis). If these planes are separated by a distance δx and if the temperature of plane A, say, is T then the temperature of plane B is $T - (\partial T/\partial x)\,\delta x$ at time t. By applying the defining relation for thermal conductivity (eqn (7.39)), calculate the net gain in the energy of the volume δV of gas lying between the two planes over a time δt. Equating this gain to $(\partial T/\partial t)$ multiplied by a suitable expression involving the molar heat capacity $C_{p,m}$ of the gas, show that

$$\frac{\partial T}{\partial t} = \left(\frac{\kappa V_m}{C_{p,m}}\right) \frac{\partial^2 T}{\partial x^2},$$

where V_m is the molar volume of the gas. The term in parentheses is usually called the thermal diffusivity.

7.17. A saturated current of 10^{-7} A is measured when a discharge tube is being irradiated with X-rays. Assuming the ions are singly charged, how many ion pairs are being produced per second by the radiation?

7.18. If a cannon ball is fired horizontally with a very high velocity it will not, in theory at least, strike the earth. At very low velocities the free path of the ball is terminated by the earth. At what speed will the ball just not collide with the earth? (This problem was first solved by Newton.) You should spot the relevance of this type of calculation to any discussion of gaseous mean free paths.

8. The solid phase

LEAVING the gas phase, we will now direct our attention to the solid phase. In this chapter we will look at some of the possible ways in which atoms are arranged in crystalline solids and at the relationship between the form of a crystal and its internal energy. Our survey of crystallography will necessarily be fairly superficial; indeed we will only look in any detail at those structures which are relevant to later discussions.

Just as we chose to define a perfect gas as one in which $\Delta E = 0$, $\frac{1}{2}\overline{mu^2} > 0$, so it would seem reasonable to define a 'thermally perfect' solid as one in which $\Delta E > 0$, $\frac{1}{2}\overline{mu^2} = 0$. Such a definition is implicit in the present chapter. In the next chapter will use the word perfect to denote the absence of those defects in the structure of a crystal in which atoms are permanently displaced from their proper sites.

8.1. Order or disorder?

We can prove by a *reductio ad absurdem* argument that the atoms of a thermally-perfect solid should be arranged in a regular rather than a haphazard fashion.

Suppose that the stationary atoms are not arranged in a regular pattern but are disposed at random as shown in Fig. 8.1. Now consider any two atoms, i and j. Because the local arrangement of atoms is, in general, different about atoms i and j the net force acting on i will differ from that on j. The only way to ensure that all the atoms are at rest (making $\frac{1}{2}\overline{mu^2} = 0$) is to have zero resultant force acting on each atom. In a random arrangement of atoms we might expect that the local arrangement of atoms about some *one* atom might indeed lead to zero resultant force on that atom. However, the resultant force acting on any other atom would, in general, be non-zero. So the random arrangement illustrated in Fig. 8.1, with all its atoms at rest, is an absurdity! To have zero resultant force on each atom demands that the local (but as yet undetermined) arrangement of atoms about each and every atom be the same. In other words the structure must be periodic.

Amorphous substances

If the arguments we have just advanced are correct, why do *amorphous* materials like glasses and certain plastics—materials with a non-periodic structure—exist? Actually the argument never said that such materials would not exist; it only asserts their structure will not be in equilibrium. Indeed over a period of centuries glass may crystallize out. If

Fig. 8.1. In a non-crystalline solid each atom i, j, etc., is subjected to a different net force due to the random arrangement of their neighbouring atoms. If i, say, is in equilibrium the odds are that j will not be in equilibrium.

the glass is kept close to its softening temperature, the crystallization is much more rapid—as will be known to anyone with experience of glass blowing.

One way of producing an amorphous material is to reduce the temperature of a melt in a time interval shorter than the time it takes the atoms to find their true equilibrium positions. An example is the formation of *plastic sulphur* by rapidly cooling molten sulphur in water. In most plastics crystallization is made difficult by having long chain molecules which can become so intertwined that the time required for crystallization to take place can be very large indeed. This may be helped by introducing a spread in the chain lengths or by introducing side chains at random. Under these conditions practically every molecule is different; the intermolecular forces will therefore be different throughout the substance and so periodicity is not required to ensure stability. Another trick is to introduce additives (plasticizers) which separate the chains and so prevents them interacting strongly with one another. Celluloid consists of nitrocellulose, which is normally crystalline, plasticized with camphor. Cellophane consists of cellulose plasticized with glycerol. The disadvantage of these now largely superseded plastics was that the plasticizers could evaporate, allowing chains to come together and so to crystallize. You may have noticed how cellophane goes brittle with age.

In *thermosetting* plastics, such as epoxy adhesives, chemical reactions produce strong cross-links between the long molecular chains so that, in essence, the piece of plastic is one giant molecule. Unlike polymers free of such strong cross-links (e.g. nylon and polyethylene) these thermosetting plastics do not soften and melt to viscous liquids as they are heated. Plastics which, on heating, soften and melt and which, on cooling, return to their original state are called *thermoplastics*.

8.2. Periodic arrays of atoms

Fig. 8.2 shows a two-dimensional *lattice*; an array of points (denoted by ·). Associated with each lattice point is a *basis*; a group of atoms (denoted by ⌒). The whole—the lattice plus its associated basis—is referred to as a *structure*. So we may describe the structure of a crystal by specifying the form of the underlying lattice *and* the basis to be associated with each point of the lattice. The basis may, as in copper for example, consist of a single atom or even, as in crystallized proteins, consist of some tens of thousands of atoms. Remember that the lattice as such does not exist; it merely tells us where to locate the basis.

In practice the lattice is normally specified by giving the form of the *primitive cell*. This is the smallest cell which, when repeated endlessly throughout space, will fill all space and reproduce the lattice. Fig. 8.3(a) shows four possible primitive cells for the lattice of Fig. 8.2. The construction of cells 1, 2, and 3 is self-evident. Cell 4—known as a Wigner–Seitz primitive cell—is constructed by drawing lines to connect a given lattice point to all neighbouring lattice points and then bisecting each such line with a perpendicular line. Fig. 8.3(b) shows how repeating just one of these cells (cell 2) throughout space does indeed reproduce the lattice. A characteristic of a primitive cell is that it contains but one lattice point. This is obviously true of cell 4 in Fig. 8.3(a). In the other cells there are lattice points at the four corners of the cells but each point is shared by the four cells which meet at each corner.

It is sometimes more convenient in describing a lattice to specify a cell which is larger than the minimum-size primitive cell. An example of such a *non-primitive* cell for the lattice of Fig. 8.2 is shown in Fig. 8.3(c).

Lattice types

By changing the magnitude of the so-called *primitive translation vectors* **a** and **b** and the magnitude of the angle ϕ in Fig. 8.2 one can

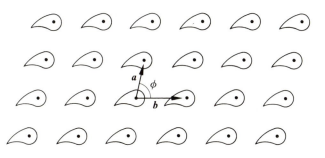

Fig. 8.2. A crystal structure results on associating a basis (⌒), which may be a single atom or a group of atoms, with each point (·) of a lattice.

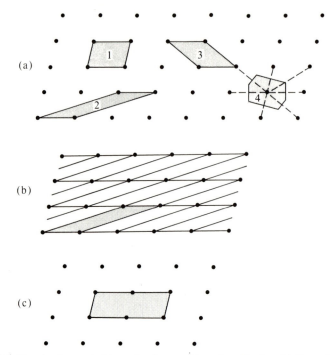

Fig. 8.3(a). Showing four primitive cells of a two-dimensional lattice. (b) Once the primitive cell is specified the lattice may be constructed by regular and endless repetition of that cell. (c) The cell shown shaded contains more than one lattice point and is called non-primitive.

generate an unlimited number of lattices which all differ from one another. We might, for example, decide to increase the magnitude of a by 5 per cent, to decrease that of b by 2 per cent, and to increase ϕ by 2°. However, this new lattice is not fundamentally different to that of Fig. 8.2. If, on the other hand, we had made $a = b$ and $\phi = 90°$ then we would have produced a quite different type of lattice (Fig. 8.4). The distinction between these two types of lattice is that, unlike that of Fig. 8.2, the

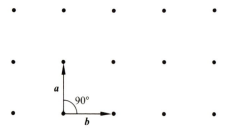

Fig. 8.4. This lattice possesses symmetry properties not present in that of Fig. 8.2. It is, for example, unchanged on rotation through 90° about a lattice point.

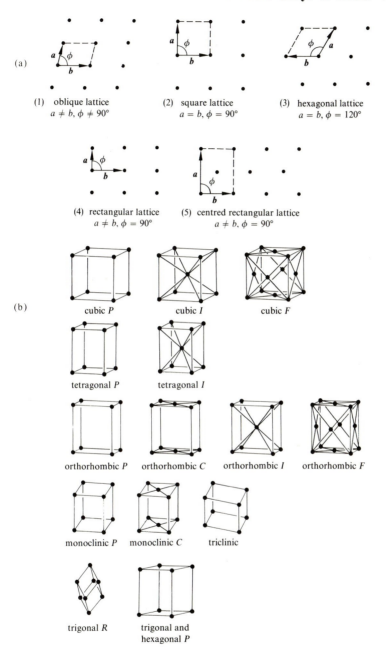

(a)

(1) oblique lattice
$a \neq b, \phi \neq 90°$

(2) square lattice
$a = b, \phi = 90°$

(3) hexagonal lattice
$a = b, \phi = 120°$

(4) rectangular lattice
$a \neq b, \phi = 90°$

(5) centred rectangular lattice
$a \neq b, \phi = 90°$

(b)

cubic *P*   cubic *I*   cubic *F*

tetragonal *P*   tetragonal *I*

orthorhombic *P*   orthorhombic *C*   orthorhombic *I*   orthorhombic *F*

monoclinic *P*   monoclinic *C*   triclinic

trigonal *R*   trigonal and
                   hexagonal *P*

Fig. 8.5. (a) All two-dimensional lattices fall into one or other of the five types shown. The cell shown in (5) is non-primitive. (b) All three-dimensional lattices fall into one or other of the fourteen types shown. The cells shown in outline are the conventional ones which are not always primitive. The letters *P*, *I*, *C*, and *F* are used to denote details of the cell; for example, *I* denotes a body-centred cell.

lattice of Fig. 8.4 is unaltered on rotating it through 90° about a lattice point, or on 'reflecting' it across a plane passing through a lattice point and parallel to **a** (or to **b**). So the lattices of Figs. 8.2 and 8.4 differ fundamentally in that they possess different lattice symmetry properties. It turns out that in two dimensions there are only the five fundamentally-different types of lattice illustrated in Fig. 8.5(a). Within any of the lattice types shown as (2), (3), (4), and (5) in Fig. 8.5(a), where ϕ is fixed, one is free to choose the magnitude of **a** and/or **b**—provided these are consistent with the type of lattice. For example, in the case of the hexagonal lattice *a* must have the same magnitude as *b* although this can be anything we wish. In the case of the oblique axis, where $a \neq b$, we must, of course, avoid making $\phi = 90°$. (To do so would generate the rectangular lattice.) By ringing all the permitted changes one can produce an infinite number of *lattices*; however, all these lattices fall into the five fundamental *types* shown in Fig. 8.5(a).

In three dimensions it turns out that there are fourteen fundamentally different lattices. These are illustrated by the cells, some of which are non-primitive, in Fig. 8.5(b). As in two dimensions, endless repetition of these cells throughout space will generate the lattice. Associating a basis with each lattice point will yield the structure of the crystal.

Describing lattice planes

The lattice points of the lattices we have just described may be thought of as lying on various sets of parallel lines in two dimensions and on various sets of parallel planes in three dimensions. Fig. 8.6 shows two different planes in a three-dimensional lattice (for clarity most of the lattice points have been omitted). The axes have been drawn in along the primitive translation vectors **a**, **b**, and **c**. Now to fix the position and orientation of a plane in three dimensions we must specify the coordinates of three non-collinear points. Thus the plane shown shaded in Fig. 8.6 is

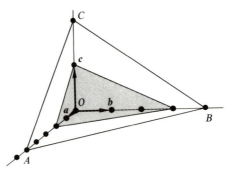

Fig. 8.6. Two different planes in a three-dimensional lattice. For clarity most of the lattice points have been omitted.

determined by the points $(2, 0, 0)$, $(0, 3, 0)$, and $(0, 0, 1)$, where the coordinates are measured in units of the lattice constants a, b, and c.

In general, if plane ABC in Fig. 8.6 makes intercepts Ha, Kb, and Lc on the three axes it follows that for this plane, or for any parallel plane,

$$OA : OB : OC = Ha : Kb : Lc$$

$$= \frac{Ha}{HKL} : \frac{Kb}{HKL} : \frac{Lc}{HKL}$$

$$\boxed{OA : OB : OC = \frac{a}{h} : \frac{b}{k} : \frac{c}{l}} \,. \tag{8.1}$$

Because H, K, L are integers we see that $h(=KL)$, $k(=HL)$ and $l(=HK)$ are integers. In the case of plane ABC illustrated in Fig. 8.6 $H = 5$, $K = 4$, and $L = 2$; so $h = 8$, $k = 10$, and $l = 20$. It is normal to reduce the three integers h, k, and l to the smallest three integers having the same ratio. The result, written in parentheses as (hkl), is referred to as the *Miller indices* of the plane. The Miller indices of plane ABC and parallel planes are therefore (4 5 10). Given the indices of a plane and, of course, the lattice vectors a, b, and c the plane is most easily constructed by appealing to eqn (8.1). For plane (hkl) one divides a by h, b by k, and c by l and then draws a plane through the three points so obtained. A negative sign (bar) placed over an index denotes that the plane cuts the axis on the negative side of the origin. Thus, for example, the plane $(4 \bar{5} 10)$ passes through the points $(a/4, 0, 0)$, $(0, -b/5, 0)$, and $(0, 0, c/10)$. Some important planes in a cubic lattice are shown in Fig. 8.7. Note that a Miller index of zero means the plane only intercepts a particular axis at infinity. Although the parallel planes $(\bar{1}00)$ and (100) in Fig. 8.7 can be described by *either* of these Miller indices if we only wish to denote the orientation of a particular plane, the two sets are required if we wish to indicate the location of the planes as seen from a particular viewpoint within the lattice.

The plane faces which bound a crystal are parallel to lattice planes. When used to describe a crystal face the symbol (hkl) denotes a face

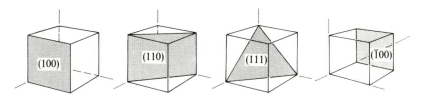

Fig. 8.7. Some important planes in a cubic lattice indicated by their Miller indices.

parallel to the plane (hkl). If there are two faces on opposite sides of the crystal, the face lying on the same side of the origin as the plane making intercepts a/h, b/k, c/l is labelled (hkl), while the opposite face is labelled ($\bar{h}\bar{k}\bar{l}$). The symbol $\{hkl\}$ means all crystal faces or sets of lattice planes which can be obtained from the face or planes (hkl) by applying the symmetry operations appropriate to the lattice in question. Thus if we were to speak of the $\{100\}$ planes in a simple cubic lattice we would actually mean the (100), (010), (001), ($\bar{1}$00), (0$\bar{1}$0), and (00$\bar{1}$) planes.

Describing directions

To specify a certain direction in a lattice one need only specify the coordinates of a line lying in this direction. If the line passes through the origin and through the lattice point with coordinates ua, vb, and wc the line is determined when u, v, and w are specified. It is usual to divide such a set of integers through by a common factor so as to give the set of smallest possible integers. If $u = 12$, $v = 8$, and $w = 6$ we would write the lattice direction in square brackets as [643]. In a cubic crystal, for example, the x-axis is the [100] direction.

Exercise 8.1

Sketch lines having Miller indices of (10) and (23) in a two-dimensional square lattice (where $a = b$ and $\phi = 90°$). Also indicate the lattice direction [25].

Solution. As applied to a two-dimensional lattice, eqn (8.1) tells us that the orientation of, say, the (23) lines is obtained by dividing a by 2 and b by 3 and joining the points $a/2$ and $b/3$. Parallel lines will have the same Miller indices; several such lines are shown in Fig. 8.8, and all are separated by the distance of the original line from the origin.

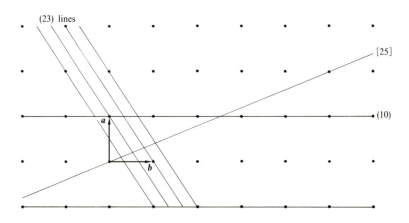

Fig. 8.8. The (23) and (10) lines and the [25] direction in a two-dimensional square lattice.

Comment. Another way of interpreting the Miller indices follows on noting that in going from one lattice point to a point a distance a away we cross 2 of the (23) lines, and that in going to the point b away we cross 3 of the (23) lines. In general, if the Miller indices of a set of planes are (hkl) we cross h planes in going a distance a, k planes in going a distance b, and l planes in going a distance c.

8.3. X-ray diffraction

To determine the structure of a crystal we must be able to ascertain the form of the underlying lattice and the basis which is associated with each lattice point. Both of these pieces of information may be obtained from X-ray diffraction measurements.

Experimental procedures

Fig. 8.9 shows one of many possible ways of studying X-ray diffraction. Here the crystal is mounted on a spindle which is concentric with a cylindrical holder containing a strip of photographic film. A narrow beam of single-wavelength (monochromatic) X-rays is directed along a radius at the crystal.

With the crystal stationary it is found that (unless one has been lucky) no diffraction spots are recorded on the developed film. It is only on rotating the crystal (either continuously through 360° or back and forth through a more limited range of angles) that these spots are normally obtained. Each spot, or groups of spots, corresponds to a unique position of the crystal relative to the film. Even when such diffraction

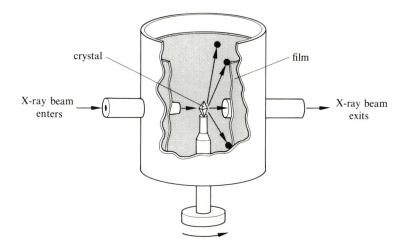

Fig. 8.9. An oscillation camera. A crystal mounted on a rotating spindle is struck by a narrow beam of monochromatic X-rays. When the crystal is in particular orientations, strong diffraction maxima are produced along particular directions.

maxima are present most of the radiation striking the crystal continues straight on and exits through a hole cut in the film (Fig. 8.9).

Obtaining the lattice

When a beam of X-rays strikes a crystal the electrons of each and every atom scatter the beam through a wide angle. We will now consider how the radiation scattered from each basis lying in plane *AB* of the cubic lattice shown in Fig. 8.10 can interfere with the radiation scattered from each basis in the adjacent parallel plane *CD*. In particular, we will consider the radiation scattered at an angle θ to these planes, equal in magnitude to the angle which the incoming beam makes with these planes (see Fig. 8.10). Should the X-rays scattered from the basis at *P* interfere constructively with the radiation scattered from the basis at *Q* it follows that the radiation scattered from all other similar pairs will also interfere constructively, and so the beam reflected in direction θ will be finite. For this constructive interference to occur the path difference between the radiation scattered at *P* and *Q* must be an integral number of wavelengths, that is

$$RQ + QS = n\lambda, \tag{8.2}$$

where n is an integer, and λ is the wavelength of the X-rays. From triangle *PQR*

$$RQ = PQ \cos \phi.$$

Also, since angle $PQS = 180° - 2\theta + \phi$,

$$QS = PQ \cos(180° - 2\theta + \phi) = -PQ \cos(2\theta - \phi).$$

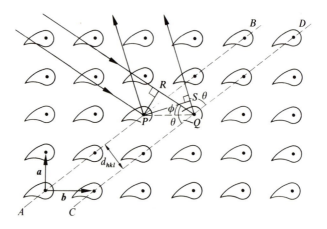

Fig. 8.10. A beam of X-rays strikes a crystal and is scattered by the basis associated with each lattice point. Here we consider the radiation scattered in a direction which makes the same angle θ with two parallel planes as does the incident beam.

Therefore,

$$RQ + QS = PQ(\cos \phi - \cos(2\theta - \phi)) = 2PQ \sin \theta \sin(\theta - \phi). \quad (8.3)$$

Now the perpendicular distance d_{hkl} between adjacent (hkl) planes is equal to $PQ \sin(\theta - \phi)$. It therefore follows from eqns (8.2) and (8.3) that the condition for a strong diffracted beam is

$$\boxed{2d_{hkl} \sin \theta = n\lambda} \quad . \quad (8.4)$$

This famous result is known as Bragg's law. Since $\sin \theta$ has a maximum value of unity, eqn (8.4) tells us that Bragg reflection can only occur for wavelengths $\lambda \leq 2d_{hkl}$. This explains why visible light cannot be used. Eqn (8.4) also explains why an arbitrarily chosen crystal setting is most unlikely to produce Bragg reflection with a monochromatic X-ray beam; the chances are that, with λ fixed, there are no planes within the lattice whose separation d_{hkl} and whose orientation θ together satisfy eqn (8.4).

Knowing the position of each diffraction spot on the developed film it is a simple matter to deduce θ. (The angle between the incident and the reflected beam in Fig. 8.10 is $180° - 2\theta$.) It is somewhat more difficult to obtain the order n which produced the spot at θ but we are helped by the knowledge that it must be an integer. We are also helped by the fact that n is low at low θ. Knowing n and θ, the spacings of various lattice planes follow from Bragg's law. By such means the lattice is obtained.

Obtaining the basis

It is important to realize that, by itself, Bragg's law only enables the *lattice* to be determined. It cannot be used to deduce the nature of the basis which is associated with each lattice point. This is because the law is only concerned with the *position* of each diffraction spot and not with its intensity.

Fig. 8.11 shows two different structures possessing the same *lattice*: in (a) the basis is a single atom; in (b) it is a molecule. It would be very surprising if the single atoms of (a) scattered X-rays in the same way as the molecules of (b). We might expect (wrongly as it happens) that the single atom would scatter X-rays equally well in all directions, so producing diffraction spots of constant intensity. We would certainly not expect the molecules in (b) to scatter X-rays equally in all directions; there is clearly the possibility of constructive or destructive interference occurring between the radiation scattered by individual atoms *within* a molecule. As a result the overall intensity of the diffraction spots will vary throughout the photograph. The basis can usually—but not always—be elucidated from the measured intensities of the spots. Knowing the lattice and the basis the structure is fully determined.

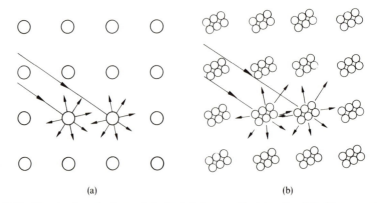

(a)   (b)

Fig. 8.11. Showing how the form of the basis influences the intensity of the X-rays scattered in different directions from a crystal. In (a) the scattering is assumed to be equal in all directions (a wrong assumption since X-rays scattered from electrons in one region of the atom can interfere with X-rays scattered from electrons in another region of the atom). In (b) Bragg reflection within the basis leads to strong scattering in particular directions.

8.4. Some simple crystal structures

We will look at just a few representative structures. You will probably find it more helpful to keep referring back to this section, as required, rather than to attempt to memorize the details of these structures.

Hexagonal close-packed (hcp) structure

This is one of the most common types of crystal structure, and occurs among some twenty-five per cent of all elements. Examples are hydrogen, magnesium, and zinc.

To understand this structure we consider first a single plane (A) of atoms which are packed together as closely as possible. Such a plane is shown by the open circles in Fig. 8.12(a). In imagination we now place another plane (B) of atoms on top of this plane so that each atom of B touches three atoms of plane A. In placing a third plane (C) of closest-packed atoms on top of plane B so that each atom of C touches three atoms of B, we have two possible choices. If we place plane C so that its atoms are directly above those of plane A we produce a hexagonal close-packed structure. For obvious reasons the packing is often referred to as $ABAB$. ... Fig. 8.12(b) is a sketch of a small portion of the hcp structure with the atoms drawn to scale. Fig. 8.12(c) shows the positions of the atoms somewhat more clearly. The lattice of the hcp structure is hexagonal and the basis consists of an atom at 000 and at $\frac{2}{3}\frac{1}{3}\frac{1}{2}$. These coordinates are given as fractions of the axial lengths a, b, and c. (In giving the coordinates of a point within a cell of the lattice it is usual to omit commas between the coordinates and also to omit brackets.)

(a)

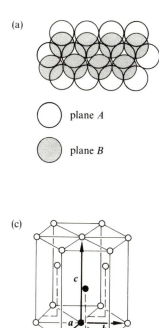

plane *A*

plane *B*

(b)

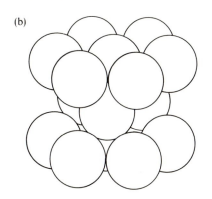

(c)

Fig. 8.12. The hcp structure in an element. (a) Two closest-packed planes of atoms are stacked on top of one another so that each atom of plane *B* makes *contact* with three atoms of plane *A*. (b) A portion of an hcp structure with the atoms drawn to size. (c) For clarity the atoms are represented by points. The lattice is hexagonal and the basis consists of an atom at 000 and at $\frac{2}{3}\frac{1}{3}\frac{1}{2}$.

Face-centred cubic (fcc) structure

This structure results when closest-packed planes of atoms are stacked in a slightly different sequence to that which leads to the hcp structure. The first two planes go in the *AB* sequence. The third plane (*C*) however is located so that the atoms in this plane are over the holes in the first plane (*A*) which are not occupied by atoms of the second plane (*B*). This is illustrated in Fig. 8.13(a). Fig. 8.13(b) shows the relative positions of the atoms in the closest-packed planes *A*, *B*, *C*, and *A*. (The packing in the fcc structure may be said to be *ABCABC*....) Although the cell which is shown in Fig. 8.13(b) is not the primitive cell its convenience is that it is cubical. A more usual way of drawing this cube is shown in Fig. 8.13(c).

Some twenty per cent of the elements crystallize with fcc structures. These include copper, silver, gold, and the inert gases (except for helium which crystallizes with a hcp structure). Sodium chloride is an example of a compound whose lattice is fcc. The basis consists of a Na$^+$ ion and a Cl$^-$

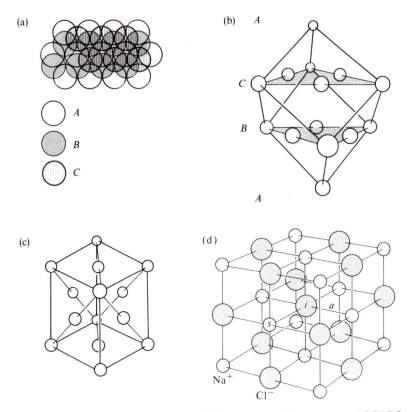

Fig. 8.13. The fcc structure in an element. (a) Shows the stacking sequence *ABCABC*... of adjacent planes. (b) The relative positions of atoms on adjacent planes may be seen. (c) On rotating the cell shown in (b) we see the fcc nature of the cell. (d) The crystal structure of sodium chloride.

ion separated by one-half the body diagonal of the fcc cell shown in Fig. 8.13(c). When this basis is associated with *each* lattice point the structure shown in Fig. 8.13(d) is obtained.

Body-centred cubic (bcc) structure

The *ABAB*... and *ABCABC*... sequences for stacking closest-packed planes exhaust the ways of repetitively stacking such planes. As we have seen, the hcp and fcc structures together account for some 45 per cent of all elements. Of the remaining elements, some 15 per cent crystallize with a body-centred cubic structure. The most notable examples are the alkali metals, Li, Na, K, Rb, and Cs. Iron and tungsten are other examples.

The body-centred cubic structure for the elements results on associating a single atom with each point of the bcc lattice. A single cell of

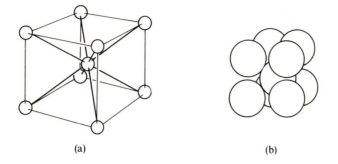

Fig. 8.14. The bcc structure of an element. In (b) the atoms have been drawn to scale.

this structure is illustrated in Fig. 8.14(a), (b). Notice that in this structure the atoms touch only along a cube diagonal and not along an edge as in the fcc structure. In other words the structure is *not* the result of stacking closest-packed planes of atoms.

Simple cubic structure

This is undoubtedly the simplest structure to visualize and to appeal to in calculating the properties of a perfect solid. However, only one *element*—polonium—crystallizes with a simple cubic lattice (and even that statement is only true within a restricted temperature range). Its structure (a single cell of which is shown in Fig. 8.15(a)) is obtained by associating a single atom with each point of the simple cubic lattice. Notice how the atoms only touch along the cube edges. If you like, this structure is formed by stacking, directly on top of one another, planes of atoms in each of which an atom only makes contact with four neighbouring atoms. Crystal structures with simple cubic lattices are by no means rare among *compounds*. Caesium chloride is an example. Here the basis consists of one Cl^- ion at 000 and one Cs^+ ion at $\frac{1}{2}\frac{1}{2}\frac{1}{2}$ (Fig. 8.15(b)).

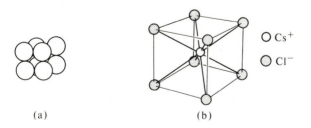

○ Cs^+

◉ Cl^-

Fig. 8.15. (a) The simple cubic structure of an element. (b) The crystal structure of caesium chloride. Here the basis to be associated with the simple cubic lattice consists of a Cs^+ ion at 000 and a Cl^- ion at $\frac{1}{2}\frac{1}{2}\frac{1}{2}$.

Zincblende structure

To obtain this structure we start with the (non-primitive) cell of the fcc lattice called cubic F in Fig. 8.5(b). The basis to be associated with each lattice point is a Zn atom at 000 and an S atom at $\frac{111}{444}$. Affixing this basis to each lattice point produces the zincblende structure—a single cell of which is shown in Fig. 8.16. If both atoms are identical in Fig. 8.16, we obtain the diamond structure. In addition to carbon in the form of diamond, silicon, germanium, and gray tin all crystallize with this structure.

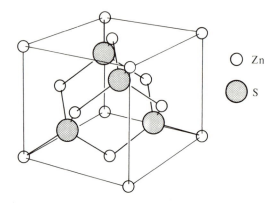

Fig. 8.16. The zincblende structure. The basis to be associated with a fcc lattice is a Zn atom at 000 and an S atom at $\frac{111}{444}$. The bonding between the Zn and S atoms is predominantly covalent.

8.5. The lattice energy of a crystal

It is natural to ask why it is that different compounds crystallize with different structures and why even a series of related compounds, such as the alkali halides, may have different structures within the series. Although these questions have a simple formal answer—namely, that the structure possessed by a crystalline solid at $T \approx 0$ will be one which minimizes the internal energy†—it is far from easy to compute the internal energies of all the possible structures that a given compound might possess. Furthermore, the true differences in the internal energies of certain structures may be less than the computational errors involved in evaluating these energies. Nevertheless, it has been possible to make fairly precise calculations of the internal energies of structures in which the bonding is due to either ionic or van der Waals forces.

† At a finite temperature it is the free energy which is minimized.

The internal energy of ionic crystals

We start by recalling that the potential energy $V(r)$ between a sodium and a chlorine ion may be written as

$$V(r) = -\frac{e^2}{4\pi\varepsilon_0 r} + \frac{B}{r^n}. \tag{8.5}$$

This is the Mie potential (eqn (2.18)) with the term $-A/r^m$ as given by eqn (2.32). Clearly the potential energy between two Na^+ ions, or between two Cl^- ions, must be written as

$$V(r) = \frac{e^2}{4\pi\varepsilon_0 r} + \frac{B}{r^n} \tag{8.6}$$

since there is now an electrostatic *repulsive* force between the two ions. (We will assume that the B/r^n term is the same for Na^+–Na^+, Cl^-–Cl^-, and Na^+–Cl^- interactions.)

We will now consider a single ion of the sodium chloride structure. Looking at Fig. 8.13(d) we see that ion i has 6 nearest neighbours of opposite sign at a distance of a. It follows from eqn (8.5) that the contribution made by these neighbours to the internal energy of i is given by

$$6\left(-\frac{e^2}{4\pi\varepsilon_0 a} + \frac{B}{a^n}\right). \tag{8.7}$$

Returning to Fig. 8.13(d) we see that the second-nearest neighbours to i are 12 ions of the same sign as i at a distance of $2^{1/2}a$. It follows from eqn (8.6) that these contribute

$$12\left(\frac{e^2}{4\pi\varepsilon_0(2^{1/2}a)} + \frac{B}{(2^{1/2}a)^n}\right) \tag{8.8}$$

to the internal energy of i. Repeating these procedures with the third-nearest neighbours, fourth-nearest neighbours, and so on, and adding together the resulting expressions (eqns (8.7), (8.8), etc.) gives the potential energy of ion i as

$$U_i(a) = -\frac{e^2 A'}{4\pi\varepsilon_0 a} + \frac{B'}{a^n}, \tag{8.9}$$

where

$$A' = \left(6 - \frac{12}{2^{1/2}} + \frac{8}{3^{1/2}} - \frac{6}{4^{1/2}} + \frac{24}{5^{1/2}} - \cdots\right), \tag{8.10}$$

$$B' = B\left(6 + \frac{12}{2^{n/2}} + \frac{8}{3^{n/2}} + \frac{6}{4^{n/2}} + \frac{24}{5^{n/2}} - \cdots\right). \tag{8.11}$$

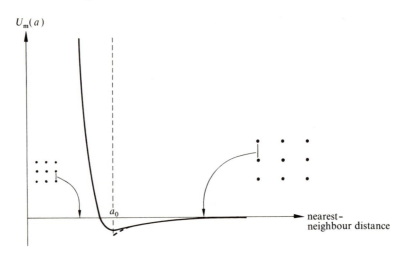

Fig. 8.17. The molar internal energy, $U_m(a)$, of a sodium chloride structure as a function of the nearest-neighbour separation a. The dotted line shows the form $U_m(a)$ would take if the ions were hard spheres.

The molar internal energy $U_m(a)$ of the sodium chloride structure (which contains $2N_A$ ions) is given by

$$U_m(a) = \tfrac{1}{2}(2N_A)U_i(a) = N_A\left(-\frac{e^2A'}{4\pi\varepsilon_0 a} + \frac{B'}{a^n}\right) \qquad (8.12)$$

where the factor of $\frac{1}{2}$ takes care of the fact that in evaluating the potential energy of i relative to j we have also evaluated the potential energy of j relative to i. Eqn (8.12) is shown graphically as the full line in Fig. 8.17. Assuming the crystal to be close to $T = 0$ the structure will be in equilibrium when $U_m(a)$ is a minimum. If this minimum occurs at $a = a_0$, the condition $dU_m(a)/da = 0$ at $a = a_0$ gives, when applied to eqn (8.12),

$$B' = (A'e^2/4\pi\varepsilon_0 n)a_0^{n-1}. \qquad (8.13)$$

Substituting eqn (8.13) back into eqn (8.12) along with $a = a_0$ gives

$$U_m(a_0) = -\frac{N_A e^2 A'}{4\pi\varepsilon_0 a_0}\left(1 - \frac{1}{n}\right) \qquad (8.14)$$

as the potential energy of an equilibrium sodium chloride structure at temperatures near $T = 0$. Although calculated for sodium chloride eqn (8.14) will, of course, apply to any other ionic compounds of the form X^+Y^- which crystallizes with the sodium chloride structure. The constant

A' which occurs in eqn (8.14) is called the *Madelung constant*. It has a value of $1\cdot7476$ for the sodium chloride structure; this being the sum of the series given in eqn (8.10). If exactly the same procedures are followed for the CsCl structure (Fig. 8.15(b)), eqn (8.14) is obtained afresh but with $A' = 1\cdot7627$. This is also true of a X^+Y^- ionic compound with the zincblende structure, but here $A' = 1\cdot6381$.

The enthalpy of sublimation

Instead of talking of the internal energy of a crystal structure we can talk of the energy $H_{m,s}$ required to dissociate one mole of the ionic crystal into its constituent *ions*. Clearly,

$$H_{m,s} = -U_m(a_0) = \frac{N_A e^2 A'}{4\pi\varepsilon_0 a_0}\left(1 - \frac{1}{n}\right). \tag{8.15}$$

Column 4 of Table 9.1 on p. 275 lists the values of $H_{m,s}$ for seven alkali halides calculated using eqn (8.15) with $A' = 1\cdot7476$, a_0 as measured in the crystal by X-ray diffraction studies, and n as given in column 3. Comparing these predicted values with the measured values given in column 5 of Table 9.1 we see they agree closely. As another example, eqn (8.15) with $A' = 1\cdot7626$, $a_0 = 3\cdot56 \times 10^{-10}$ m, and $n = 9$ predicts that $H_{m,s} = 6\cdot10 \times 10^5$ J mol^{-1} in caesium chloride; the measured value is $6\cdot60 \times 10^5$ J mol^{-1}.

A simpler, but less exact way, of arriving at the dissocation energy of an ionic crystal proceeds by imagining a single ion pair (that is, a molecule) to be removed to a large distance from the crystal, after which it is dissociated into its constituent ions. The first phase of the operation calls for little energy since the molecule has zero net charge (although it has a dipole moment). The second phase of the operation calls for the dissociation energy of the molecule as given by eqn (2.33), (where it has been assumed that $n = 9$). Since there are N_A ion pairs in a mole of the crystal, this line of argument predicts a dissociation energy of

$$H_{m,s} = \frac{N_A e^2}{4\pi\varepsilon_0 a_0}\left(1 - \frac{1}{n}\right). \tag{8.16}$$

Apart from the absence of the numerical factor A', of order unity, this is identical to eqn (8.15).

Exercise 8.2

With the exception of helium, all the inert gases crystallize with an fcc structure (Fig. 8.18). Adopting a Lennard–Jones 6–12 potential (eqn (2.20)), show that the molar internal energy $U_m(a)$ of such a structure is given by

$$U_m(a) = 2N_A\varepsilon\left[-14\cdot45\left(\frac{\sigma}{a}\right)^6 + 12\cdot13\left(\frac{\sigma}{a}\right)^{12}\right], \tag{8.17}$$

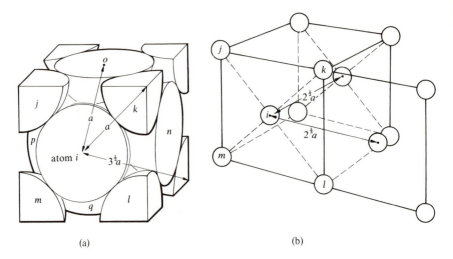

(a)   (b)

Fig. 8.18. (a) A cell of a fcc structure with the atoms drawn roughly to scale. (b) Showing part of a neighbouring cell; here the atoms are not drawn to scale.

where a is the nearest-neighbour separation. Also show that the equilibrium separation $a_0 = 1 \cdot 09\,\sigma$ and that the molar enthalpy of sublimation of solid argon is $8 \cdot 8 \times 10^3$ J mol^{-1}. Take $\varepsilon = 1 \cdot 7 \times 10^{-21}$ J (this value is deduced from the viscous behaviour of gaseous argon at high pressures).

Calculation. We start by considering a single atom i of the structure (Fig. 8.18(a)). It makes contact with 12 others; there are 4 in the corner (j, k, l, and m), another 4 on the cube faces shown (n, o, p, and q), and 4 more on the cube faces out of the plane of the paper. Adopting the Lennard–Jones $6-12$ potential (eqn (2.20)) we see that the potential energy of i due to these 12 nearest neighbours at $r = a$ is given by

$$12 \times 4\varepsilon \left[-\left(\frac{\sigma}{a}\right)^6 + \left(\frac{\sigma}{a}\right)^{12} \right]. \qquad (8.18)$$

In addition to these 12 nearest neighbours, atom i has 6 second-nearest neighbours at distance $2^{1/2}a$ between centres (Fig. 8.18(b)). Their contribution to the potential energy of i is

$$6 \times 4\varepsilon \left[-\frac{1}{2^3}\left(\frac{\sigma}{a}\right)^6 + \frac{1}{2^6}\left(\frac{\sigma}{a}\right)^{12} \right], \qquad (8.19)$$

where we have substituted $r = 2^{1/2}a$ in the Lennard–Jones potential and multiplied the result by 6.

Repeating the procedure with the third-nearest neighbours, fourth-nearest neighbours, and adding together the separate energies (eqns (8.18), (8.19), etc.) gives the total potential energy of atom i as

$$4\varepsilon \left\{ -\left[12 + \frac{6}{2^3} + \frac{24}{3^3} + \cdots \right]\left(\frac{\sigma}{a}\right)^6 + \left[12 + \frac{6}{2^6} + \frac{24}{3^6} + \cdots \right]\left(\frac{\sigma}{a}\right)^{12} \right\}.$$

Both of the series in square brackets converge; the first to a limit of 14·45392; the second (which is more obviously converging) to a limit of 12·13188. To obtain the molar internal energy of a substance which crystallizes with a fcc structure, eqn (8.21) must be multiplied by $\frac{1}{2}N_A$, where that very familiar half prevents us from counting the potential energy contribution of each pair twice. This leads immediately to eqn (8.17). The equilibrium value of a, say a_0, is obtained by setting $dU_m(a)/da = 0$ at $a = a_0$. This leads directly to $a_0 = 1·09\sigma$. Hence the molar enthalpy of sublimation $H_{m,s}$ is given by

$$H_{m,s} = -U_m(a_0) = 2N_A\varepsilon[14·45 \times (0·917)^6 - 12·13 \times (0·917)^{12}]. \quad (8.20)$$

Substituting $\varepsilon = 1·7 \times 10^{-21}$ J gives $H_{m,s} = 8·8 \times 10^3$ J mol^{-1}. The experimentally measured value if $8·36 \times 10^3$ J mol^{-1}.

Comments. Eqn (8.17) can be applied to molecular compounds (such as solid methane) which crystallize with an fcc structure. In view of the weakness of van der Waals forces compared with, say, ionic forces it is not surprising to find that the dissociation energy of a van der Waals solid is of the order of one per cent of the dissociation energy of sodium chloride. Finally, we may note that, if 14·45 and 12·13 are each approximated by 12, eqn (8.17) reduces to

$$U_m(a) = 6N_A \times 4\varepsilon\left[-\left(\frac{\sigma}{a}\right)^6 + \left(\frac{\sigma}{a}\right)^{12}\right] \quad (8.21)$$

which is precisely the result that we would have obtained had we chosen to only consider the 12 nearest-neighbour atoms to i in Fig. 8.18. Applying the equilibrium condition that $dU_m(a)/da = 0$ at $a = a_0$ to eqn (8.21) gives $a_0 = 2^{1/6}\sigma = 1·12\sigma$. Substituting this value of a_0 back into eqn (8.21) gives

$$H_{m,s} = -U_m(a_0) = 6N_A\varepsilon \quad (8.22)$$

which, as expected, agrees with the value given by eqns (2.26) and (2.27).

8.6. Factors controlling a crystal's lattice

Although we have successfully predicted the dissociation energies of sodium chloride and caesium chloride, this success has depended on assuming that NaCl has the structure shown in Fig. 8·13(d) and that CsCl has the structure shown in Fig. 8·15(b). That is, we have not really answered the question why, say, NaCl has the structure it has, rather than a structure similar to that of, say, CsCl or ZnS.

In seeking the answers to these questions we will suppose that an ionic crystal is made up of hard-sphere ions. This assumption implies that the equilibrium separation a_0 in the structure occurs when nearest neighbours touch (Fig. 8.17). Fig. 8.19(a) shows a section through the central plane of the unit cell of the sodium chloride structure shown in Fig. 8.13(d). Here although the ions have been drawn to scale (the ratio r^+/r^- being equal to the known ratio of the ionic radii in NaCl) and in their equilibrium positions (ions of opposite sign touching). What we will now do is to alter the ratio r^+/r^- while attempting to keep the ions in contact; this is tantamount to allowing other X^+Y^- ionic compounds to crystallize

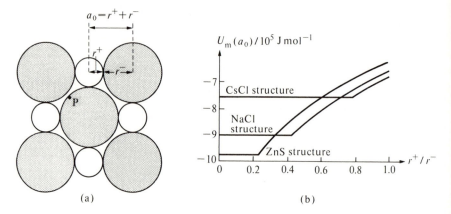

Fig. 8.19. (a) A section through the unit cell of the sodium chloride structure, as drawn parallel to the cube face. In this drawing r^+/r^- is that of NaCl. (b) Showing how the internal energy $U_m(a_0)$ of an ionic crystal in thermodynamic equilibrium at temperatures close to $T = 0$ depends on the ratio r^+/r^- of the ionic radii and on the form of the crystal structure. The equilibrium nearest-neighbour separation $a_0 = r^+ + r^-$. The ordinate values shown assume that the negative ion has a radius of 1.85×10^{-10} m (that of a Cl^- ion).

with the NaCl structure. It is immediately apparent from Fig. 8.19(a) that if we hold r^- fixed (corresponding to that of, say, the Cl^- ion) and replace the positive ion by others of progressively smaller radii we will reach a stage when the negative ions touch at P (Fig. 8.19(a)). This will occur when $2r^- = (r^+ + r^-)2^{1/2}$, that is when $r^+/r^- = 2^{1/2} - 1 = 0.414$. If we reduce the radius of the positive ion still further, there is no further drop in $U_m(a_0)$ (see Fig. 8.19(b)) and all that will happen is that it has room to 'rattle' around in the space between the negative ions.

It is a simple matter to show that, if the ionic solid has a caesium chloride structure, the value of $U_m(a_0)$ will fall with decreasing r^+/r^- in the manner shown in Fig. 8.19(b) until the two negative ions along the cell edge in Fig. 8.15(b) touch. At this point the central positive ion touches those along the body diagonals, and $r^+/r^- = (3)^{1/2} - 1 = 0.732$. At smaller ratios the postive ion has room in which to move between the negative ions. In a similar fashion $U_m(a_0)$ for the zincblende structure will fall as shown in Fig. 8.19(b) until $r^+/r^- = \frac{1}{2}(6)^{1/2} - 1 = 0.225$.

Fig. 8.19(b) tells us that for $0 < r^+/r^- < 0.33$ the ZnS structure is the most stable of the three structures considered here, that for $0.33 < r^+/r^- < 0.72$ the NaCl structure is the most stable, while for $0.72 < r^+/r^-$ it is the CsCl structure which is the most stable. This agrees with the fact that the CsCl, CsBr, and CsI, whose r^+/r^- ratios are 0.93, 0.87, and 0.78, respectively, all have the caesium chloride structure. It also agrees with the fact that all those alkali halides whose r^+/r^- lie between 0.33 (the value for LiCl) and 0·68 (the value for KBr) have the sodium chloride structure. There are, however, a number of alkali halides with $r^+/r^- > 0·72$ which possess the sodium chloride rather than the expected caesium chloride structure. These departures are not unexpected in view of the small differences in $U_m(a_0)$ of these two structures at $r^+/r^- > 0.72$ (see Fig. 8.19(b)); differences which may not, in any event, be genuine in view of our hard-sphere repulsive model.

Fig. 8.19(b) also suggests that when an X^+Y^- ionic compound has $r^+/r^- = 0·33$ we might expect to find it with either the zincblende or the sodium chloride structure. Likewise, when $r^+/r^- = 0·72$, we might expect to find either the sodium chloride or the caesium chloride structure. A substance is said to be *polymorphous* when it can exist in two or more forms with different crystal structures. Although dimorphism is not observed in the alkali halides at normal pressures and temperatures it does occur, for example, in RbCl, RbBr, and RbI at high pressures, when the sodium chloride and caesium chloride structures may be equally likely.

8.7. The surface energy of a crystal

An atom within the interior of a crystal is surrounded on all sides by other atoms. This is not true of an atom on or near the surface, where roughly half the surrounding atoms are missing. Consequently, it takes less energy to remove an atom from the surface layers of a crystal to infinity than it takes to remove an atom from the interior of the crystal to infinity. Rephrased, the internal energy of an atom in the surface layers is greater than that of an atom in the body of the crystal.

van der Waals crystal

Exercise 8.2 demonstrated that the molar internal energy of an fcc van der Waals solid as calculated assuming only nearest-neighbour interactions (eqn (8.21)) agrees, to within twenty-five per cent, with the molar internal energy calculated assuming interactions with all other atoms (eqn (8.17)). This close agreement arises because of the short range of a $1/r^7$ force. It also makes for a simple calculation of the surface energy of a van der Waals solid.

Looking at Fig. 8.18(a) we see that if atom i is located within the body of a crystal it has 12 nearest neighbours but that if it is located on a {100} face it has only 8 nearest neighbours. Denoting the number of atoms on the surface layer of the crystal by N_s, it follows that a crystal containing 1 mol of atoms has $(N_A - N_s)$ atoms within the body of the crystal. Hence

$$U_m = -\tfrac{1}{2}\varepsilon[12(N_A - N_s) + 8N_s] = -\tfrac{1}{2}\varepsilon(12N_A - 4N_s), \qquad (8.23)$$

where $-\varepsilon$ is the potential energy of an (isolated) pair of atoms (see Fig. 2.8(b)), and the factor of $\frac{1}{2}$ acknowledges that two atoms are involved in a pair. Were there to be no surface atoms ($N_s = 0$), eqn (8.23) would become

$$U_m = -\tfrac{1}{2}\varepsilon \times 12 N_A. \tag{8.24}$$

Subtracting eqn (8.24) from eqn (8.23) tells us that the molar internal energy of an actual crystal exceeds that of a 'surface free' crystal by $2N_s\varepsilon$. If the extra energy possessed per unit area of the surface by the surface atoms is denoted by γ—a quantity known as the *free surface energy*—and if the surface area of the crystal is A then

$$\gamma_{100} A = 2N_s\varepsilon, \tag{8.25}$$

where the suffix 100 reminds us that we have assumed {100} faces to the crystal. Since each atom on a {100} face occupies an area of a_0^2 (Fig. 8.18(a) shows that $1 + 4(\frac{1}{4})$ atoms occupy an area of $2a_0^2$, where a_0 is the equilibrium nearest-neighbour separation), it follows that the number of atoms per unit area of surface is $1/a_0^2$. Replacing N_s/A in eqn (8.25) by $1/a_0^2$ gives

$$\boxed{\gamma_{100} = \frac{2\varepsilon}{a_0^2}}. \tag{8.26}$$

This predicts, for example, that in solid argon $\gamma_{100} = 2 \cdot 0 \times 10^{-2}\,\text{J m}^{-2}$.

Ionic crystals

Following through similar procedures to those used in deriving the free surface energy of a van der Waals crystal leads to the conclusion that for a sodium-chloride structure

$$\gamma_{100} = \frac{0 \cdot 0145 e^2}{4\pi\varepsilon_0 a_0^3}, \tag{8.27}$$

where a_0 is the equilibrium nearest-neighbour separation. This relation (which assumes a B/r^9 in the interatomic potential) was first derived by Born and Stern in 1919. It predicts, for example, that $\gamma_{100} = 0 \cdot 41\,\text{J m}^{-2}$ in LiF; in close agreement with the measured value. Born and Stern also deduced that a {110} face of a sodium chloride structure has a free surface energy of

$$\gamma_{110} = \frac{0 \cdot 0394 e^2}{4\pi\varepsilon_0 a_0^3}. \tag{8.28}$$

Because this is 2·7 times larger than γ_{100} predicted by eqn (8.27) we would expect an alkali halide crystal to cleave more readily along a {100} plane than along a {110} plane. This agrees with the observation that alkali halides with the sodium chloride structure only cleave along {100} planes.

Experimental measurements

The most straightforward way of determining the free surface energy γ is to measure the work ΔW required to cleave a crystal. If the two surfaces so produced *each* have an area A, then, assuming that ΔW goes solely into creating these surfaces,

$$\Delta W = 2A\gamma. \tag{8.29}$$

We will now look at how this principle has been applied by Gilman to measure the free surface energies of single crystals at 77 K.

The crystals to be studied are initially cracked over part of their length using the jig shown in Fig. 8.20(a). Here a screw-driven steel wedge starts a crack which runs to the point where rubber-padded clamps apply compressional forces to the crystal. Next a small yoke is attached to

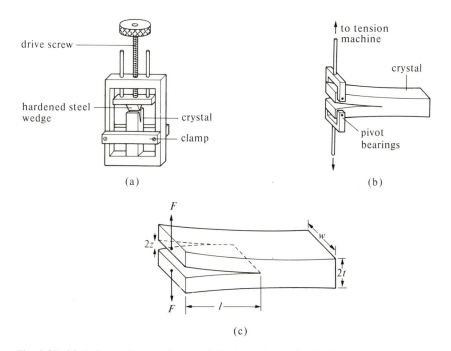

(a)

(b)

(c)

Fig. 8.20. (a) A jig used to produce partially-cleaved crystals. (b) Showing how the force is applied to the crystal. The crystal is kept immersed in liquid nitrogen. (c) The situation as the crack starts to propagate. (After Gilman, J. J. (1960). *J. appl. Phys.* **31**, 2208.)

each 'arm' of the specimen by means of pivot pins (Fig. 8.20(b)). Measured forces are applied to the partially-cracked specimen by means of a tensile-testing machine, in which the applied force is increased at a rate of about 0.08 N s^{-1}. As the force increases no movement of the crack occurs until, at a particular value F of the force, it suddenly grows in length. The initial crack length l (Fig. 8.20(c)) can be accurately measured by microscopic examination of the specimen at the conclusion of the experiment; when a crack starts and stops small steps are produced on the cleavage surface.

To see how γ is obtained we consider what happens as the crack advances through a distance dl from its starting position. Assuming the process to be thermodynamically reversible, the energy $F \times 2dz$ fed into the specimen via the tensile-testing machine must go to create two new surfaces, each of area $w \, dl$, and to increase the energy stored in each of the cantilevered arms of the crystal by dU, say (see Fig. 8.20(c)). Hence

$$2F \, dz = 2\gamma w \, dl + 2dU. \tag{8.30}$$

Standard elasticity theory can be applied to relate dz to dl and dU to dl. In fact dl cancels through eqn (8.30), leading to the relation

$$\boxed{\gamma = \frac{6F^2 l^2}{E^2 w^2 t^2}}, \tag{8.31}$$

where l, w, and t are as defined in Fig. 8.20(c), and E is the Young's modulus of the material (see problem 3.21). Thus measurements of the

TABLE 8.1
Free surface energies of solids

Substance	Surface	Environment	Temperature/K	Free surface energy/J m^{-2}
LiF	(100)	liquid N_2	77	0·34
KCl	(100)	air	298	0·11
NaCl	(100)	liquid N_2	77	0·32
CaF$_2$	(111)	liquid N_2	77	0·45
Si	(111)	liquid N_2	77	1·24
Fe(3% Si)	(100)	liquid H_2	14	1·36
MgO	(100)	air	298	1·15
		vacuum	77	1·28
Zn	(0001)	liquid N_2	77	0·11
CaCO$_3$	(10$\bar{1}$0)	liquid N_2	77	0·23
mica	(0001)	air	298	0·38
		vacuum	298	5·0

initial crack length l and the force F required to initiate crack growth allows γ to be determined. Table 8.1 lists the free surface energies for a variety of solids as determined by cleavage techniques.

PROBLEMS

8.1. Which of the cells shown in Fig. 8.5(b) are non-primitive? Remember that a primitive cell contains one lattice point.

8.2. Show that in a simple cubic lattice the distance between adjacent planes of Miller indices (hkl) is $a(h^2+k^2+l^2)^{-1/2}$, where a is the lattice constant (here the separation between neighbouring points on the lattice).

8.3. A monochromatic beam of X-rays of wavelength $1\cdot54\times10^{-10}$ m is incident on a crystal whose lattice is simple cubic with a lattice constant of $4\cdot0\times 10^{-10}$ m. At what angle θ will first-order Bragg reflection occur from (100) planes of the crystal? At what angle will third-order reflection occur?

8.4. Show that the structure of sodium chloride (Fig. 8.13(d)) can be described as a combination of a fcc lattice with a sodium ion as basis and a fcc lattice with a chlorine ion as basis. By how much is one lattice displaced relative to the other?

8.5. (a) Sketch the following planes for a simple cubic lattice: (100), (110), (120), (1$\bar{2}$0), (320), and (010). (b) When dealing with a hexagonal lattice it is usual to employ four crystal axes; three (x, y, and u) lying at 120° to each other in the basal plane, and the fourth (z) lying along the perpendicular (prism) axis. The sense of x, y, and u are such that a radius of a right-handed screw rotating about the z-axis will, in turn, pass the x, y, and u-axes. Using this information, sketch (0001) and (10$\bar{1}$0) planes in a hcp structure.

8.6. For hexagonal close packing of rigid spheres, what is the c/a ratio (see Fig. 8.12(c))?

8.7. Show that in a bcc lattice the nearest-neighbour distance is $3^{1/2}a/2$ and that in an fcc lattice it is $a/2^{1/2}$. Here a is the side of the cube edge in the cells labelled cubic I and cubic F in Fig. 8.5(b).

8.8. What is the maximum proportion of the available volume which can be filled by hard spheres when these spheres are arranged on (a) a simple cubic lattice, (b) a fcc lattice, and (c) a bcc lattice?

8.9. Show that in a monatomic hcp structure there are, about any one atom, 12 nearest neighbours (at separation a say), 6 second-nearest neighbours at separation $2^{1/2}a$, and 2 third-nearest neighbours at separation $(8/3)^{1/2}a$.

8.10. Show that the potential energy of an ion in an infinite one-dimensional sodium chloride crystal whose ions interact according to a hard-sphere coulombic model is

$$V(r) = -\frac{\alpha e^2}{4\pi\varepsilon_0 r},$$

where $\alpha = 2\ln 2$ and r is the interionic spacing. What is the internal energy for a mole of such a hypothetical crystal? You will recall that

$$\ln(1+x) = x - \frac{x^2}{2} + \frac{x^3}{3} - \frac{x^4}{4} + \cdots.$$

8.11. Starting with eqn (8.18), show that the equilibrium separation of nearest neighbours in a fcc structure whose atoms interact according to a Lennard–Jones $6-12$ potential is $1\cdot09\sigma$.

8.12. Calculate the molar internal energy of caesium chloride assuming that the potential energy between a Cs^+ and Cl^- ion is as given by eqn (8.5) and that the potential energy between two Cs^+ ions, or between two Cl^- ions, is as given by eqn (8.6). Take $n=9$ and the nearest-neighbour equilibrium separation $a_0 = 3\cdot56 \times 10^{-10}$ m. The Madelung constant has a value of $1\cdot7627$ for caesium chloride.

8.13. Verify that $U_m(a_0)$ for caesium chloride varies with the ratio r^+/r^- of the ionic radii in the manner shown in Fig. 8.19(b). Take r^- to be that of the Cl^- ion, namely $1\cdot85 \times 10^{-10}$ m. Recall that, at equilibrium, $a_0 = r^+ + r^-$ in a hard-sphere repulsive model.

8.14. When we originally calculated the internal energy of an fcc van der Waals solid in exercise 8.2 we ignored the fact that some of the atoms are located on the surface of the crystal. Assuming that we had taken this fact into account, by what percentage would the calculated value of $U_m(a_0)$ have differed from that deduced in exercise 8.2? It will be sufficient to consider only nearest-neighbour interactions. Assume the crystal has only $\{100\}$ faces.

8.15. It is possible to estimate the free surface energy γ by measuring the tensional force F required to pull a crystal apart. If the two surfaces formed each have an area A and if the force F is assumed to be constant as the two surfaces move apart by a distance s prior to separation, then $Fs = 2\gamma A$. Using these clues make a *rough* calculation of γ_{111} in sodium chloride. The nearest-neighbour separation in sodium chloride is $2\cdot8 \times 10^{-10}$ m. *Clues*: Think about the nature of a (111) plane in sodium chloride. Replace Fs by a suitable integral.

8.16. Assuming only nearest-neighbour interactions, show that the free surface energy γ_{110} in an fcc van der Waals crystal is given by

$$\gamma_{110} = \frac{5\varepsilon}{2^{3/2}a_0^2},$$

where a_0 is the nearest-neighbour separation, and ε is the dissociation energy of an isolated pair of the atoms.

8.17. One might argue that when salt dissolves in water all that happens is that the Na^+ and Cl^- ions move far apart. The energy required should come from the translational kinetic energy of the molecules in the solution, and thus the solution so formed should be at a lower temperature than the original water. Try the experiment. If you get a cooling effect, does it agree with the value you would expect from the molar internal energy of sodium chloride $(7\cdot6 \times 10^5 \text{ J mol}^{-1})$? You will need to use a table of physical constants.

8.18. Adopting a hard-sphere repulsive model, show that the free surface energy of a $\{100\}$ face of a sodium chloride crystal is given by

$$\gamma_{100} = \frac{1}{2} \frac{e^2}{4\pi\varepsilon_0 a_0^3} (A' - A''),$$

where A' is given by eqn (8.10), and where A'' is given by

$$A'' = 5 - \frac{8}{2^{1/2}} + \frac{4}{3^{1/2}} - \frac{5}{4^{1/2}} + \frac{16}{5^{1/2}} - \cdots.$$

Clues: Start by writing down an expression for the potential energy of ion s on the (100) face shown in Fig. 8.13(d). Next consider the molar internal energies of a real NaCl crystal bounded by {100} faces and of a 'surface free' crystal; the difference between these energies is equal to $\gamma_{100}A$, where A is the surface area of the real crystal.

9. Mechanical properties of solids

THERE are three distinct ways of deforming a solid. We may 'squeeze it evenly all over'—that is, subject it to a hydrostatic pressure. We may stretch it, perhaps to the point where we 'pull it apart'—that is, we may apply a tensile stress. Or we may shear it, as we might splay out a deck of cards.

We shall try to account for the behaviour of a solid under these three different stress conditions in terms of the interactions between its constituent atoms. In so doing we shall discover that most solids fail to live up to their theoretical expectations when subjected to a tensile or shear stress. Small structural imperfections can be the weak link which limits the strength of a solid.

9.1. The bulk modulus

Definition

If we wish to study how the volume of a solid changes with changing hydrostatic pressure we must clearly immerse the solid in a fluid whose pressure can be changed (Fig. 9.1(a)). A hydrostatic pressure is characterized by its constancy in all directions at any point in the fluid. If the pressure is increased from p to $p + dp$ we can expect that the volume of the solid will change by dV (which will, of course, have a negative value). Assuming the experiment to be performed at constant temperature, we define the *isothermal bulk modulus* K_T by

$$K_T = \frac{dp}{(-dV/V)} = -V_m \frac{dp}{dV_m} \tag{9.1}$$

where the last expression follows on dividing the numerator and denominator of dV/V through by the amount of the material present. The minus sign is incorporated to make the value of K_T positive. The bulk modulus may also be written as the inverse of the *compressibility*, defined as the fractional change in volume per unit change of pressure, $(-dV/V)/dp$. A solid, which normally has a low compressibility, thus has a high bulk modulus. Fig. 9.1(b) shows that the bulk modulus measures the magnitude of the slope of an isotherm multiplied by the molar volume at the point in question.

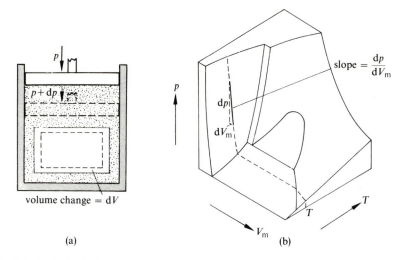

Fig. 9.1. (a) If the hydrostatic pressure is raised by dp the volume of the solid changes by dV. (b) The isothermal bulk modulus is the magnitude of the slope of the isothermal line multiplied by the molar volume of the solid at the point where the slope is measured.

Experimental techniques

As it stands the apparatus of Fig. 9.1(a) is of little practical value. To withstand high pressures the cylinder has to be made of steel, which precludes us from directly observing how the volume of the solid changes with increasing pressure.

Instead of trying to measure the fractional change in volume, $\delta V/V$, of a solid we might measure the fractional change in length, $\delta l/l$, of a rod of the solid. Assuming the substance is isotropic (has the same properties in all directions), as it is in polycrystalline samples, $\delta V/V = 3\,\delta l/l$. This result follows immediately on differentiating the expression $V = l^3$ for the volume of a cube of side l.

The leading physicist of this century in the field of high-pressure physics was P. W. Bridgman. He measured the fractional change in the length of a specimen by means of a sliding electrical contact, thereby dispensing with the need to see the specimen. Fig. 9.2(a) shows the arrangement—all of which is immersed in a high-pressure liquid contained in a compression cylinder. The specimen S, in the form of a long rod, is kept pressed against the bottom of an iron holder H by a spring O which pulls on the top of the specimen. A high-resistance wire is attached to the upper end of the specimen and this slides over a contact C which is attached to H but insulated from it. (Spring N keeps the resistance wire

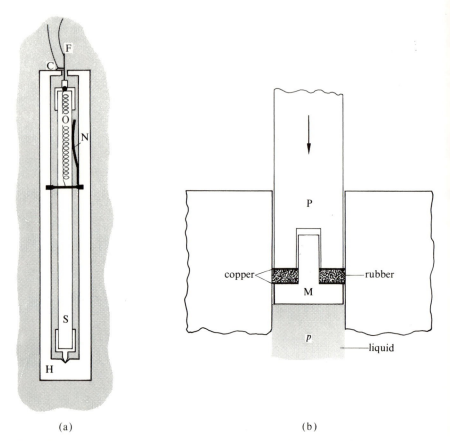

(a)   (b)

Fig. 9.2. Bridgman's equipment for measuring compressibilities. (a) The apparatus for measuring the fractional change in the length of a rod. (b) The leak-free piston for obtaining high pressures.

pressed against the contact.) By measuring the electrical resistance between C and a terminal F fixed to the wire one can deduce the change in the length of the specimen (this is proportional to the change in electrical resistance).

Early studies of high-pressure phenomena had been hampered by the fact that the high-pressure liquid tended to leak out between the piston and the cylinder to the low-pressure surroundings. Fig. 9.2(b) shows how Bridgman overcame this problem. The kernel is the mushroom-shaped plug M. The stem of the mushroom projects freely into a hollow in the end of the plunger P, made of hardened steel. Pressure is transmitted to the head of the mushroom through a soft rubber ring. At equilibrium the force acting down on the upper face of the head (which is the pressure in the rubber multiplied by the area of the ring) must equal the force acting up on the head (the pressure in the liquid multiplied by the entire area of the head). Since the area of the rubber ring is less than that of the area in contact with the liquid it follows that the hydrostatic pressure in the rubber exceeds that in the liquid. The tendency is for the rubber to leak into the liquid and not vice versa! Copper discs prevent the rubber from leaking (Fig. 9.2(b)).

Predicting K_T

We will now consider how the isothermal bulk modulus of a solid is related to the interatomic potential energy characteristic of the constituent atoms. Our line of attack will be to appeal to the first law of thermodynamics and then to relate the change in internal energy to the change in the interatomic potential energy.

If one mole of a solid, with a volume V_m, is subjected to a hydrostatic pressure p, as in Fig. 9.1(a), and if the pressure is changed slightly so that the volume of the solid changes by dV_m then the first law of thermodynamics (eqn (1.8)) tells us that (assuming the process is adiabatic) the internal energy of the solid changes by

$$dU_m = -p\, dV_m. \tag{9.2}$$

During a compression dV_m is negative, and so the internal energy will increase. Rewriting eqn (9.2) as

$$p = -\frac{dU_m}{dV_m} \tag{9.3}$$

and substituting this expression for p into the defining relation for the isothermal bulk modulus, eqn (9.1), gives

$$K_T = V_{m0}\left(\frac{d^2U_m}{dV_m^2}\right)_{V_m=V_{m0}} \tag{9.4}$$

where the suffix $V_m = V_{m0}$ reminds us that d^2U_m/dV_m^2 is to be evaluated at the equilibrium molar volume V_{m0} appropriate to the pressure at which the measurements are made. The corresponding nearest-neighbour separation is a_0. Actually eqn (9.4) is not correct. Eqn (9.1) defines the *isothermal* bulk modulus whereas eqn (9.2) assumes the compression is adiabatic. Thus eqn (9.4) really should involve the adiabatic bulk modulus K_S, rather than K_T. However K_S and K_T approach a common value as the temperature tends to $T=0$, allowing us to employ eqn (9.4) at these temperatures.

In §8.5 we saw how U_m is related to the nature of the interatomic potential characteristic, to the form of the crystal structure, and to the nearest-neighbour separation a. Before a relation like eqn (8.12) or

(8.17) can be substituted into eqn (9.4), it is necessary to rewrite it so that it involves differentials of U_m with respect to a. To introduce a into eqn (9.4) we write

$$\frac{dU_m}{dV_m} = \frac{dU_m}{da}\frac{da}{dV_m}$$

$$\frac{d^2U_m}{dV_m^2} = \frac{d}{dV_m}\left(\frac{dU_m}{da}\frac{da}{dV_m}\right)$$

$$= \frac{dU_m}{da}\frac{d^2a}{dV_m^2} + \frac{da}{dV_m}\left[\frac{da}{dV_m}\frac{d}{da}\left(\frac{dU_m}{da}\right)\right]$$

$$\frac{d^2U_m}{dV_m^2} = \frac{dU_m}{da}\frac{d^2a}{dV_m^2} + \left(\frac{da}{dV_m}\right)^2\frac{d^2U_m}{da^2}. \tag{9.5}$$

Now we can write

$$V_m = cN_A a^3 \tag{9.6}$$

$$\frac{dV_m}{da} = 3cN_A a^2, \tag{9.7}$$

where c is a constant whose value depends only on the form of the crystal structure. For example, $c = 2$ for sodium chloride; this can readily be justified by examining Fig. 8.13(d) where each ion occupies a volume of a^3, and thus $2N_A$ ions occupy a volume $V_m = 2N_A a^3$. Substituting (da/dV_m) as given by eqn (9.7) into eqn (9.5) and recalling that $dU_m/da = 0$ at $a = a_0$ gives

$$\left(\frac{d^2U_m}{dV_m^2}\right)_{V_m = V_{m0}} = \frac{1}{9c^2N_A^2 a_0^4}\left(\frac{d^2U_m}{da^2}\right)_{a = a_0}. \tag{9.8}$$

Substituting eqns (9.6) and (9.8) into eqn (9.4) gives

$$\boxed{K_T = \frac{1}{9cN_A a_0}\left(\frac{d^2U_m}{da^2}\right)_{a = a_0}.} \tag{9.9}$$

This relation was first derived by the German physicist Max Born in 1918.

The bulk modulus of an ionic crystal

Eqn (8.12) describes how the molar internal energy $U_m(a)$ of an ionic crystal made up of singly-charged ions depends on the structure of the crystal and on the nature of the interatomic potential characteristic.

When it is inserted into eqn (9.9) we obtain

$$K_T = \frac{1}{9ca_0}\left(-\frac{2e^2 A'}{4\pi\varepsilon_0 a_0^3} + \frac{n(n+1)B'}{a_0^{n+2}}\right)$$

or, substituting for B' from eqn (8.13),

$$\boxed{K_T = \frac{1}{4\pi\varepsilon_0}\frac{A'e^2(n-1)}{9ca_0^4}}. \qquad (9.10)$$

Taking sodium chloride as our example, eqn (9.10), with $A' = 1.748$, $n = 9$, $a_0 = 2.81 \times 10^{-10}$ m, and $c = 2$ predicts $K_T = 2.87 \times 10^{10}$ Pa, which agrees well with the measured value of 3.0×10^{10} Pa. In practice eqn (9.10) is normally used to deduce n—that is, the repulsive exponent which occurs in the Mie potential used to describe the interaction between the ions—from K_T, rather than the other way round. Some experimental values of n, as deduced from measurements of K_T extrapolated to $p = 0$, $T = 0$ are given in Table 9.1. It is worth pointing out that it would be very difficult to deduce n from measurements of dissociation energies; eqn (8.15) shows that changing n from, say, 9 to 10 only changes $H_{m,s}$ by 1.2 per cent. This is only $\frac{1}{8}$ of the corresponding change in K_T. The underlying reason for the strong role of n in K_T is that in compressing a solid we are reducing all the interatomic separations from r_0 in Fig. 2.6(b) over a region where the repulsive forces are dominant. The reason for the weak role of n in $H_{m,s}$ is that in removing

TABLE 9.1
Properties of ionic crystals

Substance	$K_T{}^a/10^{10}$ Pa	n^b	$H_{m,s}{}^c$ $/10^5$ J mol^{-1} (calculated)	$H_{m,s}{}^d$ 10^5 J mol^{-1} (measured)	$V_m/10^{-5}$ m^3 mol^{-1}	$(nH_{m,s}/9V_m)$ $/10^{10}$ Pa
LiF	7.0	5.9	10.04	10.04	0.98	6.72
LiCl	3.7	8.0	8.25	8.28	2.05	3.60
NaCl	3.0	9.1	7.66	7.64	2.70	2.58
NaBr	2.6	9.5	7.25	7.24	3.21	2.38
KBr	1.8	10.0	6.60	6.53	4.33	1.68
KI	1.4	10.5	6.22	6.33	5.30	1.39
RbBr	1.5	10.0	6.37	6.41	4.94	1.44

[a] Values obtained by P. W. Bridgman, as quoted by Slater, J. C. (1924). *Phys. Rev.* **23**, 488.
[b] As deduced by J. C. Slater (*loc. sit.*) from K_T values.
[c] Calculated using values of n listed in column 3.
[d] When solid alkali halides sublime they do not normally dissociate into widely separated ions. Hence $H_{m,s}$ does not directly measure the required dissociation energy and allowance must be made for the energy required to form the ions. If, for example, the alkali halide dissociates to X^+Y^- molecules we must add $N_A\varepsilon$ (see eqn (2.26)) to $H_{m,s}$ to obtain the required dissociation energy. There is no one single pattern of behaviour at dissociation throughout the alkali halides.

an ion to infinity we must pull against the attractive force of all the other ions of opposite sign; the only $1/r^n$ forces which offer much help are those arising from the few nearest-neighbour ions to the one being removed.

Exercise 9.1

Show that the isothermal bulk modulus of a crystallized inert gas (fcc structure) is given by

$$K_T = \frac{69\varepsilon}{ca_0^3}. \tag{9.11}$$

Calculate the value of K_T for solid argon, given that $\varepsilon = 1 \cdot 7 \times 10^{-21}$ J and $a_0 = 3 \cdot 7 \times 10^{-10}$ m.

Proof. The molar internal energy $U_m(a)$ expressed as a function of the nearest-neighbour separation a is given by eqn (8.17). Differentiating this expression with respect to a gives

$$\frac{dU_m}{da} = 2N_A\varepsilon \left[\frac{6 \times 14 \cdot 45\sigma^6}{a^7} - \frac{12 \times 12 \cdot 13\sigma^{12}}{a^{13}} \right] \tag{9.12}$$

$$\frac{d^2U_m}{da^2} = 2N_A\varepsilon \left[\frac{-7 \times 6 \times 14 \cdot 45\sigma^6}{a^8} + \frac{13 \times 12 \times 12 \cdot 13\sigma^{12}}{a^{14}} \right]. \tag{9.13}$$

It follows from eqn (9.12) that the equilibrium nearest-neighbour separation a_0, at which $dU_m/da = 0$, is given by $a_0 = 1 \cdot 090\sigma$. Substituting eqn (9.13) with $a = a_0 = 1 \cdot 090\sigma$ into eqn (9.9) leads to eqn (9.11).

To evaluate K_T we must know the value of c. It follows from the defining relation (eqn (9.6)) that c is equal to the volume occupied per atom of the structure divided by the cube of the nearest-neighbour separation. Looking at Fig. 8.18(a) we see that $6(\frac{1}{2}) + 8(\frac{1}{8}) = 4$ atoms occupy a volume of $2^{3/2}a_0^3$. Hence $c = (2^{3/2}a_0^3/4)a_0^{-3} = 2^{-1/2}$. Substituting this result, along with the given values of a_0 and ε, into eqn (9.11) predicts $K_T = 3 \cdot 3 \times 10^9$ Pa; in fair agreement with the measured value of $2 \cdot 5 \times 10^9$ Pa.

Comment. The bulk moduli of the solid inert gases are substantially less than the bulk moduli of ionic salts. That is, the solid inert gas are *more* compressible than the ionic salts (which contain $1/r^2$ repulsive forces in addition to the $1/r^n$ repulsive forces).

Relationship of K_T to $H_{m,s}$

Eliminating A' between eqns (8.15) and (9.10) and using eqn (9.6) shows that for an ionic crystal the isothermal bulk modulus K_T and the molar enthalpy of sublimation $H_{m,s}$ (meaning dissociation to widely separated ions) are related by

$$\boxed{K_T = \frac{n}{9} \frac{H_{m,s}}{V_m}}. \tag{9.14}$$

Comparing the measured values of K_T given in column 2 of Table 9.1 with the ratio $nH_{m,s}/9V_m$, as calculated using experimentally determined values of n, $H_{m,s}$, and V_m and given in column 7, we see that eqn (9.14) is obeyed, or nearly obeyed, by the alkali halides.

Likewise, if we eliminate ε between eqns (8.20) and (9.11), we obtain the following relation for a crystallized inert gas (fcc structure);

$$K_T = 8 \frac{H_{m,s}}{V_m}. \tag{9.15}$$

For solid argon $H_{m,s} = 7 \cdot 7 \times 10^3 \, \text{J mol}^{-1}$ and $V_m = 2 \cdot 27 \times 10^{-5}$ $\text{m}^3 \, \text{mol}^{-1}$. Substituting these values into eqn (9.15) predicts an isothermal bulk modulus of $2 \cdot 7 \times 10^9 \, \text{Pa}$. The experimental value is $2 \cdot 5 \times 10^9 \, \text{Pa}$.

Finally, it is worth emphasizing that we have only been able to make quantitative predictions as to the behaviour of the two 'extreme' types of solids; namely solids in which the bonding is predominantly ionic or predominantly van der Waals in character. Between these two extremes lie the metals and the covalently-bonded solids (such as diamond). Because a lump of such solids is really one giant molecule it takes sophisticated quantum mechanics to arrive at $H_{m,s}$ and K_T in these cases. However, in view of the fact that eqns (9.14) and (9.15) only differ in the numerical factor which multiplies their right-hand side we might hazard the guess that, say, a metal would obey a relation of the form

$$K_T = \frac{CH_{m,s}}{V_m}, \tag{9.16}$$

where the constant C lies somewhere in the range 1 to 8. In fact, a value of $C = 3$ allows eqn (9.16) to be satisfied by a wide variety of metals.

9.2. Young's modulus

If a tensional force F, applied to a specimen of constant cross-sectional area A (Fig. 9.3(a)) and length l, causes the length to change by δl we say that the *tensile stress* of F/A has caused a *strain* of $\delta l/l$. Note that stress is measured in units of N m^{-2} (i.e. Pa) and that strain, being a ratio, is a pure number. Tensile stresses can act in rather unexpected ways. Fig. 9.3(b) shows, in exaggerated fashion, what happens when we walk across a wooden floor. As a plank bends under our weight, the upper surface AB is compressed and the lower surface CD is extended, relative to the length PQ they had before we walked on the plank. Decreasing the interatomic separation in a solid below the equilibrium value brings repulsive restoring forces into play; increasing the interatomic separation beyond the equilibrium value brings attractive forces into play. If it were not for these forces we would fall through the floor.

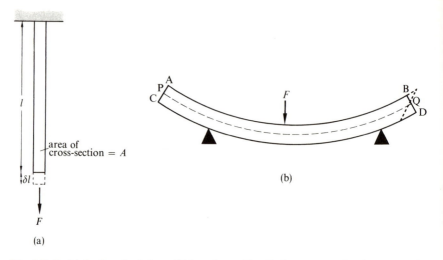

Fig. 9.3. In (a) the length of the solid has changed by δl, the same at all points across the solid, when a force F is applied. In (b) the top surface AB has contracted, and the bottom surface CD has expanded, relative to the unchanged central portion PQ.

Young's modulus

Fig. 9.4(a) shows the general form of the stress–strain curve of a *ductile* material. It would apply to a polycrystalline metal like stainless steel or to certain fcc and bcc single crystals; examples being copper (fcc) and iron (bcc). In this plot the stress is defined as the tensile force divided by the *original* cross-sectional area of the specimen. Up to point P, the *proportionality limit*, the strain is proportional to the applied stress and the strain disappears on removing the stress. This reversible (or near-reversible) behaviour continues to a point Q known as the *yield point*, usually defined as the point beyond which a permanent strain of greater than 0·2 per cent is produced. Fig. 9.4(a) shows the effect of removing— and then reapplying—the tensile stress from the specimen. Notice how the elastic (that is, reversible) behaviour continues right up to the point from which the stress was originally reduced. As a consequence the yield point is higher when measured the second time round. (The 0·2 per cent permanent strain defining the yield point must obviously be calculated using the specimen's length at the conclusion of the first cycle.) This procedure for raising the yield stress of a material is called *work-hardening*. With ever-increasing stress (Fig. 9.4(a)) the material continues to flow plastically, reaching a maximum stress at R, at which point the stress is called the *tensile strength* of the material, and ultimately to fracture at S. If the stress is defined as the tensile force divided by the *actual*, rather than the original cross-sectional area, the maximum at R

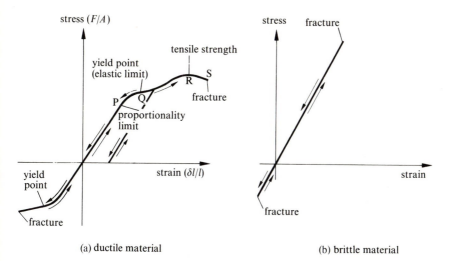

(a) ductile material   (b) brittle material

Fig. 9.4. The stress–strain curves of (a) a ductile material and (b) a brittle material. The exact form of the curves depends, of course, on the material in question. These curves are purely a qualitative indication of how these two classes of material behave.

usually disappears and the stress continues to increase until fracture occurs. Under compression (Fig. 9.4(a)), an initial elastic region is followed by plastic flow and finally fracture. When performed with great care the form of the stress–strain curves obtained under compression and extension are normally identical.

   Fig. 9.4(b) shows the behaviour of a *brittle* material like cast iron or glass. In such materials the elastic region is abruptly terminated when the specimen suddenly snaps. Some thermosetting plastics, notably the phenolics and melamines, behave in a completely brittle fashion. Yet other plastics, for example polythene, have a stress–strain curve which is qualitatively similar to Fig. 9.4(a). However, the vast majority of commercial plastics fall between the two extremes. These have an initial linear, or near-linear, stress–strain characteristic which gradually turns over towards a plateau.

   Whether a material is ductile or brittle the stress–strain characteristic always has an initial rise which is at least approximately linear. The slope of this portion is known as the *elastic*, or *Young's, modulus E*. Thus,

$$E = \frac{\text{stress}}{\text{strain}} = \frac{F/A}{\delta l/l}.$$

(9.17)

TABLE 9.2
Mechanical properties of solid materials

Material (at 293 K)	$K_T/10^9$ Pa	$E/10^9$ Pa	$G/10^9$ Pa	Tensile strength[a]$/10^6$ Pa
Aluminium	74	71	26	150–450
Copper	138	130	48	300–500
Iron (pure)	170	211	82	400–600
Lead	46	17	5·5	10–15
Platinum	230	160	60	300–400
Silver	104	76	30	300–350
Tin	58	45	18	20–40
Zinc	72	109	43	100–150
Brass	112	100	37	350–550
Steel (mild)	169	212	82	1000–1200
Glass (crown)	40	70	30	30–100

[a] The quoted tensile fracture stresses are for specimens in the form of rods (or wires).

Values of Young's modulus for a variety of different materials are given in Table 9.2. We will now try to predict E for a single crystal in terms of the interatomic force characteristic of the constituent atoms. We start by considering sodium chloride.

Young's modulus derived

Fig. 9.5(a) shows a section through a crystal of sodium chloride which is being subjected to a force F. If there are n rows of ions the force acting along each row is, of course, F/n. Denoting the row spacing by r_0 (which we will assume to be constant, independent of F) and the cross-sectional area by A, we have $n = A/r_0^2$ and so the force acting along each row is Fr_0^2/A. For the crystal to be at rest the support must provide an equal upward force F, that is, an upward force Fr_0^2/A along each row of ions (Fig. 9.5(a)).

We will now isolate a single pair of ions (Fig. 9.5(b)) and ask how their separation varies with Fr_0^2/A. In so doing we are deliberately ignoring forces from ions on neighbouring rows and from forces due to other than nearest neighbours along the row. The unstrained length of the crystal is, of course, $l = Nr_0$, where N (or rather $N+1$) is the number of ions along the row. If the interionic separation changes by δr under the influence of the force Fr_0^2/A—and it will change until, as shown in Fig. 9.5(b), the restoring force between the ions reaches Fr_0^2/A— the overall length of the crystal will change by $\delta l = N\,\delta r$. Therefore

$$\text{strain} = \frac{\delta l}{l} = \frac{N\,\delta r}{Nr_0} = \frac{\delta r}{r_0}. \tag{9.18}$$

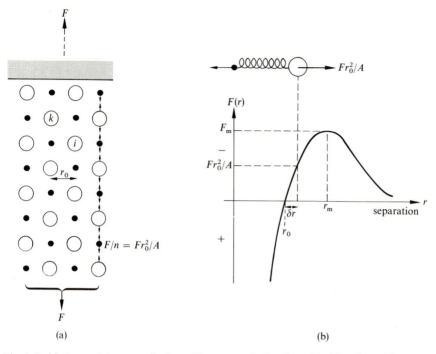

Fig. 9.5. (a) On applying a tensile force F to a crystal of sodium chloride a force F/n acts along each row of ions, where n is the number of rows of ions across the area A of the crystal. (b) At equilibrium this force will equal the restoring force between a pair of ions.

For small strains we can write $\delta r = (-dr/dF(r))\,\delta F$. The negative sign is necessary since the applied force acts in the opposite sense to the interionic force $F(r)$. Here $\delta F = F r_0^2/A$, so that eqn (9.18) becomes

$$\text{strain} = -\frac{F r_0}{A}\left(\frac{dr}{dF(r)}\right)_{r=r_0}.$$

Substituting this expression for strain along with that for the applied stress (F/A) into eqn (9.17) gives

$$E = -\frac{1}{r_0}\left(\frac{dF(r)}{dr}\right)_{r=r_0} \tag{9.19}$$

or, since $F(r) = -d\mathcal{V}(r)/dr$ (eqn (2.17)),

$$\boxed{E = \frac{1}{r_0}\left(\frac{d^2\mathcal{V}(r)}{dr^2}\right)_{r=r_0}.} \tag{9.20}$$

Adopting a Mie potential model for the interaction between a Na^+ and a Cl^- ion (eqn (2.29) with $A = e^2/4\pi\varepsilon_0$, $m = 1$, and $n = 9$) gives

$$\mathcal{V}(r) = \frac{e^2}{4\pi\varepsilon_0 r_0}\left(-\frac{r_0}{r} + \frac{1}{9}\frac{r_0^9}{r^9}\right) \tag{9.21}$$

$$\frac{d^2\mathcal{V}(r)}{dr^2} = \frac{e^2}{4\pi\varepsilon_0 r_0}\left(-\frac{2r_0}{r^3} + \frac{10r_0^9}{r^{11}}\right).$$

Substituting this equation with $r = r_0$ into eqn (9.20) leads to

$$\boxed{E = \frac{8e^2}{4\pi\varepsilon_0 r_0^4}} \ . \tag{9.22}$$

Since the nearest-neighbour separation is $2\cdot81\times10^{-10}$ m in sodium chloride, eqn (9.22) predicts $E = 2\cdot95\times10^{11}$ Pa. The measured value at temperatures close to $T = 0$ is $5\cdot8\times10^{10}$ Pa. The reason for the pronounced disagreement is that we have ignored cross-linking in the crystal, such as between ions i and k in Fig. 9.5(a). When these cross-links are considered the theoretical and measured values agree to within a factor of two. Had we chosen to consider, say, an fcc or hcp structure in Fig. 9.5(a) the need to consider cross-links would have been obvious from the start.

Experiments show that the value of Young's modulus of almost all materials decreases with increasing temperature. For example, it decreases in sodium chloride linearly with rising temperature from $5\cdot5\times10^{10}$ Pa at 140 K to $3\cdot0\times10^{10}$ Pa at 810 K. This decrease in E can be readily understood from Fig. 9.5(b). As a solid expands with increasing temperature so the mean interatomic separation r_0 must increase. Hence the point at which the slope $dF(r)/dr$ is evaluated in eqn (9.19) increases, moving to a region where the slope has a lower value.

Exercise 9.2

Make a rough estimate of the theoretical value of Young's modulus of solid argon. Assume a Lennard-Jones 6–12 potential with $\varepsilon = 1\cdot7\times10^{-21}$ J. The nearest-neighbour separation is $3\cdot7\times10^{-10}$ m.

Calculation. Proceeding in the same way as we did in calculating the Young's modulus of sodium chloride we may take over eqn (9.19) with $F(r)$ as given by eqn (2.24). This leads to

$$E = \frac{12\varepsilon}{r_0}\left(\frac{-7r_0^6}{r^8} + \frac{13r_0^{12}}{r^{14}}\right)_{r=r_0} = \frac{72\varepsilon}{r_0^3} \ .$$

Substituting $\varepsilon = 1\cdot7\times10^{-21}$ J and $r_0 = 3\cdot7\times10^{-10}$ m (comparing Fig. 8.18(a) with Fig. 9.5(a) may lead you to opt for a slightly different value of r_0) predicts $E = 2\cdot4\times10^9$ Pa. The measured value is $4\cdot8\times10^9$ Pa.

Comments. The reason why the theoretical and experimental values of E agree more closely in solid argon than in sodium chloride is that the interatomic forces are shorter-ranged in solid argon. Consequently, simplistic models which consider only nearest-neighbour interactions are likely to be more successful when used with short-range forces than with long-range forces. Finally, it is worth noting that the measured values of K_T and E are often of the same magnitude. Thus for argon $K_T = 2\cdot5 \times 10^9$ Pa and $E = 4\cdot8 \times 10^9$ Pa. Some indication of the generality of this similarity of K_T and E may be seen from Table 9.2. These similarities are to be expected since both moduli involve changes in the interatomic separations.

Speed of sound in a solid

Fig. 9.6(a) shows a view through a simple-cubic crystal. If we suppose the crystal to be close to $T = 0$, the atoms will (classically speaking) be at rest. If, in imagination, we now displace an entire plane AB of atoms sideways until it 'touches' a neighbouring plane CD of atoms and then release it, we know from eqns (3.44) and (3.46) that every atom of the displaced plane will oscillate back and forth with a period T_v given by

$$T_v = 2\pi \left(\frac{m}{2k_s}\right)^{1/2}, \qquad (9.23)$$

where m is the mass of each atom, and k_s is the force constant between an isolated pair of atoms. Consequently, the time t that it takes for the oscillating plane AB to move from a position midway between planes CD and EF to, say, the right-hand extremity of its motion is given by

$$t = \tfrac{1}{4}T_v = \frac{2\pi}{4}\left(\frac{m}{2k_s}\right)^{1/2}. \qquad (9.24)$$

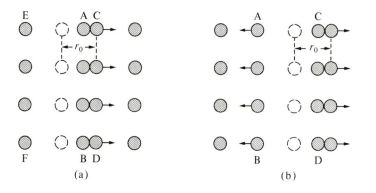

(a)   (b)

Fig. 9.6. Illustrating the mechanism whereby a longitudinal wave propagates in an elastic solid. (a) Plane AB oscillates back and forth, touching planes CD and EF at the extremities of its motion. Here planes CD and EF are fixed. (b) If plane CD is free to move, it will travel to the right after being struck by plane AB. This figure shows the situation at a time $\tfrac{1}{4}T_v$ after that shown in (a).

Up to now we have assumed that planes CD and EF are fixed (this is implicit in eqn (9.23); see §3.9). If we now relax this assumption while, rather inconsistently, retaining eqn (9.23), we see that after being struck by plane AB, plane CD will be set moving to the right. Thus in a time t given by eqn (9.24) the disturbance will have moved a distance r_0 equal to the mean separation between two adjacent planes (Fig. 9.6(a)). After a further interval of $\frac{1}{4}T_v$ plane CD will have struck the next plane (Fig. 9.6(b)), at a distance r_0 away, setting it into motion, and so on. Clearly, the speed v at which a disturbance propagates along the solid is given by

$$v = \frac{r_0}{t} = \frac{4r_0}{2\pi}\left(\frac{2k_s}{m}\right)^{1/2}.$$

(9.25)

Now it follows from eqn (9.19), with $-dF(r)/dr$ at $r = r_0$ replaced by the force constant k_s, that $E = k_s/r_0$. In addition, the mean density ρ of the solid shown in Fig. 9.6 is given by $\rho = m/r_0^3$. Hence eqn (9.25) can be written as

$$v = \frac{2^{3/2}}{\pi}\left(\frac{k_s/r_0}{m/r_0^3}\right)^{1/2}$$

or, approximating $2^{3/2}/\pi$ ($= 0{\cdot}90$) by unity,

$$\boxed{v = \left(\frac{E}{\rho}\right)^{1/2}}.$$

(9.26)

Although derived here by somewhat unsatisfactory arguments at the atomic level, eqns (9.26) can be derived using macroscopic elasticity theory. It predicts, for example, that the speed of sound along a copper bar ($\rho = 8{\cdot}9 \times 10^3 \text{ kg m}^{-3}$, $E = 1{\cdot}3 \times 10^{11}$ Pa) is $3{\cdot}8 \times 10^3 \text{ m s}^{-1}$, which is in close agreement with the speed measured at low frequencies. Eqn (9.26) is often used to deduce E from the speed at which sound propagates down a rod of the material; the speed can be found by measuring the frequencies required to set up longitudinal standing waves along the rod (see problem 9.6).

9.3. The tensile strength of a solid

The theoretical strength

Looking at Fig. 9.5 we see that a crystal should only break under tension when the force Fr_0^2/A along a row of atoms exceeds F_m, the maximum restoring force which can be provided by a pair of atoms (Fig. 9.5(b)). Equating Fr_0^2/A with $-F_m$ gives the tensile stress σ_{tc} ($= F/A$) at which fracture should occur as

$$\sigma_{tc} = -F_m/r_0^2. \tag{9.27}$$

Taking ubiquitous sodium chloride as our example we start by finding $F(r) = -d\mathscr{V}(r)/dr$. Since $\mathscr{V}(r)$ is given by eqn (9.21) we obtain

$$F(r) = -\frac{e^2}{4\pi\varepsilon_0 r_0}\left(\frac{r_0}{r^2} - \frac{r_0^9}{r^{10}}\right). \tag{9.28}$$

The maximum restoring force F_m (see Fig. 9.5(b)) occurs at $r = r_m$ where $dF(r)/dr = 0$. Applying this condition to eqn (9.28) gives

$$r_m = 5^{1/8}r_0 = 1\cdot22r_0. \tag{9.29}$$

The theoretical strain at which fracture should occur is therefore given by $(r_m - r_0)/r_0 = 0\cdot22$. Substituting $r = r_m = 5^{1/8}r_0$ into eqn (9.28) yields

$$F_m = -\frac{e^2}{4\pi\varepsilon_0 r_0^2}\frac{4}{5^{5/4}}. \tag{9.30}$$

Eqn (9.27) therefore predicts that

$$\sigma_{tc} = \frac{e^2}{4\pi\varepsilon_0 r_0^4}\frac{4}{5^{5/4}} \tag{9.31}$$

which, on substituting $r_0 = 2\cdot81 \times 10^{-10}$ m, works out at $1\cdot97 \times 10^{10}$ Pa. Alternatively, we may substitute for $e^2/4\pi\varepsilon_0 r_0^4$ as given by eqn (9.22) into eqn (9.31). This predicts that

$$\sigma_{tc} \approx E/15. \tag{9.32}$$

Inserting the measured value of E for sodium chloride predicts a fracture stress of 4×10^9 Pa. Other interatomic force characteristics give different constants in eqn (9.32). For example, a Lennard-Jones 6–12 potential model gives a fracture stress of approximately $E/27$ (see problem 9.9). In fact, one often assumes the following 'universal' relation:

$$\boxed{\text{fracture stress} \approx E/10} \tag{9.33}$$

for the theoretical fracture stress of a material.

The measured strength

The actual tensile fracture stress of a sodium chloride crystal as 'taken off the shelf' is around 1×10^8 Pa. This is approximately two orders of magnitude down on the theoretical value. The strain at which fracture occurs is correspondingly down on the theoretical value of $0\cdot22$. Most crystals as found in a bottle in a chemistry laboratory will break at a

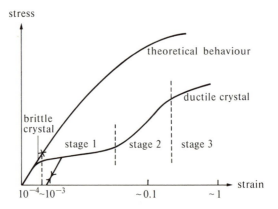

Fig. 9.7. In theory a crystalline solid should withstand a tensile strain of around 10^{-1} before fracture occurs. In practice a brittle crystal will fracture at a strain of about 10^{-4} to 10^{-3}; a ductile crystal at a strain of about 10^{-1} to 1. The strain is here defined as the tensile force divided by the true cross-sectional area. The ductile behaviour shown here is typical of that found with single crystals of fcc metals.

tensile stress which lies between about a hundredth and a thousandth of their theoretical value. This sort of discrepancy would seem to suggest that something is *drastically* wrong with our model; yet the level of agreement between the theoretical and experimental values of E would argue otherwise. Fig. 9.7 highlights the problem: the experimental and theoretical slopes of the stress–strain curve more or less agree, but the experimental and theoretical breaking stresses wildly disagree. This disagreement exists irrespective of whether the material is brittle or ductile. The irreversible behaviour of ductile materials poses an additional problem, which we shall presently consider.

9.4. Whiskers

Under appropriate conditions it is possible to grow whisker-like crystals. These are long filamentary single crystals of uniform cross-section whose length to diameter ratio may be a thousand or more. Sodium chloride, for example, normally crystallizes as small imperfect cubes; the familiar table salt. Slowly cooling the solution will produce a few whiskers. Adding a few per cent of polyvinyl alcohol (a large organic molecule) prior to crystallization induces the salt to crystallize exclusively as whiskers. Lengths of a centimetre or more are easily obtained.

Experimental techniques

It was found by Gordon and others in the 1950s that whiskers could be bent into very tight circles before they snapped. As Fig. 9.3(b) reminds

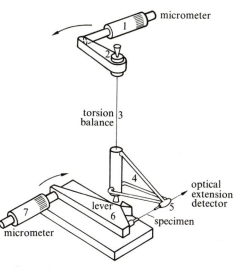

Fig. 9.8. A micro-tensile testing machine for obtaining the stress–strain characteristic of crystalline whiskers.

us, this implies a large breaking strain. Although Gordon was able to measure the strain by manipulating the whiskers under the microscope with a dissecting needle he was unable to measure the stress. This problem was overcome by Marsh who built a micro-tensile testing machine which enabled the stress–strain curve of a whisker to be measured directly. Fig. 9.8 shows the basic principles of the machine. A torsion head arm 2 transmits a torque via torsion wire 3 to the support 4, so applying a tensile force to the specimen. The tensile stress depends on the setting of the head arm 2 and this is accurately controlled by means of micrometer 1. As the specimen strains, the position of the specimen end-mount 5 will, of course, shift. Rather than attempt to measure the extension directly the mount 5 is restored to its original position by means of the lever 6 which is rotated by micrometer 7. An optical detector is used to ascertain that the mount is in its original position. Such is the versatility and the precision of the micro-tensile testing machine that it can apply loads of from 10^{-5} to about 4 N to specimens having cross-sections from 10^{-13} to $10^{-8}\,\mathrm{m}^2$, and lengths from 5×10^{-4} to $1 \times 10^{-2}\,\mathrm{m}$. It can measure extensions of from 5×10^{-10} to $1 \cdot 5 \times 10^{-2}\,\mathrm{m}$.

The properties of whiskers

Fig. 9.9 shows the stress–strain characteristic of a $2 \times 10^{-6}\,\mathrm{m}$ diameter whisker of silicon as measured with a micro-tensile testing machine. You will notice that although the crystal is elastic, Hooke's law is not

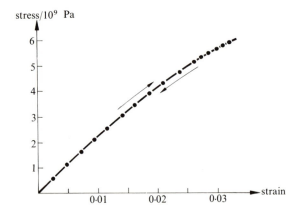

Fig. 9.9. Stress–strain characteristic of a silicon whisker. (Data: J. E. Gordon.)

obeyed. This non-linear behaviour is to be expected when we recall that, except for small displacements about the equilibrium separation, the interatomic force characteristic is non-linear (Fig. 2.6(b)). The measured strain of 0·05 at which fracture occurred is gratifyingly close to the theoretical fracture strain of 0·15 for silicon. Further evidence that when a whisker fractures the interatomic separation approaches the value at which the interatomic force is at its maximum value comes from the observation that the whisker will often fracture explosively, disappearing into a cloud of fine dust. As another example of their strength, a whisker of sodium chloride may withstand stresses of up to one-third of the theoretical fracture stress. This makes it considerably stronger than commercial steels, whose tensile strengths lie mainly in the range 2×10^8 to 7×10^8 Pa.

Steps and cracks

The strength of most whiskers depends on their diameter, decreasing from near theoretical values at diameters of less than about 10^{-6} m to the characteristic bulk value at diameters of about 10^{-4} m. Gordon and Marsh have explained this dependence in terms of stress concentrations which arise at steps on the surface of these near-perfect crystals. The first stage in the growth of a whisker is the formation of a fine smooth filament. Different layers, each of which may be a single layer of atoms thick, will advance at different rates; the exact rate of advance of a layer depends on how fast it is fed with atoms from the surrounding liquid or vapour. Those layers which are fed faster may catch up on more slowly moving lower layers, resulting in the formation of steps on the surface of the whisker. You might think that these steps would only lead to an increase in stress by virtue of a reduction in the cross-sectional area of the whisker. However, as Fig. 9.10(a) shows, the situation in which atoms throughout region ABCD are unstrained is inadmissible since it leads to mismatch between atoms in planes BC and EF. Such a mismatch would in fact lead to interatomic forces between atoms in planes BC and EF which would attempt to align the atoms in these planes. It is clear that much of the stress which is necessary to maintain region ABCD in a strained condition must be provided by those atoms near the root B of the step. This is illustrated very schematically in Fig. 9.10(b). Only in the vicinity of A is the stress relieved, with interatomic separations close to their unstrained value. Because of the stress concentration which occurs near the root B of the step, the stress in this region will attain the value at which fracture occurs long before the average stress throughout the whisker reaches this value. So steps can lead to premature failure of a whisker during a tensile test. It is also plausible that the greater the step height h (Fig. 9.10(b)) the greater will be the stress concentration at the tip; as h increases, the stress in more and more rows of atoms in region ABCD must be fed through a few atoms in the root. Marsh found that, in the case of silicon whiskers, the step height h was a constant *fraction*

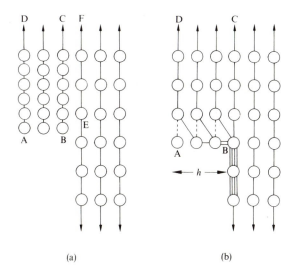

(a)   (b)

Fig. 9.10 (a) The presence of a step AB cannot lead to the complete removal of strain throughout region ABCD since this would introduce a shear strain between BC and EF (caused by the atoms in these planes being out of step). (b) In practice the stress necessary to maintain much of ABCD in a state of strain is provided by a relatively few atoms at the root B of the step.

$(\frac{1}{40})$ of the whisker diameter. Using these ideas he was able to account for the observation that the thicker the whisker the lower its strength.

Surface steps and cracks—a crack is also effective in concentrating stress—are not unique to whiskers. Most brittle fracture, as for example happens when a glass is dropped on the floor, can be traced to the spreading of such cracks. On a larger scale, many a ship has broken in two because the designer was unaware of the stress concentrating nature of such small apertures as hatchways. We will return to the role of cracks in §9.7.

9.5. The shear modulus

Definition

If a force F acts along the surface (area A) of a solid as shown in Fig. 9.11 we say that the solid is subjected to a shear stress, defined as F/A. The resulting shear strain is defined as $\delta x/h$ which, for small deformations, is approximately equal to the angle θ. The ratio of stress to strain, so defined, is called the *shear modulus G*:

$$G = \frac{\text{stress}}{\text{strain}} = \frac{F/A}{\delta x/h} \, . \tag{9.34}$$

As can be seen from Table 9.2, G is usually a factor of two or three smaller than K_T or E.

Fig. 9.11 shows the system of forces necessary to keep the sheared solid in equilibrium. To stop the solid from moving the mount must provide a force equal and opposite force to F. However, if these were the only two forces acting on the solid it would rotate. To prevent this an equal and opposite couple must be applied (Fig. 9.11).

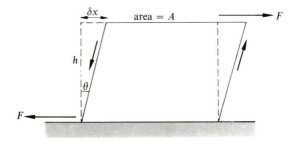

Fig. 9.11. Showing the shear strain $\delta x/h$ which results on applying a shear stress F/A to a solid.

Shear stresses produced by tensile stresses

Shear stresses can arise in rather unexpected ways. If we subject a solid to a tensile stress we necessarily alter the separation between cross-linked atoms (such as i and k in Fig. 9.5(a)). This, of course, introduces new stresses. At the macroscopic level we may resolve a tensile force F into components acting tangential and normal to the slice shown in Fig. 9.12. We see that the slice, which has a cross-sectional area of $A/\sin \phi$, where A is the cross-sectional area of the solid, is subject to a shear force of $F \cos \phi$, and thus to a shear stress σ_s, given by

$$\sigma_s = \frac{F}{A} \cos \phi \sin \phi = \frac{F}{2A} \sin 2\phi. \qquad (9.35)$$

This has a maximum when $\phi = 45°$. In fact whiskers of isotropic substances like glass do tend to fail along planes at 45° to the tensile stress. More generally, we can expect a single crystal subjected to an increasing axial tensile stress to fail along whichever plane the resolved shear stress first reaches the critical shear stress appropriate to that plane—provided the crystal does not first fail by cleaving along a plane perpendicular to the axial tensile stress, as discussed in §9.3. Most large metal whiskers fail in shear; whiskers of normally brittle materials (such as silicon) undergo tensile failure.

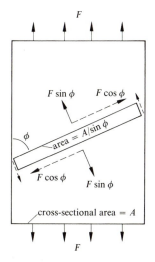

Fig. 9.12. A tensile force F resolves into a shear force $F \cos \phi$ acting parallel to the surface of the slice and into a tensile force $F \sin \phi$ acting normal to the slice.

Calculating the critical shear stress

In shearing a crystal we move, or rather we attempt to move, two planes of atoms past one another. If we assume that the planes remain individually undistorted the discussion becomes essentially that of how an atom (i in Fig. 9.13(a)) moves from one equilibrium position A to an adjacent equilibrium position B. Clearly the net restoring force F in the shear plane is zero at A and B since these are equilibrium positions. It is also clear, by symmetry, that F will be zero at the midpoint between A and B. (At this point the component of the force which j exerts on i in the shear plane will be equal and opposite to that of l on i.) Symmetry also demands that whatever be the form of F from $x = 0$ to $x = b/2$ the form from $x = b/2$ to $x = b$ will be equal in magnitude but opposite in direction. Here b is the interatomic separation in the shear direction. So the restoring force must be a periodic function of x with period b. Following Frenkel, we will assume F to be sinusoidal, this being the simplest function with the desired properties. As a consequence the shear stress σ_s will be related to x by

$$\sigma_s = C \sin(2\pi x/b), \tag{9.36}$$

where C is a constant whose value can be determined by noting that, at

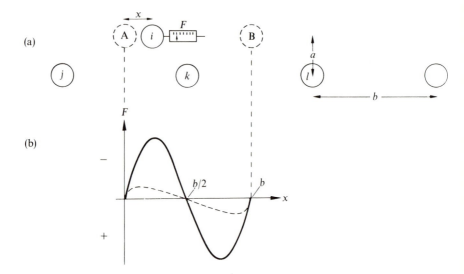

Fig. 9.13. (a) In shearing a solid we displace a plane of atoms (of which only one atom i is shown) by an amount x relative to the underlying plane of atoms (j, k, l, etc). (b) A thought experiment in which we displace atom i from one equilibrium site, A, to an adjacent equilibrium site, B, recording the restoring force on i as we shift it. The full curve is a sinusoidal approximation. The dotted curve is a more realistic approximation.

small values of x, eqn (9.36) reduces to

$$\sigma_s = C\frac{2\pi x}{b} = \frac{C2\pi a}{b}\left(\frac{x}{a}\right).$$

(9.37)

Since x/a is the strain, eqn (9.34) gives $G = C2\pi a/b$, that is $C = Gb/2\pi a$. Hence eqn (9.36) becomes

$$\sigma_s = \frac{Gb}{2\pi a}\sin\left(\frac{2\pi x}{b}\right).$$

(9.38)

The critical shear stress σ_{sc} at which one plane of atoms will slip over the adjacent plane of atoms is given by the maximum value of eqn (9.38), namely

$$\sigma_{sc} = \frac{Gb}{2\pi a}.$$

(9.39)

Since a and b have approximately the same value in most crystal structures a crystal should fracture when the shear stress reaches $G/2\pi$, where G is the shear modulus measured at small strains. We could equally well conclude that a solid should fracture when the shear strain (given by σ_{sc}/G; see eqn (9.34)) reaches $(G/2\pi)/G = 1/2\pi = 0.16$. This is several orders of magnitude greater than the shear strain at which most ordinary materials fail.

A small part of the discrepancy between the theoretical and experimental shear strengths of a solid lies in our assuming a sinusoidal force characteristic (Fig. 9.13(b)). Introducing the true interatomic characteristic between atom i and atoms j, k, l, etc. leads to the dotted curve shown in Fig. 9.13(b). A Lennard-Jones 6–12 potential, for example, predicts a critical shear strain of 0.03. Although a step in the right direction, the theoretical critical strains are still orders of magnitude greater than the measured values. Certain whiskers alone live up to expectations.

Slip

We have seen (Fig. 9.4(a)) that a ductile material subjected to a tensile stress undergoes an irreversible increase in length before it fractures. We have also seen that a tensile stress has a shear component (Fig. 9.12). These two observations are closely related.

When a single crystal undergoes plastic deformation (as does copper, for example) it is found that the crystal divides itself into regions which have slipped with respect to one another (Fig. 9.14(a)). If the surface of the deformed crystal is now examined with a low-power optical microscope it will be seen to be crossed by what are known as *slip lines* (Fig.

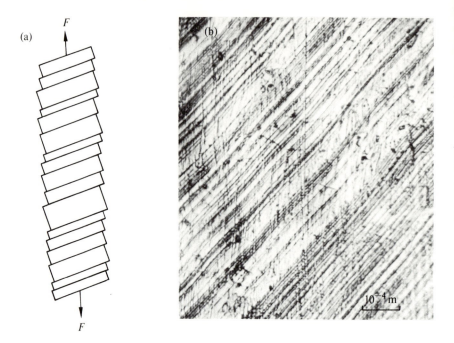

Fig. 9.14. (a) On stretching a single crystal rod of a ductile material it divides itself up into bands which have slipped with respect to one another. (b) Slip lines on the surface of a Cu-10 per cent Al single crystal which has been strained by approximately ten per cent in a tensile test. (Photograph by courtesy of I. Woodgate and Dr G. D. W. Smith.)

9.14(b)). What appears as a slip line is in reality a *slip band* made up of several *slip planes*, a few hundred atomic distances apart. Slip (or glide) planes are often densely-packed crystal planes; this might be expected since the denser the packing the farther apart will adjacent planes be. Face-centred cubic metals, for example, slip on {111} planes. Sodium chloride slips on {110} planes. When all planes are equally densely packed, as they are in isotropic materials, slip occurs in a plane at an angle of 45° to the tensile stress. In view of the arguments summarized in Fig. 9.12, slip *per se* is not unexpected. What is unexpected is that it normally occurs at strains many times lower than the predicted value.

9.6. Dislocations

The edge dislocation

In imagination, let us apply a shear stress to a simple cubic crystal, as shown in Fig. 9.15(a), such that the row *AB* of atoms at the left-hand end of the crystal have slipped by one lattice period *a* and therefore lie

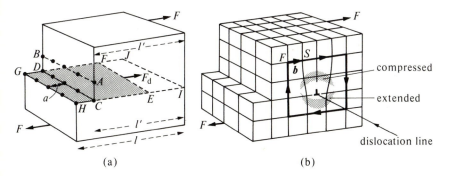

Fig. 9.15. (a) The applied shear stress has displaced the atoms in row *AB* so that they lie over the atoms in row *CD*. The shaded plane is called the slipped plane; line *EF* the dislocation line. (b) Showing a circuit used to define the Burger's vector *b* of the dislocation.

directly over the row *CD* of atoms. Since the atoms at the right-hand end of the crystal are unslipped it follows that there must exist some line *EF* such that the slip occurring to the left of *EF* is greater than $\frac{1}{2}a$ while the slip occurring to the right of *EF* is less than $\frac{1}{2}a$. Line *EF* is called the *dislocation line*; plane *EFGH* the *slipped plane*. If we move a distance *l'* through the upper portion of the crystal (Fig. 9.15(a)) we must clearly pass through one more plane of atoms than if we move the same distance through the lower portion of the crystal. In an edge dislocation the dislocation line marks the location at which this extra half-plane of atoms is squeezed into the upper portion of the crystal. The nature of the dislocation (denoted by a T-shaped symbol) is brought out more clearly in Fig. 9.15(b).

In describing a dislocation it is customary to follow a set of agreed procedures. One first defines a *sense* to the dislocation line. The sense is arbitrary; we shall suppose that the line runs into the paper in Fig. 9.15(b). Applying the right-hand screw convention we now make an atom-to-atom path which encloses the dislocation but which passes entirely through 'good' regions of the crystal. The path consists of equal numbers of atomic spacings along opposite sides of the circuit (Fig. 9.15(b)). Because of the dislocation the path will not close. A vector drawn from the finishing point *F* to the starting point *S* is called the *Burger's vector b*. (To be precise, the vector so-defined is called the *local Burger's vector*; the so-called *true Burger's vector* is the subject of problem 9.14.) Since we have taken a path whose sense is determined by that of a right-handed screw advancing along the dislocation line and have joined the finishing point to the starting point we may say that we have followed a *FS/RH* convention. (Some writers define *b* via a *SF/RH*

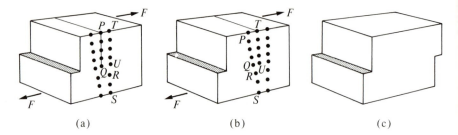

Fig. 9.16. Showing how an edge dislocation moves. (a) *PQ* labels one end of the extra half-plane of atoms. (b) On applying a shear stress the continuous plane *PQRS* forms and *TU* becomes the extra half-plane. (c) By such means the half-plane is displaced to the surface.

convention.) For an edge dislocation **b** lies perpendicular to the dislocation line (Fig. 9.15(b)).

Fig. 9.16 shows how an edge dislocation moves under the influence of an applied shear stress. It takes only a small stress to cause the extra half-plane of atoms labelled *PQ* in Fig. 9.16(a) to line up with *RS*, making *TU* the extra half-plane (Fig. 9.16(b)). This process continues until the extra half-plane arrives at the surface of the crystal (Fig. 9.16(c)); slip has occurred, and in a direction parallel to that of the Burger's vector. Returning to Fig. 9.15, we usually call plane *GHIJ* the *slip plane*. It indicates the plane on which slip *will* occur; not to be confused with the slipped plane *EFGH*, where slip has already taken place. The motion of an edge dislocation has been likened to the motion of a ruck or wrinkle across a carpet; it takes a lower force to move the ruck than it takes to move the entire carpet.

The screw dislocation

Fig. 9.17 shows how a screw dislocation can be formed. In imagination, we first draw in the plane shown in Fig. 9.17(a). Next we distort the crystal so that the portion lying on one side of the plane moves up while that lying on the opposite side of the plane moves down. When the slip in regions remote from the dislocation line reaches one lattice spacing the atoms will once more be in registration (Fig. 9.17(b)) and we say that a screw dislocation has been created. Assuming the sense of the dislocation to run from the top to the bottom of the crystal in Fig. 9.17(b) we can readily determine the Burger's vector **b** (using the *FS/RH* convention); in the particular case shown in Fig. 9.17(b) it has a magnitude equal to the lattice period *a*. Note that **b** lies parallel to the dislocation line. The circular symbol used to denote a screw dislocation indicates the direction

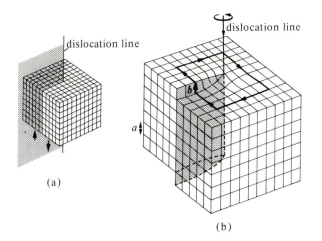

Fig. 9.17. (a) Showing how a screw dislocation may be formed. The shaded plane is drawn as shown in a perfect crystal and the crystal is then distorted in the directions given by the arrows until the atomic planes remote from the dislocation line are again in register. (b) Showing the Burger's vector **b**, the dislocation line, and the slipped plane of the dislocation.

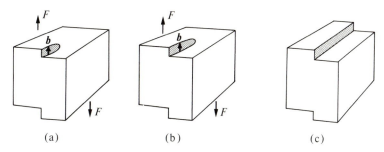

Fig. 9.18. (a) A screw dislocation with Burger's vector **b**. (b) Under the influence of the applied shear stress the disclocation moves in a direction perpendicular to **b**. (c) Continued movement of the dislocation procedures slip in the crystal.

that a right-handed screw will rotate when advancing along the Burger's vector.

Fig. 9.18 illustrates how a screw dislocation moves under the influence of a shear stress to produce a step which extends the length of the surface and which represents slip in the crystal. It is clear from this figure that the dislocation line of the screw dislocation moves in a direction perpendicular to that of the Burger's vector.

The mixed dislocation

Dislocations are not restricted to the edge and screw varieties. A so-called *mixed dislocation* has some edge character and some screw character. To see how a mixed dislocation can arise we imagine ourselves stressing a crystal until it has slipped over the plane shown shaded in Fig. 9.19(a). The vector **b** denotes the magnitude of the slip. As before, the dislocation line separates the slipped from the unslipped region of the crystal. It is clear that the dislocation has pure screw-character at only one point (where the dislocation line emerges from the left-hand face) and has pure edge-character at only one point (on the adjoining face). At all other points on the line the dislocation is part edge and part screw.

Up to now we have only considered dislocations which terminate on a crystal's surfaces (Figs. 9.15(b), 9.17(b), and 9.19(a)). However, it is possible to form a dislocation loop by 'joining together' four crystals, each containing a quarter loop. Fig. 9.19(b) shows how this can be done. The resulting single crystal will contain a closed loop which can be shown shaded to indicate that—except very close to the dislocation line—the atoms above this plane have slipped by an amount **b** relative to the atoms below this plane. Clearly, the dislocation loop so formed will have pure edge character at only two points and pure screw character at only two points. Fig. 9.20(a) and (b) shows a plan-view and a cross-sectional view, respectively, of dislocation loops in silicon, as taken with a transmission electron microscope. The vertical segments of the loops shown in Fig. 9.20(a) are pure screw dislocations while the remaining segments are of mixed character, with the Burger's vector making an angle of 60° to the dislocation line. This follows immediately since **b** (which lies along the screw direction) is constant along the slipped plane. It is not difficult to

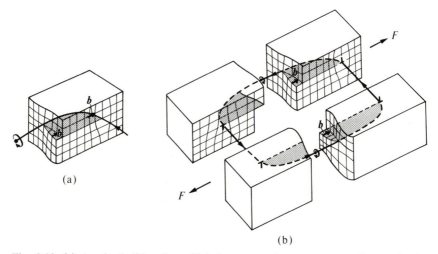

(a)

(b)

Fig. 9.19. (a) A mixed dislocation which has screw character on one face and edge character on the adjoining face. The arrow on the dislocation line indicates its sense. (b) Showing how a planar dislocation loop can be formed out of four quarter loops.

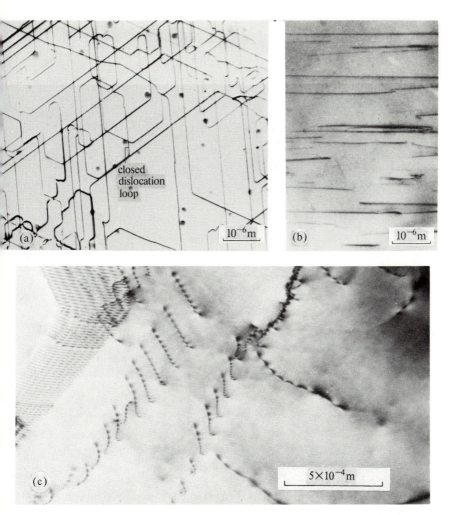

closed
dislocation
loop

(a) 10^{-6}m

(b) 10^{-6}m

(c) 5×10^{-4}m

Fig. 9.20. (a) A plan view and (b) a cross-sectional view of dislocations in silicon (doped with 5×10^{18} parts m^{-3} of boron) which had been subjected to high stress deformation. One complete dislocation loop is shown arrowed in (a). The larger loops are completed outside the field of view. (Transmission electron micrographs by courtesy of Drs. A. Ourmazd and H. Alexander.) (c) Dislocation arrays in a thin aluminium foil. The lines appear short since they terminate on the upper and lower surfaces of the foil. The oscillating contrast seen along a line is an interference effect arising from diffraction of the transmission microscope's electron beam by the crystal lattice. Dislocations can, as here, lower their energy by arranging themselves in a regular array. (Transmission electron micrograph by courtesy of Dr. C. Rae.)

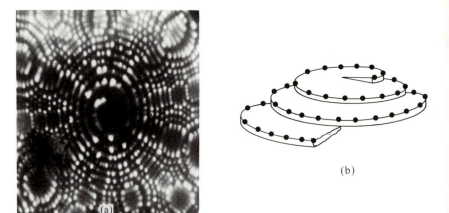

Fig. 9.21. (a) A field ion micrograph showing the surface of a single crystal of tungsten. The white dots show individual tungsten atoms. (Photograph courtesy of Drs. T. J. Godfrey and G. D. W. Smith). (b) Showing the nature of the crystal's surface in the vicinity of the dislocation (which emerges from the surface of the crystal at the centre of the spiral). The locations of those atoms which show up in the micrograph are also indicated. In field ion microscopy a very fine metal needle (here of tungsten) with a hemispherical tip is cooled with liquid helium and held at a high positive potential relative to a distant fluorescent screen. Very low-pressure helium gas is present. The high electric field gradients which exist in the vicinity of the *exposed* atoms (such as those shown in Fig. 9.21(b)) lead to local ionization of the helium gas atoms. The resulting positive helium ions are then accelerated towards the fluorescent screen. On striking this screen they produce the type of image shown in Fig. 9.21(a).

argue that the effect of applying the shear stress indicated in Fig. 9.19(b) will be to cause the dislocation loop to expand until it disappears from the crystal. When this happens the crystal will have slipped by an amount b. It is worth pointing out that a hydrostatic pressure will not cause dislocations to move. This explains why materials which undoubtedly contain dislocations nevertheless have values of K_T which lie close to the predicted values.

Electron microscope studies require thin specimens and thus electron micrographs frequently show dislocations as short lines terminating on the upper and lower surfaces of the specimen. Fig. 9.20(c) shows such dislocations in aluminium. Dislocations also show up in studies made using the field ion microscope (an instrument capable of resolving individual atoms). Fig. 9.21(a) shows a field ion micrograph of the surface of a single crystal of tungsten. This crystal contains a dislocation line which emerges from the crystal at the centre of the spiral pattern. Fig. 9.21(b) shows how the image seen in Fig. 9.21(a) may be produced.

Energy of dislocations

As can be seen from Fig. 9.15(b), part of the crystal near an edge dislocation is extended, and part is compressed, from its normal equilibrium condition. Since energy is required both to increase and to decrease the interatomic separation from its equilibrium value (recall Fig. 8.17) it follows that more energy is associated with the edge dislocation than is associated with the same volume of a perfect crystal. The same is true of a screw dislocation. Fig. 9.22(a) shows a crystal of radius R containing a single screw dislocation of Burger's vector b. Straightening out the region of the crystal lying between radii r and $r + dr$ (Fig. 9.22(b)) we see that it is strained by an amount $b/2\pi r$. Now the elastic energy present per unit volume is given by $\frac{1}{2}$stress $\times$ strain (see problem 9.8) or, since stress $=$ $G \times$ strain (eqn (9.34)), by $\frac{1}{2}G$(strain)$^2 = Gb^2/8\pi^2 r^2$. Therefore the energy stored within the cylindrical shell of radii r and $r + dr$ and of length L, whose volume is $L2\pi r\,dr$, is given by $LGb^2\,dr/4\pi r$. Integrating from $r = b$ (which we shall take to be equal to the radius of the 'core' of the dislocation) to $r = R$ gives the elastic energy $\mathscr{E}_1$ stored per unit length of screw dislocation as

$$\mathscr{E}_1 = \int_b^R \frac{Gb^2}{4\pi r}\,dr$$

$$\boxed{\mathscr{E}_1 = \frac{Gb^2}{4\pi} \ln\left(\frac{R}{b}\right)}.$$

(9.40)

If many dislocations are present, R can reasonably be taken to be half the

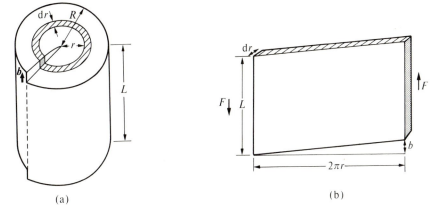

(a)

(b)

Fig. 9.22. (a) A screw dislocation of length L lying along the axis of a crystal of radius R. (b) The region of the crystal lying between radii r and $r + dr$ shown straightened out.

distance between neighbouring dislocations. Of course, had we chosen to integrate from $r = 0$ to $r = R$ the energy would have become infinite! It is difficult to give a rigorous defence of our choice of $r = b$ as the lower limit of integration. However it does seem reasonable since the strain falls towards zero at $r < b$ (see Fig. 9.22(a)); suggesting that the energy stored in the core is much less than the energy stored between $r = b$ and $r = R$.

The calculation of the energy of an edge dislocation is more complicated. When the calculations are made an expression similar to eqn (9.40) is obtained (but multiplied by a factor of order unity). This is not unexpected since, were $\mathscr{E}_1$ quite different in the two cases, one type of dislocation would—contrary to observation—occur overwhelmingly more frequently than the other type.

The quantity R which occurs in eqn (9.40) can be readily found from the *dislocation density* ρ, which may be defined as the average number of dislocation lines passing through a randomly oriented plane of unit area lying within the material (see also problem 9.15). The value of ρ can be directly determined from electron micrographs, such as that shown in Fig. 9.20(c). In annealed crystals ρ is of order $10^{10} \, \mathrm{m}^{-2}$; in work-hardened materials it may rise to $10^{16} \, \mathrm{m}^{-2}$. Adopting a mean value of $10^{13} \, \mathrm{m}^{-2}$ we see that each line passes through an area of $10^{-13} \, \mathrm{m}^2$ and hence is separated from its nearest neighbour by $(10^{-13})^{1/2} \, \mathrm{m} = 3 \cdot 2 \times 10^{-7} \, \mathrm{m}$. Substituting $R = \frac{1}{2}(3 \cdot 2 \times 10^{-7} \, \mathrm{m})$ and $b = 2 \times 10^{-10} \, \mathrm{m}$ (a typical value) into eqn (9.40) gives $\mathscr{E}_1 = 0 \cdot 53 G b^2$. Adopting $\rho = 10^{10} \, \mathrm{m}^{-2}$ leads to $\mathscr{E}_1 = 0 \cdot 80 G b^2$, whereas $\rho = 10^{16} \, \mathrm{m}^{-2}$ gives $\mathscr{E}_1 = 0 \cdot 26 G b^2$. If we therefore write

$$\boxed{\mathscr{E}_1 = G b^2} \, , \qquad (9.41)$$

we shall obtain order-of-magnitude correct estimates of the energy per unit length of a dislocation. By way of example, when $G = 4 \times 10^{10} \, \mathrm{Pa}$ (a typical value for a metal; see Table 9.2) and $b = 2 \cdot 5 \times 10^{-10} \, \mathrm{m}$, then $\mathscr{E}_1 = 2 \cdot 5 \times 10^{-9} \, \mathrm{J \, m}^{-1}$.

Because of the extra elastic energy associated with unit length of a dislocation there will always be a tendency for a dislocation to minimize its overall length. The analogy of a stretched piece of rubber readily comes to mind. To find the *line tension T* of a dislocation we imagine ourselves stretching the dislocation so as to increase its length by δl. The energy thereby transferred to the dislocation during this operation is $T \, \delta l$. Equating this to the increase in the stored energy, namely $\mathscr{E}_1 \, \delta l$, gives $T = \mathscr{E}_1$. Thus, from eqn (9.41),

$$\boxed{T = G b^2} \, . \qquad (9.42)$$

The force on a dislocation

Fig. 9.15 reminds us how a shear force F can cause a dislocation to move. As a consequence, there must be some force F_d acting on the dislocation proper. To find the relationship between F_d and F we look, from two different viewpoints, at what happens as the crystal slips by an amount b across the slip plane $GHIJ$ shown in Fig. 9.15(a). Here b is the magnitude of the Burger's vector of the single dislocation lying in the crystal (Fig. 9.15(b)). From a microscopic viewpoint we may state that a force F_d moved the dislocation through a distance l, equal to the length of the crystal, thereby doing work $F_d l$. From a macroscopic viewpoint we may state that a shear force F caused the crystal to slip by an amount b, so doing work Fb. Since the same work must be done in both cases we have

$$F_d = Fb/l$$

or, in terms of the shear stress $\sigma_s = F/lw$, where w is the width of the crystal (equal to the length L of the dislocation line in Fig. 9.15(a)),

$$\boxed{F_d = \sigma_s bL} \; . \tag{9.43}$$

This tells us the force experienced by an edge dislocation of length L, with a Burger's vector b, when an external shear stress σ_s is applied to the crystal.

Sources of dislocations

A well-annealed single crystal contains a dislocation density ρ of around $10^{10} \, \mathrm{m}^{-2}$. Assuming that all these dislocations are free to move under the influence of an applied stress we can readily make a rough calculation of the maximum strain to be expected in such a crystal. To do so we shall suppose that slip occurs on a plane making an angle ϕ to the crystal's axis (as defined in Fig. 9.12) and that the crystal has a square cross-section of side l. It follows from this that the slip plane contains $\rho l^2/\sin \phi$ dislocations, where ρ is the dislocation density. If each dislocation has a Burger's vector of magnitude b then the slip will be limited to $\rho l^2 b/\sin \phi$, corresponding to a tensile strain of $\rho l^2 b(\cos \phi/\sin \phi)/\mathscr{L}$, where $\mathscr{L}$ is the length of the specimen. Taking $\rho = 10^{10} \, \mathrm{m}^{-2}$, $l = 3 \times 10^{-3} \, \mathrm{m}$, $b = 2{\cdot}5 \times 10^{-10} \, \mathrm{m}$, $\phi = 45°$, and $\mathscr{L} = 1 \times 10^{-2} \, \mathrm{m}$ gives a strain of $0{\cdot}2$ per cent. Since single crystals of ductile materials can slip by some tens of per cent before failing, we are forced to conclude that extra dislocations must be produced during deformation.

Fig. 9.23 shows a source of extra dislocations in a crystal. It is known as a Frank–Read source and consists of a dislocation which is pinned at

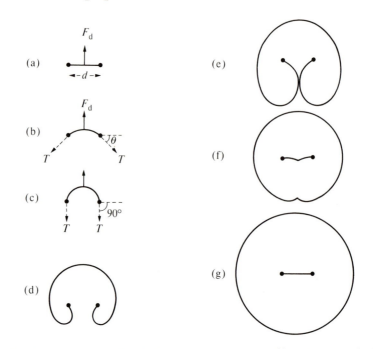

Fig. 9.23. A planar view of a Frank–Read source in operation. (a) A dislocation is pinned at two sites a distance d apart. The applied shear stress lies in the plane of the paper normal to d. In (b) to (c) the dislocation bows out under the influence of F_d but remains in equilibrium. If the stress is further increased, the dislocation spontaneously passes through the sequence shown in (d)–(g) leading to the formation of a dislocation loop and a new pinned dislocation.

two points a distance d apart. When an external shear stress σ_s is applied perpendicular to the dislocation, it will (see eqn (9.43)) experience a force of F_d (Fig. 9.23(a)) causing the dislocation to bow out as shown in Fig. 9.23(b). For the dislocation to be in equilibrium we must have

$$F_d = 2T \sin \theta$$

or, substituting for F_d from eqn (9.43) with $L = d$ and for T from eqn (9.42),

$$\sigma_s = \left(\frac{2Gb}{d}\right) \sin \theta. \qquad (9.44)$$

This relation implicitly assumes that eqn (9.43), with L given by the *separation of the pinning sites*, applies even though the dislocation has bowed out. This assumption can be justified by redrawing the bowed dislocation in stepwise fashion so that it consists of short lengths of edge dislocation (lying parallel to the original unbowed edge dislocation),

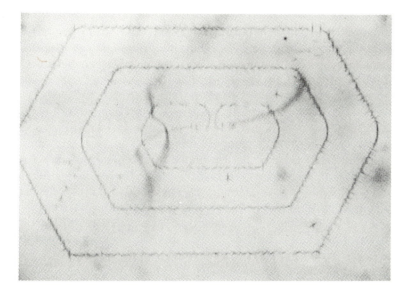

Fig. 9.24. A Frank–Read source operating in a single crystal of silicon.

alternating with short lengths of screw dislocation (lying perpendicular to the original dislocation). Now the forces on the screw elements all lie perpendicular to the applied stress, and thus need not concern us. We are therefore left with the forces on the edge elements, and these add up to F_d as given by eqn (9.43).

Eqn (9.44) tells us that the stress will be a maximum when $\theta = 90°$ (Fig. 9.23(c)). Once this stage is reached the dislocation will *spontaneously* pass through the stages shown in Fig. 9.23(d)–(g) leading to the formation of one dislocation loop *plus* a dislocation which is again anchored at the pinning sites. The applied stress will then cause this loop to expand outwards in the manner already discussed on p. 300, while the source produces yet another loop, and so on. A beautiful example of a Frank–Reid source in action in silicon is shown in Fig. 9.24.

The stress–strain characteristic

Returning to Fig. 9.7 we can now account for the main features of the stress–strain characteristic of a single crystal of a ductile material like copper. At the very lowest stresses (those that produce strains of up to about 2×10^{-6} in the case of a copper single crystal) the existing dislocations do not move, and the crystal behaves in a strictly elastic manner. Once these dislocations start to move out of the crystal a small permanent strain will be produced. The yield point roughly corresponds to the point at which dislocation sources become operative. During stage 1 of the deformation, characterized by a quasi-linear stress–strain characteristic (see Fig. 9.7), the sources produce an abundant supply of dislocations. Some of these become interlocked with one another, so that the stress required for further deformation gradually increases. Throughout stage 1 slip occurs on one set of parallel planes.

At the end of stage 1 the applied stress is high enough for the critical resolved shear stress to be exceeded on other potential slip planes. With slip planes crossing there is plenty of opportunity for dislocations to pile up and to intersect to form hard-to-move tangles. This accounts for the rising slope which characterizes stage 2 of the deformation (Fig. 9.7). During stage 3 the crystal continues to work-harden but at a decreasing rate. The decreasing slope is attributable to the fact that some of the obstacles to dislocation slip can now be bypassed. Thus a dislocation at the head of a pile-up can avoid any obstacle which is impeding its progress by 'cross-slipping' on to another plane. This cross-slipping is only significant at large stresses.

Hardening materials

An important part of a metallurgist's work is to know how to convert the relative weakness of a single crystal of a metallic element (with a tensile strength lying in the range 10^6 to 10^7 Pa) into the strength required of engineering materials (where tensile strengths of 10^8 to 10^9 Pa may be demanded). These techniques rely for their success on reducing the number of dislocations present and/or restricting their movement. The first of these criteria is satisfied by whiskers, which are normally either dislocation-free or which contain a single screw dislocation along their central axis (but which is unable to move under an axial tension). Whiskers have found limited use in isolation, although they are used as fibrous reinforcements of composite materials. In certain solids (known as *inherently strong solids* and typified by diamond and tungsten carbide) the dislocations present cannot be moved at stresses much below the theoretical shear strength.

With mass-produced metals one must normally look for ways of restricting the movement of dislocations. As we have just seen this may be done by tangling the dislocations by the process of work-hardening. This can be accomplished by, for example, cold-drawing the metal through a die. A more everyday example of a work-hardening process is the repeated bending back and forth of a paper clip. An obvious way of impeding the motion of dislocations is to add alien ('impurity') atoms to the metal. These will serve as pinning centres; eqn (9.44) reminds us that the smaller the separation d of the centres the greater will be force required to move the dislocation. One way of introducing these pinning centres is to add a controlled amount of impurity and then to disperse it throughout the material. This process is known as *dispersion hardening*, and it may be brought about in several ways. As an example, aluminium oxide particles may be dispersed in aluminium by heating particles of the oxide and the metal together at temperatures close to, but below, the melting point. In the process known as *precipitation hardening*, an alloy is rapidly cooled (quenched) from a high temperature to a temperature at which it would, given time, separate out into its separate solid phases. This supersaturated solid solution will over a long period partially separate out, thereby introducing pinning centres. The process may be accelerated by gently reheating the solid solution. Aluminium–copper alloys may be treated in this way. A final technique for strengthening a metal is to produce it in a polycrystalline form rather than as a single crystal. Grain boundaries can be effective barriers to dislocation movement. As an example, a polycrystalline copper rod can have a tensile strength some five times that of the same material in the form of a single crystal.

9.7. Brittle versus ductile failure

The surfaces of virtually all materials—other than of certain types of whisker—contain many small steps and cracks. As we saw in §9.4 these defects act as stress concentrators (Fig. 9.10). It is therefore natural to ask why most materials do not fail in brittle fashion by the growth of these defects across the body of the specimen. The first person to offer a convincing answer to this question was A. A. Griffith. He pointed out (in 1921) that as a crack grows so its surface area, and therefore its surface energy, increases but that the elastic energy stored in the vicinity of the enlarging crack decreases. If the magnitude of the decrease in the elastic energy exceeds the energy required to create the new surface, we can expect the crack to grow, leading to brittle failure. If the converse is true, there should be no tendency for the crack to grow, leaving the way open for ductile failure.

In quantifying these ideas we start by considering a strained slab of material held between fixed grips (Fig. 9.25(a)). Denoting the tensile stress of the material by σ_t ($=F/A$), the total elastic strain energy stored within the volume Al of the slab is given by $(\frac{1}{2}\sigma_t^2/E)Al$. This is obtained by evaluating the area under the force-extension curve shown in Fig. 9.25(b); the extension $\delta l = (\sigma_t/E)l$ being related to the tensile force $A\sigma_t$ via the definition of E (eqn (9.17)). Hence the elastic energy $\mathscr{E}_v$ stored per unit volume of the solid is given by

$$\mathscr{E}_v = \tfrac{1}{2}\sigma_t^2/E. \tag{9.45}$$

Next we introduce an elliptical-shaped crack of major axis $2a$ (which

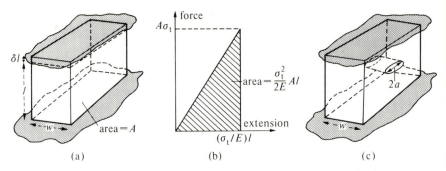

Fig. 9.25. (a) An elastic solid of cross-sectional area A is held between fixed supports so that it is strained to a value of $\delta l/l$. (b) Showing how the elastic-strain energy is calculated. (c) Introducing the crack shown lowers the total elastic-strain energy stored in the solid.

we shall take to be very much greater than the minor axis) running the full width w of the slab. Its effect will be to lower the strain energy stored within the vicinity of the crack. The strain energy stored remote from the crack is unchanged. We might reasonably guess that the effect of the crack is totally to remove the elastic strain from within a cylinder of radius a and length w. Since this cylinder has a volume of $\pi a^2 w$ it follows from eqn (9.45) that the effect of introducing the crack is to reduce the elastic energy stored in the solid by an amount $(\frac{1}{2}\sigma_t^2/E)\pi a^2 w$. An exact

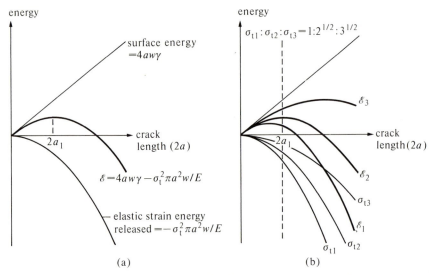

Fig. 9.26. Showing how the decrease in the elastic-strain energy and the increase in the surface energy, separately and summed, depend on the length $2a$ of a crack in the solid. In (a) a single stress is considered. In (b) the effect of three different stresses, σ_{t1}, σ_{t2}, and σ_{t3} is considered. Here $\mathscr{E}_1$, $\mathscr{E}_2$, and $\mathscr{E}_3$ represent the corresponding total energies. The material is brittle under stress σ_{t1} and ductile under stress σ_{t3}.

calculation made by integrating the energy stored throughout the strain field gives double this value for the energy reduction, namely $\sigma_t^2 \pi a^2 w/E$.

Finally we calculate the energy associated with the new surfaces created by introducing the crack. Since these surfaces have an area of close to $2 \times 2aw$ it follows that this surface energy is $4aw\gamma$, where γ is the free surface energy (see §8.7).

Fig. 9.26(a) shows the decrease in the elastic-strain energy and the increase in the surface energy, plotted separately and combined, as a function of the crack length $2a$. We see immediately that short-length cracks only serve to increase the total energy $\mathscr{E}$. Any pre-existing short cracks will therefore have no tendency to grow. On the other hand it is energetically favourable for pre-existing long cracks to grow. The critical crack length $2a_1$ at which instability sets in occurs when the rate of decrease of elastic-strain energy equals the rate of increase of surface energy. This occurs at the maximum in $\mathscr{E}$, where

$$\frac{d\mathscr{E}}{da} = \frac{d}{da}\left(4aw\gamma - \frac{\sigma_t^2 \pi a^2 w}{E}\right) = 0$$

giving

$$2a_1 = \frac{4\gamma E}{\pi \sigma_t^2} \tag{9.46}$$

as the critical crack length for a given stress σ_t in a material of Young's modulus E and free surface energy γ. Alternatively, if we consider a particular crack, of length say $2a_1$, and examine the effect of changing the applied tensile stress σ_t, as in Fig. 9.26(b), we see that when σ_t is such that $2a_1$ lies below the maximum in $\mathscr{E}$ the crack is stable, but that instability sets in when the maximum in $\mathscr{E}$ occurs at $2a_1$. Rearranging eqn (9.46) we obtain

$$\sigma_{tc} = \left(\frac{4\gamma E}{\pi 2a_1}\right)^{1/2} \tag{9.47}$$

as the critical stress for a crack of length $2a_1$. At lower stresses this crack will not grow and the material may fail in a ductile manner. At higher stresses the material will undergo brittle fracture. Eqn (9.47) is known as the Griffith stress for the growth of a sharp crack. Although derived for an interior crack this same result applies to a surface crack of length $2a_1$.

Griffith's stress relation has been checked by measuring the tensile stress at which glass, containing sharp cracks of known length, fractures.

As predicted by eqn (9.47) the fracture stress is proportional to $(2a_1)^{-1/2}$. Ordinary window glass, for example, usually has surface cracks of some 10^{-6} m deep. Introducing a 1-cm crack therefore lowers the fracture stress by a factor of $(10^{-2}/10^{-6})^{1/2} = 100$. As another example, rock salt (NaCl) is normally brittle, it thus has cracks longer than the value given by eqn (9.46). When these cracks are dissolved away by immersing the crystals in warm water they become ductile.

Although isolated edge dislocations do not act as Griffith's cracks they can do so if a sufficient number of them pile up at some obstruction thus creating a local void which then grows in brittle fashion. Solids (such as hcp metals) which fail in this manner are often called semi-brittle solids. Those solids which, like fcc metals, simply continue to deform plastically until the atomic planes literally slide apart can, if one wishes, be called non-brittle solids. Finally, it should be pointed out that eqns (9.46) and (9.47) need to be modified before they can be applied to metals. The reason is that when a crack grows in a metal a sizeable part of the strain energy goes to increase the kinetic energy of the metal's atoms, leaving little energy available to form new surfaces.

PROBLEMS

9.1. Given that the isothermal bulk modulus of KBr is $1 \cdot 8 \times 10^{10}$ Pa, deduce the value of the repulsive exponent n which occurs in the Mie potential model describing the interionic potential energy between a K^+ and a Br^- ion. Potassium bromide crystallizes with the sodium chloride structure and has a nearest-neighbour separation of $3 \cdot 3 \times 10^{-10}$ m.

9.2. Predict the isothermal bulk modulus of beryllium oxide (BeO) given that the bonding in the solid between the Be^{2+} and the O^{2-} ions is wholly ionic. Beryllium oxide has a zincblende structure (Madelung constant $1 \cdot 638$) and a nearest-neighbour separation of $1 \cdot 7 \times 10^{-10}$ m. Assume a value of $n = 9$ in the Mie potential model.

9.3. Making the simplifying assumptions that the molar internal energy of a monatomic solid can be written as $U_m = \frac{1}{2} N_A \mathcal{V}(r)$, where n is the coordination number (that is, number of nearest neighbours) of an atom of the solid, and that the molar volume of the solid can be written as $V_m = N_A r^3$ (thus imagining that each atom occupies a cubical volume r^3), show that the isothermal bulk modulus K_T of the solid is given by

$$K_T = \frac{n}{18 r_0} \left(\frac{d^2 \mathcal{V}(r)}{dr^2} \right)_{r=r_0},$$

where r_0 is the equilibrium separation of an *isolated* pair of atoms. Hence show that $K_T = 4n\varepsilon/r_0^3$ for a substance whose interatomic potential energy $\mathcal{V}(r)$ is represented by a Lennard-Jones 6–12 potential. Taking $\varepsilon = 1 \cdot 7 \times 10^{-21}$ J and $r_0 = 3 \cdot 7 \times 10^{-10}$ m, deduce K_T for solid argon.

9.4. Show that the molar dissociation energy $H_{m,s}$ of solid sodium chloride, where the dissociation is into widely separated Na^+ and Cl^- ions, exceeds the molar dissociation energy, where dissociation is to isolated NaCl molecules, by approximately $\frac{1}{3}H_{m,s}$.

9.5. Show that the periodic time T_v of atomic vibrations in a solid of nearest-neighbour separation a is approximately given by $T_v = 2^{1/2}\pi a(\rho/E)^{1/2}$, where ρ is the density of the solid, and E is its Young's modulus. Calculate T_v in iron, given that $E = 2 \times 10^{11}$ Pa, $\rho = 7 \cdot 8 \times 10^3$ kg m^{-3} and $a = 2 \cdot 5 \times 10^{-10}$ m.

9.6. It is proposed to measure Young's modulus in steels (in which E is around 2×10^{11} Pa and ρ is $7 \cdot 8 \times 10^3$ kg m^{-3}) by setting up longitudinal standing waves in a 50-cm long rod of the steel under investigation. This can be done by wrapping a coil of wire connected to a sine-wave signal generator around the rod. The coil lies between the poles of a permanent magnet which produces a field perpendicular to the axis of the rod. The longitudinal oscillations of the end of the rod remote from the coil can be picked up by means of a piezoelectric transducer connected to a cathode-ray oscilloscope. Calculate the minimum frequency at which longitudinal standing waves can be established.

9.7. Starting with eqns (9.9) and (9.20), show that the bulk modulus and Young's modulus agree, to within about one order of magnitude, in sodium chloride. Because of the 'two-ion' approximation implicit in eqn (9.20) you should replace U_m in eqn (9.9) by $N_A V(r)$, where $V(r)$ is the potential energy between a Na$^+$ ion and a Cl$^-$ ion. Also check out the agreement starting with eqns (9.10) and (9.20).

9.8. Show that the elastic-strain energy stored per unit volume in a strained body is $\frac{1}{2}$(stress × strain) if the stress–strain characteristic is linear. An inventor proposes to replace the 4-kg 'weight' in his grandfather clock by a 60-m long steel wire of circular cross-section: The gravitational potential energy normally possessed by the raised weight is to be replaced by elastic-strain energy in the stretched wire. Assuming the Young's modulus of the steel is 2×10^{11} Pa, what diameter wire should be used to ensure that, when stressed, it has the same energy as the weight in the fully-wound clock? The weight can fall through a distance of $0 \cdot 8$ m; equal to the stretch of the strained wire.

9.9. Adopting the procedures (and approximations) introduced in §9.3, calculate the stress and strain at which a single crystal of argon should fail in a tensile test. Assume a Lennard-Jones 6–12 potential with $\varepsilon = 1 \cdot 7 \times 10^{-21}$ J and take the nearest-neighbour separation to be $3 \cdot 7 \times 10^{-10}$ m in the unstrained crystal. By suitably extending the arguments show that the theoretical fracture stress is $E7^{7/6}/13^{13/6}$.

9.10. By suitably combining eqns (9.31) and (8.27) obtain the relationship between the free surface energy of a {100} face in sodium chloride and the stress at which tensile failure should occur along a {100} plane. Explain, in words, why this sort of relation is to be expected on macroscopic grounds.

9.11. A rectangular slab of material is maintained in equilibrium by a pair of equal and opposite stresses τ_1 acting along the top and bottom surfaces, respectively, and by another equal and opposite pair of stresses τ_2 acting on both end faces. What is the relation between τ_1 and τ_2?

9.12. Fig. 9.27 show the equilibrium positions of atom i relative to two other atoms j and k; we may take it that r_0 is the equilibrium separation of an isolated pair of atoms. Considering only the interaction of atoms i and j, and assuming a Lennard-Jones potential model (eqn (2.19)), show that the

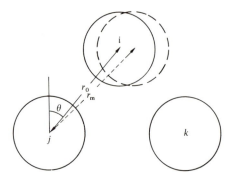

Fig. 9.27. A simple two-atom model of shear.

separation r_m at which the restoring force is a maximum is given by $r_m = r_0[(q+1)/(p+1)]^{1/(q-p)}$. Since $r_m \approx r_0$, the shear strain when the restoring force is a maximum is approximately $(r_m \sin \theta - r_0 \sin \theta)/r_0 \cos \theta$. So what is the theoretical critical shear strain on the basis of this simple two-atom model if these two atoms interact with a Lennard-Jones 6–12 potential?

9.13. Show that if a couple of moment Γ is applied about the axis of a solid straight cylinder of length l and radius R so that the cylinder twists through an angle ϕ then

$$\Gamma = G\pi R^4 \phi/2l,$$

where G is the shear modulus of the material out of which the cylinder is constructed. The twist ϕ is measured at the end at which the couple is applied; the other end of the cylinder is fixed. What couple is required to twist a $1 \cdot 5$ m length of solid brass rod ($G = 1 \cdot 0 \times 10^{11}$ Pa) of diameter 5×10^{-4} m through an angle of 45°? *Clues*: A section of radii r and $r+dr$ taken from the cylinder 'straightens out' to a slab of dimensions $l \times 2\pi r \times dr$. This slab experiences a shear strain of $r\phi/l$. Work out the force, multiply by r to convert to a couple, and integrate.

9.14. The true Burger's vector may be defined operationally in the following manner. Jumping from one lattice point to a neighbouring lattice point form a *closed* circuit enclosing the dislocation line. The circuit must be made in 'good' material. Now make a corresponding set of jumps in an entirely perfect region of the crystal. The lattice vector which is needed to complete the circuit in the *ideal* crystal is defined as the true Burger's vector of the dislocation. Show that the true Burger's vector thus defined (adopting a *FS/RH* convention) is equal to the local Burger's vector as defined on p. 295 (adopting a *SF/RH* convention). Proceed by drawing a crystal structure containing (a) an edge dislocation, and (b) a screw dislocation.

9.15. An alternative definition to that given on p. 302 defines the dislocation density ρ in a crystal as the total length of dislocation line passing through unit volume of the material. Show that these two definitions are equivalent. What is the dislocation density in the aluminium foil whose transmission electron micrograph is shown in Fig. 9.20(c)?

9.16. A specimen of molybdenum which has been cold-worked is found to have a dislocation density of 2×10^{15} m^{-2}. Make a rough estimate of the strain energy present per unit volume of the material. Molybdenum has a bcc structure with a nearest-neighbour separation of $2\cdot72\times10^{-10}$ m. It slips along $\{101\}$ planes. Take $G = 1\cdot33\times10^{11}$ Pa.

9.17. A particular crystal is found to yield plastically at a shear stress of 4×10^{8} Pa. Assuming that the Frank–Read sources are all identical, calculate the length of each source (the distance between pinning sites), given that the shear modulus of the crystal is 3×10^{10} Pa and the Burger's vector is 2×10^{-10} m.

9.18. Glass fractures in brittle fashion when subjected to a tensile stress. If a slab of glass with a Young's modulus of $6\cdot4\times10^{10}$ Pa, and containing a sharp crack normal to the length of the slab, fractures under a tensile stress of $2\cdot6\times10^{6}$ Pa, what is the length of the crack? The glass has a free surface energy of $1\cdot75$ J m^{-2}.

9.19. In 1949, Orowan suggested that the attractive portion of the interatomic force characteristic between a pair of atoms could be written as

$$F(u) = -F_{m}\sin(2\pi u/\lambda);\qquad 0\le2u\le\lambda$$

where $2u = r - b_{0}$ is the displacement of an atom from its equilibrium separation $r = b_{0}$, and λ is a 'range' parameter for the bond. (a) Sketch out the characteristic proposed by Orowan. (b) Ignoring cross-links, what is the tensile fracture stress σ_{tc} of a simple cubic crystal? (c) Again ignoring cross-links, what is Young's modulus of the crystal? (d) By considering the process of fracturing a solid through the growth of a crack deduce an expression for the $\{100\}$ free surface energy of a simple cubic crystal. (e) By eliminating λ from your answers to parts (c) and (d) obtain a relationship between σ_{tc} and E.

9.20. Use arguments at the atomic level to explain the following facts: (a) The speed of sound is greater in solids than in gases. (b) The bulk modulus of solids increases with increasing pressures. (c) The bulk modulus of a solid is approximately equal to the Young's modulus of the solid. (d) The Young's modulus of a solid decreases with increasing temperature.

9.21. Review the factors which determine whether a given material will fail in brittle or in ductile fashion.

10. Thermal properties of solids

IN discussing the structural and mechanical properties of solids we implicitly assumed that $\frac{1}{2}m\overline{u^2} = 0$. This was akin to assuming that $\Delta E = 0$ in a gas. Just as removing the restriction $\Delta E = 0$ allowed us to discuss the properties of real (imperfect) gases so we can expect that acknowledging that $\frac{1}{2}m\overline{u^2} > 0$ in a solid will allow us to discuss a wider range of processes than hitherto. Thus we shall examine the thermal properties of solids. We will start by discussing their heat capacity at constant volume; a discussion which will force us to conclude that the atoms of a solid do not obey Newtonian mechanics. We will also look at why solids expand on heating and at a mechanical model of melting.

10.1. Heat capacity of non-metals: classical theory

In §3.9 we showed how the laws of classical mechanics predict that an atom of a monatomic crystal at a temperature T has a mean energy (kinetic plus potential) of $3kT$ (eqn (3.60)). Therefore the total energy of 1 mol of the crystal is $N_A(3kT) = 3RT$. To this we should properly add the internal energy $U_m(a_0)$ of the crystal structure (as discussed in §8.5). Hence the internal energy U_m of 1 mol of a monatomic crystal is given by

$$U_m = 3RT + U_m(a_0). \tag{10.1}$$

At a fixed volume the value of a_0, and therefore of $U_m(a_0)$, will be independent of T. Therefore,

$$\boxed{C_{V,m} = \left(\frac{\partial U_m}{\partial T}\right)_{V_m} = 3R} \ . \tag{10.2}$$

Classical physics thus predicts that $C_{V,m}$ for all solid elements should have the same value $3R$, which is constant *independent of temperature*. Experiments show that at sufficiently high temperatures $C_{V,m}$ is indeed constant and with a value $3R$ $(= 24 \cdot 9 \, \text{J K}^{-1} \, \text{mol}^{-1})$; an observation known as Dulong and Petit's law. However, as the temperature is lowered $C_{V,m}$ is found to fall away, tending towards zero as T tends to zero. Fig. 10.1(a) shows results for diamond and Fig. 10.1(b) for copper. These measurements were actually made at constant pressure but because solids expand so little on heating them at constant pressure the difference between $C_{p,m}$ and $C_{V,m}$ (see eqn (1.15)) is usually negligible and may be ignored.

We are clearly witnessing the same sort of failure of theory to match

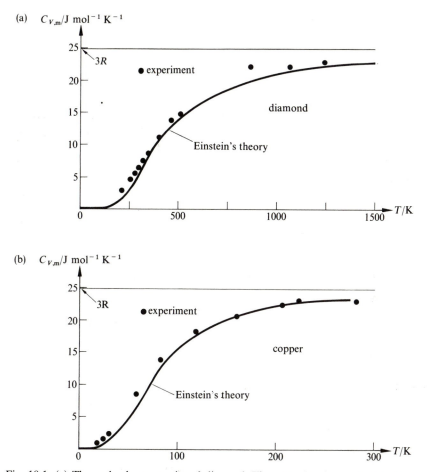

Fig. 10.1. (a) The molar heat capacity of diamond. The curve has been calculated from Einstein's theory (§10.2) assuming $\theta_E = 1320$ K. (b) The molar heat capacity of copper. Here $\theta_E = 240$ K has been assumed.

the experimental findings as we saw in our discussions of the heat capacity of diatomic gases (§5.6). Once again the failure stems from the inability of classical mechanics to describe correctly the motions of bound atoms.

10.2. Heat capacity of non-metals: Einstein's theory

The model

The first explanation of the breakdown of Dulong and Petit's law came from Einstein in 1906, who capitalized on an earlier result of Planck. In 1900 Planck had assumed that the energy E_x of an atom vibrating in, say, the x-direction could only have certain discrete values, namely those values given by the relation

$$E_x = n\frac{h}{2\pi}\omega = n\hbar\omega \; ,$$ (10.3)

where n is an integer, 0, 1, 2, 3, etc., h is a constant known as Planck's constant (for brevity we write $\hbar$ for $h/2\pi$), and ω is the circular frequency of vibration of an atom as calculated classically.

The average energy of an atom

Our immediate goal is to deduce the average energy of an atom of a solid which shares out its total energy according to Planck's rules.

As a preliminary we consider how, say, four 'Planck oscillators' A, B, C, and D share out a total of, say, five units of energy. One possible way is shown in Fig. 10.2. Here oscillator B has an energy $E_x = 3\hbar\omega$; oscillators C and A each have energy $1\hbar\omega$; and oscillator D has zero energy. What we must now do is to draw out all the possible 'pictures' which correspond to physically distinct situations and then find the average number of oscillators with energy 0, $1\hbar\omega$, $2\hbar\omega$, etc. This problem is formally equivalent to one we discussed in §4.4 in considering a two-dimensional gas. In the present context Fig. 4.13 shows all the ways in which four oscillators can share out five units of energy. To find the mean number of oscillators with a particular energy, say $E_x = 3\hbar\omega$, we count up the total number of appearances at this level (24) and divide by the total number of pictures (56). On carrying out this operation at each level we obtain the histogram shown in Fig. 10.3 (which is, of course, similar to Fig. 4.14).

Because of these formal similarities between the energy distribution

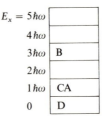

Fig. 10.2. One possible way in which four Planck oscillators, A, B, C, and D can share out a total of five units of energy. It would make no difference had we written AC in the cell at $E_x = 1\hbar\omega$ since this would still tell us that A has energy $1\hbar\omega$ and C has energy $1\hbar\omega$. Thus rearranging letters within a cell does not represent new ways of sharing out the energy.

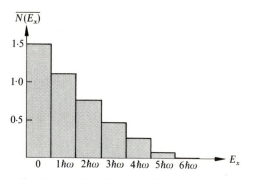

Fig. 10.3. The energy distribution of four Planck oscillators sharing out five units of energy.

of a two-dimensional gas and of a collection of Planck oscillators we may immediately take over eqn (4.34) with g_i having the same value g at all energy levels (it is unity in Fig. 10.2). Thus, on average, the number of oscillators, $N(E_i)$, with an energy E_i is given by

$$N(E_i) = A\, e^{-\beta E_i} \qquad (10.4)$$

where A $(= g\exp(-\alpha))$ and β are constants, as yet undetermined.

To find the mean energy $\bar{E}$ of a Planck oscillator we start with the definition

$$\bar{E} = \frac{\sum_i N(E_i) \times E_i}{\sum_i N(E_i)}. \qquad (10.5)$$

Substituting for $N(E_i)$ from eqn (10.4) and for E_i from eqn (10.3) gives

$$\bar{E} = \frac{A(0\, e^{-\beta 0} + \hbar\omega\, e^{-\beta\hbar\omega} + 2\hbar\omega\, e^{-2\beta\hbar\omega} + 3\hbar\omega\, e^{-3\beta\hbar\omega} + \cdots)}{A(e^{-\beta 0} + e^{-\beta\hbar\omega} + e^{-2\beta\hbar\omega} + e^{-3\beta\hbar\omega} + \cdots)}.$$

Because of the standard form

$$\frac{d}{dx}(\ln u) = \frac{1}{u}\frac{du}{dx} \qquad (10.6)$$

we may write

$$\bar{E} = -\frac{d}{d\beta}\ln(1 + e^{-\beta\hbar\omega} + e^{-2\beta\hbar\omega} + e^{-3\beta\hbar\omega} + \cdots).$$

Now the expression inside the brackets is a geometrical series whose ratio r of any term to the preceding term has a value $e^{-\beta\hbar\omega}$. When $|r| < 1$, as it is here, the infinite series $1 + r + r^2 + \cdots$ has a value $1/(1-r)$. Therefore

$$\bar{E} = -\frac{d}{d\beta} \ln\left(\frac{1}{1-e^{-\beta\hbar\omega}}\right).$$

Applying eqn (10.6) with $u = (1 - e^{-\beta\hbar\omega})^{-1}$ leads to

$$\bar{E}_x = \frac{\hbar\omega}{e^{\beta\hbar\omega} - 1} \tag{10.7}$$

where we have reinserted the suffix x to emphasize that we have only considered oscillations in the x-direction. However, similar results hold in the y- and z-directions. Thus the total mean energy $\bar{E}$ of atom in the solid is

$$\bar{E} = \frac{3\hbar\omega}{e^{\beta\hbar\omega} - 1}. \tag{10.8}$$

Although undetermined by us we shall tentatively assume that, as in a gas, the constant $\beta = 1/kT$. The justification will come presently when we find that, in accord with experiment, the predicted value of $C_{V,m}$ tends towards $3R$ at high temperatures.

The molar heat capacity predicted

Since $U_m = N_A \bar{E}$ it follows from eqn (10.8) with $\beta = 1/kT$ that

$$C_{V,m} = \left(\frac{\partial U_m}{\partial T}\right)_{V_m} = 3R\left(\frac{\hbar\omega}{kT}\right)^2 \frac{e^{\hbar\omega/kT}}{(e^{\hbar\omega/kT} - 1)^2}.$$

This may be written as

$$C_{V,m} = 3R\left(\frac{\theta_E}{T}\right)^2 \frac{e^{\theta_E/T}}{(e^{\theta_E/T} - 1)^2} \tag{10.9}$$

where

$$\theta_E = \hbar\omega/k \tag{10.10}$$

is known as the *Einstein temperature*. At high temperatures $(T \gg \theta_E)$ eqn (10.9) becomes

$$C_{V,m} = 3R\left(\frac{\theta_E}{T}\right)^2 \frac{1 + (\theta_E/T) + \cdots}{(1 + (\theta_E/T) + \cdots - 1)^2}$$

which tends to $3R$ as T tends to infinity. At low temperatures $(T \ll \theta_E)$ we have $\exp(\theta_E/T) \gg 1$, so that eqn (10.9) becomes

$$C_{V,m} = 3R\left(\frac{\theta_E}{T}\right)^2 e^{-\theta_E/T}. \tag{10.11}$$

With decreasing temperature $\exp(-\theta_E/T)$ in eqn (10.11) decreases more rapidly than (θ_E/T) increases. As a consequence, the predicted heat capacity falls. However, as Fig. 10.1 demonstrates, it falls rather too rapidly. The values assigned to θ_E in Fig. 10.1 were chosen to best fit the experimental data to eqn (10.9); values of θ_E for other materials are given in Table 10.2. Of course, a value of $\theta_E = 240$ K for copper implies a vibrational period $T_v = 2\pi/\omega = h/k\theta_E$ of $2\cdot0\times10^{-13}$ s. To see whether this is at all reasonable we recall (p. 284) that the speed v of sound in a solid is approximately $r_0/\frac{1}{4}T_v$, where r_0 is the nearest-neighbour separation. In copper, $v = 3\cdot81\times10^3$ m s^{-1} and $r_0 = 2\cdot6\times10^{-10}$ m, predicting $T_v = 2\cdot7\times10^{-13}$ s.

Einstein's model fails to fit the experimental data at low temperatures because it assumes that all the atoms have a common vibrational frequency ω. This common frequency can only be justified if each atom vibrates independently of its neighbours (§3.9). Once we remove this fiction we shall discover that the atoms of a solid possess a range of frequencies. It is this range of frequencies which lies at the heart of a correct explanation of the temperature dependence of $C_{V,m}$.

Exercise 10.1

Before leaving Einstein's model it is worth demonstrating that it can be used to discuss the rise in $C_{V,m}$ of molecular hydrogen which starts to occur at around 500 K (Fig. 5.19), when $C_{V,m}$ rises from $\frac{5}{2}R$ toward $\frac{7}{2}R$. You are asked to show that this rise is consistent with the molecular vibrational period of $3\cdot8\times10^{-15}$ s (the value deduced from spectroscopic data).

Calculation. Since a hydrogen molecule only vibrates along a line joining the centres of the two hydrogen atoms the factor of three on the right-hand side of eqn (10.9) must be replaced by unity. Hence the molar heat capacity at temperatures greater than about 500 K takes the form

$$C_{V,m} = R\left(\frac{\theta_E}{T}\right)^2 \frac{e^{\theta_E/T}}{(e^{\theta_E/T}-1)^2} + \frac{5}{2}R. \tag{10.12}$$

Rather than plotting out eqn (10.12) we shall simply check that it is consistent with Fig. 5.19 at, say, $T = 0\cdot2\theta_E$, $T = 0\cdot5\theta_E$, and $T = \theta_E$. By definition (eqn (10.10)), $\theta_E = \hbar\omega/k = h/kT_v = 6\cdot6\times10^{-34}$ J s/$(1\cdot38\times10^{-23}$ J K$^{-1}\times3\cdot8\times10^{-15}$ s)$ = 1\cdot26\times10^4$ K. Substituting $T/\theta_E = 0\cdot2$ into eqn (10.12) gives $C_{V,m} = 2\cdot7R$ at $T = 2500$ K. Likewise, eqn (10.12) predicts that $C_{V,m} = 3\cdot2R$ at 6300 K and $3\cdot4R$ at 12 600 K. These values fit Fig. 5.19.

Comment. Quantum mechanics shows that the vibrational energies of a diatomic molecule are given by $(n+\frac{1}{2})\hbar\omega$, rather than by $n\hbar\omega$ (eqn (10.3)). The only effect of the additional $\frac{1}{2}\hbar\omega$ term is to add $\frac{1}{2}N_A\hbar\omega$ to the molar internal energy. Since ω is independent of T this extra internal energy makes no contribution to $C_{V,m} = (\partial U_m/\partial T)_{V_m}$. Similarly a term $\frac{1}{2}\hbar\omega$ should be added to the right-hand side of eqn (10.7). It too has no effect on $C_{V,m}$.

10.3. The one-dimensional coupled-oscillator model

A simulation study

Before discussing the behaviour of a real three-dimensional solid we shall consider a one-dimensional solid made up of a line of identical atoms, each one coupled only to its nearest neighbours. Adopting a strategy we have employed on previous occasions, we shall begin with a simulation study.

A suitable simulator consists of 'gliders' connected together by springs on a linear air-track (a frictionless mount) as shown in Fig. 10.4(a). When one of the outer gliders is moved back and forth a compressional wave travels down the row of gliders-plus-springs to the far end where it is reflected back. In carrying out this experiment we find that *provided* the vibrational period T_v (or, equivalently, the circular frequency $\omega = 2\pi/T_v$) is correctly chosen, standing wave patterns are established. Fig. 10.4(a) shows one such standing wave pattern, in which two of the gliders remain permanently at rest. The nature of the standing wave pattern is more clearly brought out in Fig. 10.4(b) which gives the

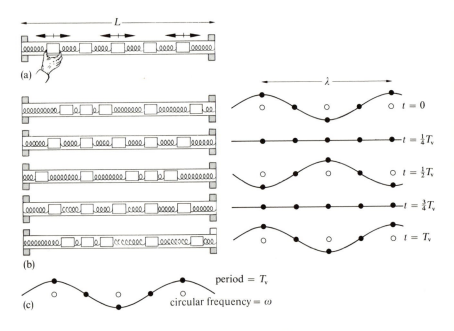

Fig. 10.4. Five identical gliders are connected together by identical springs. (a) Showing one of the standing wave patterns which can be set up. (b) The standing wave pattern examined in detail. Open circles indicate the equilibrium positions of the gliders. The ordinates show the displacements of each glider at the stated times. (c) A shorthand description of the standing wave pattern.

displacement of each glider at times $t = 0, \frac{1}{4}T_v, \frac{1}{2}T_v, \frac{3}{4}T_v$, and T_v. Here the horizontal displacement of a glider from its equilibrium position is represented by a graph in which the positive ordinate is used for displacements to the right and the negative ordinate for displacements to the left. Fig. 10.4(c) shows a shorthand representation of the standing wave pattern of Fig. 10.4(b). To obtain the 'envelope' which indicates the limits to the displacement of each glider this curve must be reflected about the abscissa.

In the course of experimenting with the simulator one soon discovers that several more standing wave patterns can be established. Fig. 10.5(a) shows the full set of patterns for the five gliders of Fig. 10.4(a). This figure also gives the wavelength λ of each of the various vibrational patterns—or *modes* for short—which can be set up. Instead of quoting λ we can instead quote the *circular wavenumber* q defined by $q = 2\pi/\lambda$. If

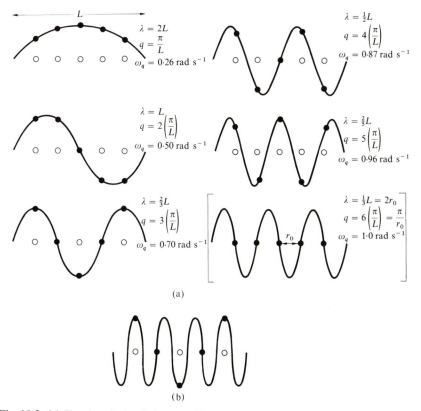

(a)

(b)

Fig. 10.5. (a) Showing all the distinct standing wave patterns which can be set up with the glider-plus-spring apparatus of Fig. 10.4(a). (b) So far as the gliders are concerned this mode is the same as that labelled $q = 3(\pi/L)$ in (a).

we disregard the springs' motion (which we can do since we are only interested in how the gliders, and hence the atoms, behave) we see that there are five distinct modes; equal to the number of the gliders. (In the last mode shown in Fig. 10.5(a) all the gliders are at rest and only the springs move.) While experimenting with the simulator one may find what, at first sight, appears to be additional modes. One such mode is shown in Fig. 10.5(b). However, forgetting the springs, we see that this mode is the same as that labelled $\lambda = \frac{2}{3}L$ (that is, $q = 3\pi/L$) in Fig. 10.5(a).

In setting up all the modes shown in Fig. 10.5(a) one discovers that, in any one mode (for example, that with $q = 5(\pi/L)$), all the atoms vibrate with the same frequency but that this vibrational frequency is different in each mode. Fig. 10.5(a) lists the values of ω_q for each mode when every glider has a mass of $0{\cdot}2$ kg and every spring a force constant of 5×10^{-2} N m^{-1}.

A one-dimensional solid

At $T = 0$ the atoms of a one-dimensional solid would occupy the lattice sites as shown in Fig. 10.6(a). At $T > 0$ the atoms will, of course, oscillate about these sites. Fig. 10.6(b) is a snapshot of how the structure

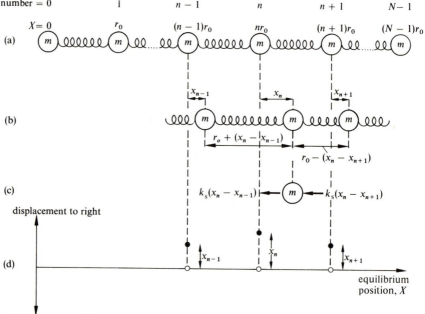

Fig. 10.6. A one-dimensional crystal. (a) At $T = 0$ the N atoms would occupy the lattice sites. (b) At $T > 0$ the atoms will be displaced from their lattice sites. (c) Each displaced atom will experience a restoring force. (d) The displacements represented graphically.

might appear at a particular instant. The displacement of the atoms from their lattice sites can be represented graphically as shown in Fig. 10.6(d).

If we denote the horizontal displacement of each atom by x_0, $x_1, \ldots, x_{n-1}, x_n, x_{n+1} \cdots$ and the interatomic force constant by k_s, we have, on applying Newton's second law to atom n (see Fig. 10.6(c)):

$$m \frac{d^2 x_n}{dt^2} = -k_s(x_n - x_{n-1}) - k_s(x_n - x_{n+1})$$

$$m \frac{d^2 x_n}{dt^2} = k_s(x_{n-1} + x_{n+1} - 2x_n). \tag{10.13}$$

You may find it helpful to compare atom n with atom i of Fig. 3.10. This will remind you that what matters is the *change* in the spring lengths from their equilibrium lengths r_0.

Drawing on our knowledge of how a simulated solid behaves we shall look for standing wave solutions to eqn (10.13). A standing wave can be written as

$$x = A \cos 2\pi \frac{X}{\lambda} \cos 2\pi \frac{t}{T_v} = A \cos qX \cos \omega t.$$

Since atom n of Fig. 10.6 is at $X = nr_0$ its displacement at time t is thus given by

$$x_n = A \cos qnr_0 \cos \omega_q t. \tag{10.14}$$

This equation correctly expresses the fact that the amplitude of vibration of an atom ($= A \cos qnr_0$) depends on which atom it is (that is, on n). It also expresses the fact that in any particular mode (that is, q) the value of ω_q is independent of n.

Substituting eqn (10.14) into eqn (10.13) gives

$$mA \cos qnr_0 \frac{d^2}{dt^2} (\cos \omega_q t)$$

$$= k_s A[\cos q(n-1)r_0 + \cos q(n+1)r_0 - 2 \cos qnr_0] \cos \omega_q t$$

which becomes

$$-m\omega_q^2 = 2k_s(\cos qr_0 - 1)$$

or, since $\cos \theta = 1 - 2 \sin^2(\theta/2)$,

$$m\omega_q^2 = 4k_s \sin^2 \tfrac{1}{2} qr_0. \tag{10.15}$$

Therefore

$$\boxed{\omega_q = 2\left(\frac{k_s}{m}\right)^{1/2} |\sin \tfrac{1}{2} qr_0|} . \tag{10.16}$$

Because the negative root of eqn (10.15) implies a negative frequency, which is without physical significance, we only take the positive root into account. The equation relating ω_q and q (here eqn (10.16)) is known as the *dispersion relation*.

We have yet to make use of the fact that the one-dimensional crystal (Fig. 10.6(a)) has a finite length L with an antinode (a maximum amplitude of vibration) at each end, that is at $X = 0$ and $X = L$. This demands that L is an integral number of the half-wavelengths (see Fig. 10.9), or in terms of q ($= 2\pi/\lambda$), that

$$q = \left(\frac{\pi}{L}\right), 2\left(\frac{\pi}{L}\right), 3\left(\frac{\pi}{L}\right), \dots, N\left(\frac{\pi}{L}\right), \qquad (10.17)$$

where N is the total number of atoms. In terminating the sequence at $N(\pi/L)$ we are expressing the reality that beyond $N(\pi/L)$ the standing wave patterns repeat earlier ones (recall Fig. 10.5). Ignoring the difference between N and $N-1$ we can write $L = Nr_0$ (see Fig. 10.6(a)) so that the maximum value of q is given by

$$q_{\text{max}} = \frac{\pi}{r_0}. \qquad (10.18)$$

Fig. 10.7 shows how to represent the information contained in eqns

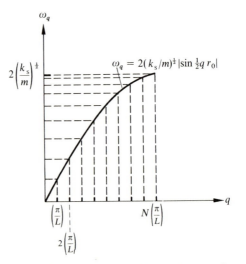

Fig. 10.7. The abscissa records the values of q that produce standing wave patterns in a one-dimensional structure. Each of the modes has a unique circular frequency, indicated by the heights of the dashed vertical lines. The tops of these lines fit a smooth curve, given by eqn (10.16).

(10.16) and (10.17). Along the abscissa we mark off the permitted values of q as given by eqn (10.17). To find the values of ω_q—that is, the circular frequencies of the atoms in the mode of circular wavenumber q—we merely substitute each permitted value of q into eqn (10.16) and plot the result.

In a one-dimensional crystal the motion of an individual atom will be the resultant of the motion which it executes in sustaining all possible vibrational patterns in the crystal. A casual observer would see each atom behaving in a highly complex, seemingly erratic, fashion. What interests us, of course, is not the motion of an atom *per se* but its mean kinetic and potential energies. To find the total internal energy of our solid we must first evaluate the energy of a single mode; of a single vibrational pattern in which *all* the atoms participate. We will then sum the energies of all the modes that are present. It is important to keep in mind throughout the following discussions that in looking at a single mode we are only looking at a *part* of the motions of these atoms; it takes the other modes to complete the picture.

The energy of a single mode

In evaluating the total energy associated with a vibrational mode of circular frequency ω_q we will consider first the kinetic energy of all the atoms in the structure when this mode is present. Since

$$\text{K.E.} = \tfrac{1}{2}m \sum_{\text{atoms}} \left(\frac{\mathrm{d}x_n}{\mathrm{d}t}\right)^2$$

we have, on substituting for x_n from eqn (10.14),

$$\text{K.E.} = \tfrac{1}{2}mA^2\omega_q^2 \sin^2\omega_q t \sum_{n=0}^{N-1} \cos^2 qnr_0. \tag{10.19}$$

We must now find the total potential energy associated with this particular vibrational mode of circular wavenumber q. We begin by considering the energy V_n required to displace atom n from its lattice site nr_0. If, as shown in Fig. 10.8, this atom is displaced an amount x_n from its lattice site it will, according to eqn (10.13), experience a (restoring) force of value $k_s(x_{n-1}+x_{n+1}-2x_n)$. In displacing atom n through a further distance $\mathrm{d}x_n$ energy of amount $-k_s(x_{n-1}+x_{n+1}-2x_n)\,\mathrm{d}x_n = k_s(2x_n-x_{n-1}-x_{n+1})\,\mathrm{d}x_n$ is fed into the structure, where it is stored as potential energy. The potential energy V_n of atom n, when at some displacement x_n, is therefore given by

$$V_n = \int_0^{x_n} k_s(2x_n - x_{n-1} - x_{n+1})\,\mathrm{d}x_n$$

$$V_n = k_s(x_n^2 - x_{n-1}x_n - x_{n+1}x_n).$$

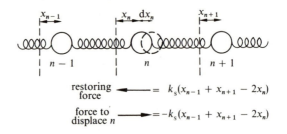

$$\text{restoring force} \longleftarrow = k_s(x_{n-1} + x_{n+1} - 2x_n)$$

$$\text{force to displace } n \longrightarrow = -k_s(x_{n-1} + x_{n+1} - 2x_n)$$

Fig. 10.8. As atom n of a one-dimensional crystal is displaced through dx_n its potential energy increases.

In performing this calculation we have taken the potential energy of atom n as zero when it is on its lattice site nr_0. An exactly similar calculation can be carried out for each and every atom in the structure, giving the total potential energy of the disturbed structure as

$$\text{P.E.} = k_s \sum_{n=0}^{N-1} x_n^2 - \tfrac{1}{2}k_s \sum_{n=0}^{N-1} (x_{n-1}x_n + x_{n+1}x_n), \qquad (10.20)$$

where the summation is over all N atoms. The factor of $\tfrac{1}{2}$ has been introduced into the summation involving mixed terms to ensure that $-\partial(\text{P.E.})/\partial x_n$ correctly gives the restoring force on atom n as $k_s(x_{n-1} + x_{n+1} - 2x_n)$, this being the value arrived at on p. 325. Substituting for x_n as given by eqn (10.14) into eqn (10.20) leads to

$$\text{P.E.} = 2A^2 k_s \cos^2\omega_q t \, \sin^2\tfrac{1}{2}qr_0 \sum_{n=0}^{N-1} \cos^2 qnr_0$$

or, eliminating k_s via eqn (10.16),

$$\text{P.E.} = \tfrac{1}{2}mA^2\omega_q^2 \cos^2\omega_q t \sum_{n=0}^{N-1} \cos^2 qnr_0. \qquad (10.21)$$

To find the total energy E_q associated with this one mode of circular wavelength q we add the kinetic and potential energies as given by eqns (10.19) and (10.21), respectively. This gives

$$\boxed{E_q = \tfrac{1}{2}mA_1^2\omega_q^2}. \qquad (10.22)$$

where

$$A_1 = A\left(\sum_{n=0}^{N-1} \cos^2 qnr_0\right)^{1/2}. \qquad (10.23)$$

Comparing eqn (10.22) with eqn (3.52) we see that E_q—which involves all N atoms—is equal to the energy of a simple harmonic oscillator with an oscillating mass m equal to that of an atom of the structure, a circular frequency equal to that of the mode, and an amplitude A_1 of oscillation which is related to the amplitude A of the mode by eqn (10.23). Having demonstrated the energy equivalence of a mode and a harmonic oscillator, we next assume that the oscillator is a Planck oscillator whose mean

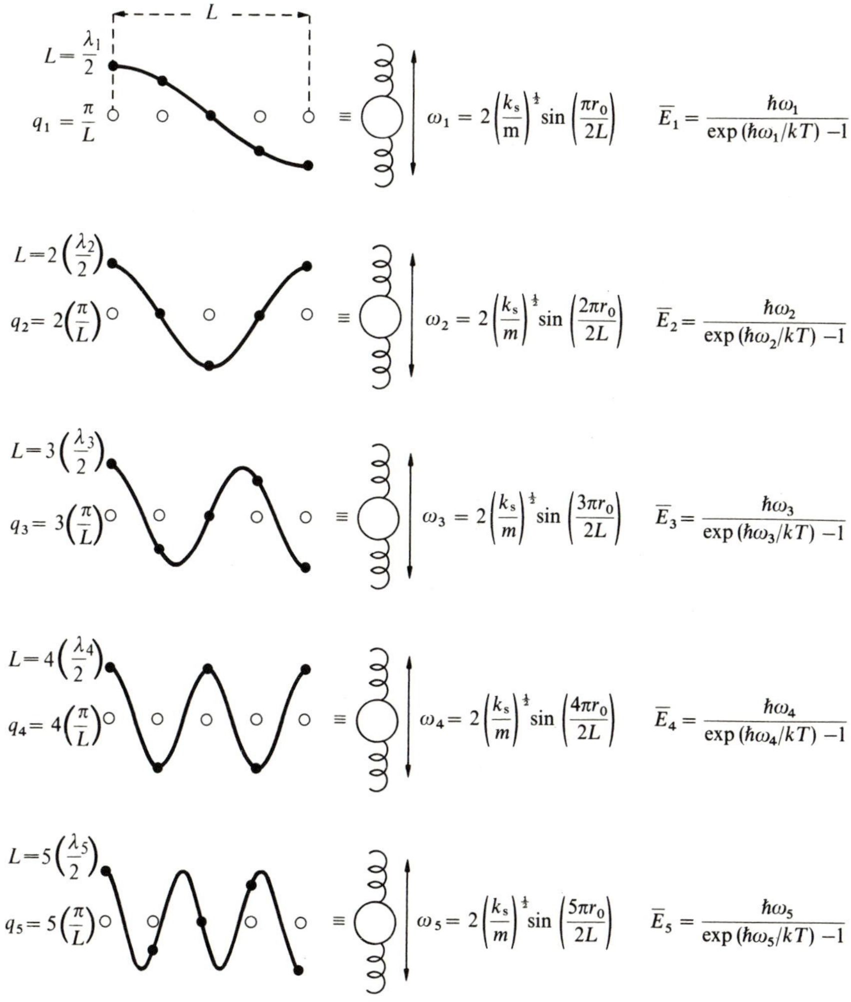

$$L = \frac{\lambda_1}{2}$$

$$q_1 = \frac{\pi}{L}$$

$$\omega_1 = 2\left(\frac{k_s}{m}\right)^{\frac{1}{2}} \sin\left(\frac{\pi r_0}{2L}\right) \qquad \bar{E}_1 = \frac{\hbar\omega_1}{\exp(\hbar\omega_1/kT) - 1}$$

$$L = 2\left(\frac{\lambda_2}{2}\right)$$

$$q_2 = 2\left(\frac{\pi}{L}\right)$$

$$\omega_2 = 2\left(\frac{k_s}{m}\right)^{\frac{1}{2}} \sin\left(\frac{2\pi r_0}{2L}\right) \qquad \bar{E}_2 = \frac{\hbar\omega_2}{\exp(\hbar\omega_2/kT) - 1}$$

$$L = 3\left(\frac{\lambda_3}{2}\right)$$

$$q_3 = 3\left(\frac{\pi}{L}\right)$$

$$\omega_3 = 2\left(\frac{k_s}{m}\right)^{\frac{1}{2}} \sin\left(\frac{3\pi r_0}{2L}\right) \qquad \bar{E}_3 = \frac{\hbar\omega_3}{\exp(\hbar\omega_3/kT) - 1}$$

$$L = 4\left(\frac{\lambda_4}{2}\right)$$

$$q_4 = 4\left(\frac{\pi}{L}\right)$$

$$\omega_4 = 2\left(\frac{k_s}{m}\right)^{\frac{1}{2}} \sin\left(\frac{4\pi r_0}{2L}\right) \qquad \bar{E}_4 = \frac{\hbar\omega_4}{\exp(\hbar\omega_4/kT) - 1}$$

$$L = 5\left(\frac{\lambda_5}{2}\right)$$

$$q_5 = 5\left(\frac{\pi}{L}\right)$$

$$\omega_5 = 2\left(\frac{k_s}{m}\right)^{\frac{1}{2}} \sin\left(\frac{5\pi r_0}{2L}\right) \qquad \bar{E}_5 = \frac{\hbar\omega_5}{\exp(\hbar\omega_5/kT) - 1}$$

Fig. 10.9. Each vibrational mode of a one-dimensional crystal has an energy equal to that of a Planck oscillator vibrating with the same circular frequency as the mode. Here ω_q and q are related by eqn (10.16).

energy $\bar{E}_q$ is given by eqn (10.7) with $\beta = 1/kT$. Thus

$$\bar{E}_q = \frac{\hbar\omega_q}{e^{\hbar\omega_q/kT} - 1}.$$
(10.24)

Fig. 10.9 summarizes the arguments which we have advanced in the case of a five-atom one-dimensional solid.

The total internal energy

To obtain the molar internal energy U_{m1} of our one-dimensional crystal we must sum the energies of all the distinct modes which the crystal can sustain; if you like, we must sum the entries in the last column of Fig. 10.9. Since the number of distinct modes is equal to the number of atoms present we can thus write

$$U_{m1} = \sum_{q = \pi/L}^{N_A(\pi/L)} \frac{\hbar\omega_q}{e^{\hbar\omega_q/kT} - 1}.$$
(10.25)

Now according to eqn (10.17) each mode is separated by π/L in circular wavenumber. It therefore follows that there are L/π modes per unit range of q. (Go a unit distance along the abscissa of Fig. 10.7 and you will pass L/π points denoting modes.) In a range dq there are therefore $(L/\pi)\,dq$ modes, allowing eqn (10.25) to be rewritten as

$$U_{m1} = \frac{L}{\pi} \int_0^{\pi/r_0} \frac{\hbar\omega_q}{e^{\hbar\omega_q/kT} - 1}\,dq.$$
(10.26)

The lower limit of the integral has been taken as zero rather than π/L (clearly justified since $L = (N_A - 1)r_0 = 1\cdot2 \times 10^{14}$ m in a crystal with $r_0 = 2 \times 10^{-10}$ m) and the upper limit as π/r_0 rather than $N_A\pi/(N_A - 1)r_0$.

Differentiating eqn (10.26) with respect to T gives the molar heat capacity at constant volume of a one-dimensional crystal as

$$C_{V,m} = \frac{L}{\pi kT^2} \int_0^{\pi/r_0} \frac{(\hbar\omega_q)^2\, e^{\hbar\omega_q/kT}}{(e^{\hbar\omega_q/kT} - 1)^2}\,dq,$$
(10.27)

where ω_q is determined by the appropriate dispersion relation, eqn (10.16). You should be able to confirm that $C_{V,m}$ tends to a constant value at high temperatures and falls towards zero as $T \to 0$. This being so we can therefore seek to extend eqn (10.27) so that it applies to a three-dimensional crystal.

Exercise 10.2

Five gliders, each of mass $0\cdot2$ kg, are connected together by springs as shown in Fig. 10.4(a) on a track of total length $1\cdot8$ m. The force constant of each spring is $5 \times 10^{-2}\, \text{N m}^{-1}$. Without looking back at Fig. 10.5, draw out all the physically distinct vibrational patterns. Calculate the circular frequency of each mode and check your answers against the values given in Fig. 10.5.

Calculation. The vibrational frequencies are, of course, found from the dispersion relation, eqn (10.16), with $k_s = 5 \times 10^{-2}\, \text{N m}^{-1}$, $m = 0 \cdot 2\, \text{kg}$, and $r_0 = 1 \cdot 8\, \text{m}/6 = 0 \cdot 3\, \text{m}$. Eqn (10.16) therefore becomes $\omega_q = \sin(0 \cdot 15q)$. Using the values of q as deduced from the standing wave patterns shown in Fig. 10.5 (namely $q = 2\pi/\lambda$, where $\lambda = 3 \cdot 6$, $1 \cdot 8$, $1 \cdot 2$, $0 \cdot 9$, and $0 \cdot 72\, \text{m}$) yields the values of ω_q given in Fig. 10.5.

Comments. To simplify the analysis the simulation deliberately included outer springs to the first and last gliders (Fig. 10.4(a)). This ensured that every glider was connected to two springs. Such outer 'springs' are clearly absent in a one-dimensional crystal. Their absence will, however, have an utterly insignificant effect on the behaviour of a crystal containing N_A atoms.

10.4. The three-dimensional coupled-oscillator model

For simplicity we will only consider the vibrational patterns which can be set up in a simple cubic structure. We will further suppose that the crystal is in the form of a rectangular block with {100} faces.

Fig. 10.10 illustrates how entire planes of atoms can vibrate so as to set up standing wave patterns characterized by circular wavenumbers which are integral multiples of π/L_x, π/L_y, and π/L_z, where L_x, L_y, and L_z are the lengths of the block's edges. The values of q_x range from π/L_x to $N_x(\pi/L_x)$, where N_x is the number of atoms along the x-edge; q_y ranges from π/L_y to $N_y(\pi/L_y)$; q_z from π/L_z to $N_z(\pi/L_z)$. In general, any permitted value of q_x may be present with any permitted values of q_y and q_z; a mode may therefore be characterized by giving the coordinates (q_x, q_y, q_z) of a point in q_x-q_y-q_z space. Fig. 10.11 shows all the permitted points—all the permitted modes—which have a common value of q_z. It is clear that the total number of permitted modes is $N_x \times N_y \times N_z$, equal to total number of atoms N in the crystal.

So far we have considered longitudinal vibrations in a crystal. However, transverse vibrations can also occur (Fig. 10.12). Each of these types of transverse vibration will lead to standing wave patterns in which q_x, q_y, and q_z are, respectively, integral multiples of π/L_x, π/L_y, and π/L_z. So in our crystal containing N_A atoms there are a total of $3N_A$ possible modes; N_A due to longitudinal vibrations and $2N_A$ due to transverse vibrations. We may therefore write down that the molar internal energy of our three-dimensional crystal is (cf. eqn (10.25))

$$U_{m3} = \sum_q \frac{\hbar\omega_q}{e^{\hbar\omega_q/kT} - 1}, \tag{10.28}$$

where the summation is over all $3N_A$ modes.

Without the dispersion relations—the relations between ω_q and q—to accompany it, eqn (10.28) will not tell us U_{m3} and so $C_{V,m}$. As with

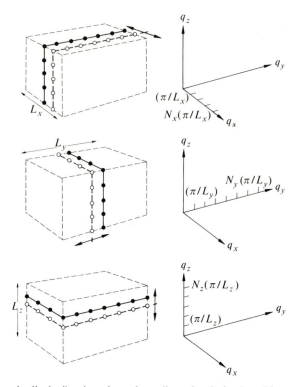

Fig. 10.10. Longitudinal vibrations in a three-dimensional simple cubic crystal produce standing wave patterns characterized by the values of their circular wavenumbers q_x, q_y, and q_z.

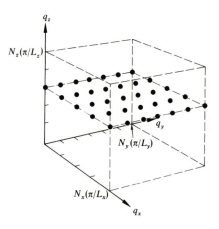

Fig. 10.11. The longitudinal modes in a three-dimensional crystal are characterized by the triplet (q_x, q_y, q_z). For clarity only those modes having $q_z = 3(\pi/L_z)$ are shown.

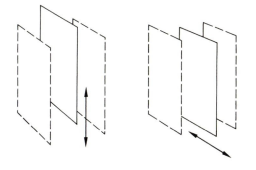

Fig. 10.12. Showing how transverse vibrations can exist in a three-dimensional crystal.

a one-dimensional crystal, these relations depend on the form of the crystal and on the nature of the interatomic forces. However, they are much more tedious to derive than was the dispersion relation for a one-dimensional crystal (eqn (10.16)) and we shall not attempt their derivation here.

10.5. The Debye approximation

Having arrived at the procedure for evaluating the heat capacity of a real crystal we have been forced to abandon the calculation because of the problems likely to be encountered in deriving the dispersion relations. Now is the time to introduce a number of judiciously chosen approximations. In so doing we shall assume that the crystal has a simple cubic structure, as in §10.4, and is in the form of a cube of side L. If there are N atoms along each edge then $L = Na$, where a is the nearest-neighbour separation.

The approximations

The first approximation we shall make is to replace the three-dimensional array of points which is bounded by planes at $q_x = N_x \pi / L_x = \pi / a$, at $q_y = \pi / a$, and at $q_z = \pi / a$ by the same array of points but now bounded by an octant of the same volume in $q_x - q_y - q_z$ space (Fig. 10.13(a)). This demands that

$$\left(\frac{\pi}{a}\right)^3 = \frac{1}{8}\left(\frac{4}{3}\pi q_D^3\right)$$

$$q_D = \left(\frac{6\pi^2}{a^3}\right)^{1/3} = 1 \cdot 2\left(\frac{\pi}{a}\right), \qquad (10.29)$$

where q_D is the radius of the replacement octant, and is often called the Debye circular wavenumber.

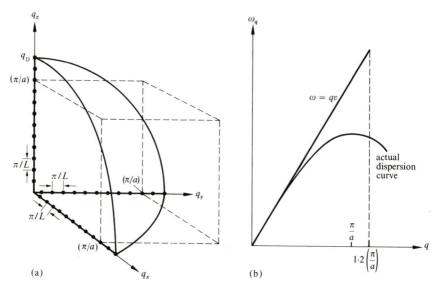

Fig. 10.13. The nature of the Debye approximations. (a) The lattice of points in q_x–q_y–q_z space is taken as bounded by an octant of radius q_D such that the total number of points within the octant equals the number within the cube of side π/a. (b) The linear dispersion relation $\omega = qv$ replaces the actual dispersion relation.

The second approximation we shall make use of was introduced by Debye in 1912. He assumed that the velocity of both longitudinal and transverse waves in a solid share a common value v which is independent of frequency. Since $v = \lambda/T_v = 2\pi\lambda/2\pi T_v = \omega/q$, this implies

$$\omega = qv. \tag{10.30}$$

Fig. 10.13(b) shows the nature of this approximation. Only at high q (for example, at $q = \pi/a$, where $\lambda = 2a$) does the true dispersion curve differ significantly from that assumed by Debye. We shall now use both these approximations in evaluating the molar internal energy of the crystal via eqn (10.28).

The heat capacity

The first step in evaluating eqn (10.28) is to replace the summation by an integral. Since each point in q-space describing the various modes which can exist is separated from its nearest-neighbour by π/L (see Fig. 10.13(a)) it follows that each point occupies a volume of $(\pi/L)^3$, as illustrated in Fig. 10.14(a). The number of points which lie within the shell of radius q and thickness dq shown in Fig. 10.14(b) is therefore equal to the volume of the shell divided by the volume occupied by each

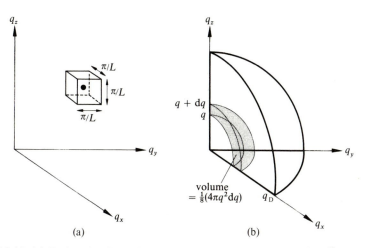

Fig. 10.14. (a) Each point, indicating a mode, occupies a volume of $(\pi/L)^3$ in q_x-q_y-q_z space. (b) The number of points within the shaded shell is equal to its volume divided by $(\pi/L)^3$.

point, that is $\frac{1}{8}(4\pi q^2\,dq)/(\pi/L)^3$. Eqn (10.28) thus becomes

$$U_{m3} = \frac{3}{8}\left(\frac{L}{\pi}\right)^3 \int_0^{q_D} \frac{\hbar\omega_q}{e^{\hbar\omega_q/kT}-1}\, 4\pi q^2\,dq \qquad (10.31)$$

where the factor of three arises because there is one longitudinal and two transverse modes at each value of (q_x, q_y, q_z).

The next step is to introduce the dispersion relation (eqn (10.30)). This tells us that $q = \omega/v$ and $dq = d\omega/v$. Substituting these expressions into eqn (10.31) gives

$$U_{m3} = \frac{3}{8}\left(\frac{L}{\pi}\right)^3 \frac{1}{v^3} \int_0^{\omega_D} \frac{\hbar\omega}{e^{\hbar\omega/kT}-1}\, 4\pi\omega^2\,d\omega \qquad (10.32)$$

where the upper limit of integration is

$$\omega_D = q_D v = \left(\frac{6\pi^2}{a^3}\right)^{1/3} v \qquad (10.33)$$

as follows from eqn (10.29).

Differentiating eqn (10.32) with respect to T gives

$$C_{V,m} = \frac{3}{8}\left(\frac{L}{\pi}\right)^3 \frac{1}{v^3} \int_0^{\omega_D} \frac{\hbar\omega(\hbar\omega/kT^2)\,e^{\hbar\omega/kT}}{(e^{\hbar\omega/kT}-1)^2}\, 4\pi\omega^2\,d\omega$$

$$C_{V,m} = \frac{3V_m k^4 T^3}{2\pi^2 v^3 \hbar^3} \int_0^{\omega_D} \frac{(\hbar\omega/kT)^2\,e^{\hbar\omega/kT}}{(e^{\hbar\omega/kT}-1)^2}\left(\frac{\hbar\omega}{kT}\right)^2 d\left(\frac{\hbar\omega}{kT}\right) \qquad (10.34)$$

where the molar volume $V_m = L^3$. For neatness let us write $x = (\hbar\omega/kT)$. This means, of course, that the upper limit of integration becomes $x_D = \hbar\omega_D/kT = \theta_D/T$, where θ_D, known as the *Debye temperature*, is given by

$$\theta_D = \frac{\hbar\omega_D}{k} = \frac{\hbar}{k}\left(\frac{6\pi^2}{a^3}\right)^{1/3}v. \tag{10.35}$$

On making these substitutions, along with $R = N_A k = (V_m/a^3)k$, into eqn (10.34) we obtain

$$C_{V,m} = 9R\left(\frac{T}{\theta_D}\right)^3 \int_0^{\theta_D/T} \frac{x^4 e^x}{(e^x - 1)^2}\,dx \tag{10.36}$$

as Debye's famous relation for the heat capacity of a monatomic solid.

At high temperatures θ_D/T is small, so that the values of x which the integrand takes are small. This allows us to approximate e^x in the numerator by unity and to write $e^x = 1 + x$ in the denominator. Eqn (10.36) then approximates to

$$C_{V,m} = 9R\left(\frac{T}{\theta_D}\right)^3 \int_0^{\theta_D/T} \frac{x^4}{(1+x-1)^2}\,dx$$

$$C_{V,m} = 3R = 24\cdot9 \text{ J K}^{-1}\text{ mol}^{-1}$$

at $T \gg \theta_D$ which is, of course, in accord with experiment (Fig. 10.15).

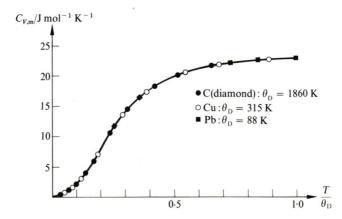

Fig. 10.15. The curve shows the molar heat capacity of diamond, copper, and lead calculated from Debye's relation (eqn (10.36)) assuming $\theta_D = 1860$ K for diamond, $\theta_D = 315$ K for copper, and $\theta_D = 88$ K for lead. These values of θ_D compare favourably with those given by eqn (10.35).

TABLE 10.1
The Debye temperature of selected solids

Substance	θ_D/K measured[a]	θ_D/K predicted[b]
Pb	88	135
Na	150	165
Ag	215	220
Zn	250	310
Cu	315	365
Al	390	440
CaF_2	475	540
NaCl	280	290

[a] These values best fit eqn (10.36) to the experimental $C_{V,m}$ data.
[b] Deduced from eqn (10.35).

At low temperatures the upper limit of integration in eqn (10.36) may be taken as infinity for all practical purposes. This being so the integral tends to a constant value $(4\pi^4/15)$. Therefore

$$C_{V,m} = \frac{12\pi^4}{5} R\left(\frac{T}{\theta_D}\right)^3 \qquad (10.37)$$

at $T \ll \theta_D$. This T^3 variation in heat capacity is very closely obeyed by a wide variety of substances.

Although the integral in eqn (10.36) can only be expressed analytically at the extremes of high and low temperatures it can be evaluated numerically at any temperature. This leads to the curve shown in Fig. 10.15. Comparing Fig. 10.15 with Fig. 10.1 we see that Debye's theory fits the experimental data much more closely than does Einstein's theory. To improve on Debye's theory one must work with eqn (10.28) and the calculated dispersion relations.

Table 10.1 records the Debye temperatures of a variety of materials; column 2 gives the values of θ_D which best fit eqn (10.36) to the measured values of $C_{V,m}$; column 3 gives θ_D as calculated from eqn (10.35).

10.6. Heat capacity of metals

The heat capacities of metals are particularly enigmatic. To explain their high electrical conductivity it is necessary to suppose that at least some of the metal's electrons are free to move. However, classical physics dictates that each of these free electrons will (as a result of collisions with the metal ions) come into thermal equilibrium with the structure and so have a kinetic energy of $\frac{1}{2}m\overline{u^2} = \frac{3}{2}kT$. Consequently, each free electron should contribute $\frac{3}{2}k$ to the heat capacity. By way of example, the observed electrical conductivity of copper can be explained on the assumption that every atom in the metal gives up one electron to the common pool. Hence the molar heat capacity of copper should exceed that of a monatomic insulator by $1 \cdot 5R$, and thus reach a value of $4 \cdot 5R$ at high temperatures. In fact, as Fig. 10.1(b) shows, it never exceeds $3R$. Clearly, the electrons do not obey the laws of classical mechanics. A new model is required.

The free-electron model

This model starts by replacing the localized positive ions (such as the Cu^+ ions) by a uniform background of positive charge—a kind of jelly—within which the electron gas can move freely.

According to Newton's second law, if a free electon (mass m) residing in an isolated rectangular block of copper (Fig. 10.16(a)) is given a momentum mu_x it will proceed to bounce back and forth between opposite faces of the block. The electron is normally prevented from escaping by the net positive charge it leaves behind as it attempts to escape (the block plus electron are electrically neutral). Only if the speed of the electron is such that $\frac{1}{2}mu_x^2$ exceeds the energy required to remove the electron to infinity will it escape. Classical mechanics assures us that provided u_x is less than the escape speed it—and thus mu_x—can be given any value we choose. Wave mechanics says otherwise.

According to wave mechanics a free electron with a specified momentum mu_x must be thought of as a group of waves whose wavelength is $\lambda = h/mu_x$ (eqn (2.11)). So what then is to replace our classical picture of a particle bouncing back and forth between opposite faces of the block? Recalling how transverse standing wave patterns can be established on a rope we might guess that the electron's wave functions $\psi(x)$ will have the form shown in Fig. 10.16(b). Here an integral number n of half-wavelengths are shown fitted into the length L_x of the block. Thus the wave functions are characterized by $n(\lambda_x/2) = L_x$ or, in terms of the circular wavenumber k_x ($= 2\pi/\lambda_x$), by

$$k_x = \left(\frac{\pi}{L_x}\right), 2\left(\frac{\pi}{L_x}\right), 3\left(\frac{\pi}{L_x}\right), \dots \tag{10.38}$$

A necessary condition to be satisfied by any acceptable wave function is that $|\psi(x)|^2 = 0$ at $0 > x > L_x$. In words, this condition states that there is no probability of finding the electron outside the block. Fig. 10.16(c)

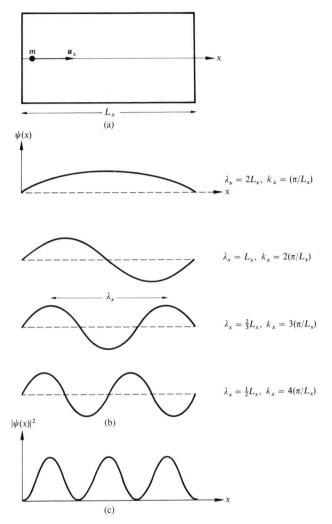

Fig. 10.16. A free electron in a rectangular block of metal as described (a) by classical mechanics, and (b) by wave mechanics. (c) Showing how $|\psi(x)|^2$ varies along the x-axis for the wave function with $k_x = 3(\pi/L_x)$.

shows that $|\psi(x)|^2 = 0$ at $0 > x > L_x$ for the wave function with $k_x = 3(\pi/L_x)$. The same is obviously true of the other wave functions shown in Fig. 10.16(b).†

† In suggesting that the wave functions have the form shown in Fig. 10.16(b) we are *not* saying that $\psi(x)$ has the same time-dependence as standing waves on a rope. If it did, $\psi(x, t)$ would be zero at certain times (when the rope is straight). At such times $|\psi(x, t)|^2$ would vanish everywhere—the electron would disappear! In fact $\psi(x, t) = \psi(x)[\cos(Et/\hbar) - i \sin(Et/\hbar)]$, where $i = (-1)^{1/2}$ and E is the energy of our free electron. All we are doing in Fig. 10.16(b) is to sketch the form of $\psi(x)$. Although $\psi(x, t)$ is complex, $|\psi(x, t)|^2$ is always real. In the present problem $|\psi(x, t)|^2 = |\psi(x)|^2$.

So far we have really considered the one-dimensional problem of an electron confined between $x = 0$ and $x = L_x$. Extending the arguments we might reasonably expect that the wave function of a free electron in a rectangular block of metal of length L_x, width L_y, and depth L_z is characterized by the values of (k_x, k_y, k_z), where k_x, k_y, and k_z must be chosen from the following list. (We cannot choose $k_x = k_y = k_z = 0$, that is, $\lambda_x = \lambda_y = \lambda_z = \infty$, since this says the electron is not in the metal.)

$$
\begin{array}{l}
k_x = 0, (\pi/L_x), 2(\pi/L_x), 3(\pi/L_x), \ldots \\
k_y = 0, (\pi/L_y), 2(\pi/L_y), 3(\pi/L_y), \ldots \\
k_z = 0, (\pi/L_z), 2(\pi/L_z), 3(\pi/L_z), \ldots
\end{array}
\qquad (10.39)
$$

From now on it will suit our purposes best to take the block of metal to be a cube of side $L = L_x = L_y = L_z$. It will also help provide a neat graphical method of recording the (k_x, k_y, k_z) value of an electron's wave function. Fig. 10.17 shows the method we shall adopt. What we do is to construct a simple cubic lattice of points in k_x–k_y–k_z space in which the nearest-neighbour separation is (π/L). Because of the form of eqn (10.39) (with

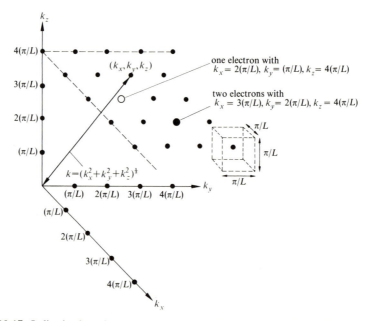

Fig. 10.17. Indicating how the quantum numbers of the free electrons in a cube of metal (of side L) may be recorded. An open circle at (k_x, k_y, k_z) shows that only one electron has a wave function with these circular wavenumbers; a filled circle shows that two electrons have the same (k_x, k_y, k_z).

$L = L_x = L_y = L_z$) it follows that each point may be used to characterize the wave function of an electron. Suppose an electron has a wave function with $k_x = 2(\pi/L)$, $k_y = (\pi/L)$, $k_z = 4(\pi/L)$. To denote this we may draw in an open circle around this lattice point as in Fig. 10.17. If a second electron has the same value of (k_x, k_y, k_z) we may fill in the circle. Fig. 10.17 thus shows that two electrons have $k_x = 3(\pi/L)$, $k_y = 2(\pi/L)$, $k_z = 4(\pi/L)$. According to quantum mechanics a maximum of two electrons— of opposite 'spin'—can have the same values of (k_x, k_y, k_z). (In the language of classical physics, we would picture the electron as a spinning sphere whose axis may point in two different directions, often called 'up' and 'down'.) The values of (k_x, k_y, k_z) and a statement of whether the electron is 'spin up' or 'spin down'—all four pieces of information—may be said to constitute the *quantum numbers* of our free electron.

The energy of a free electron

We have argued that the potential energy of a free electron is constant inside a metal. Taking this constant value as defining the zero of potential energy allows us to equate the total energy E of an electron to its kinetic energy. Thus

$$E = \tfrac{1}{2}mu^2 = \frac{1}{2m}[(mu_x)^2 + (mu_y)^2 + (mu_z)^2]. \qquad (10.40)$$

Since $k_x = 2\pi/\lambda_x$ and $\lambda_x = h/mu_x$, that is $k_x = mu_x/\hbar$, with similar expressions for k_y and k_z, eqn (10.40) becomes

$$\boxed{E = \frac{\hbar^2}{2m}(k_x^2 + k_y^2 + k_z^2) = \frac{\hbar^2 k^2}{2m}} \qquad (10.41)$$

where k is the distance of the point (k_x, k_y, k_z) from the origin in Fig. 10.17. In other words, a sphere, or rather an octant, drawn in k_x–k_y–k_z space with the origin as centre and of radius $k = (k_x^2 + k_y^2 + k_z^2)^{1/2}$ is a surface of constant energy on which $E = \hbar^2 k^2/2m$.

If the metal is supposed to be at $T = 0$, then the quantum numbers must be assigned to the free electrons in such a way as to minimize the mean energy of an electron. Because E and (k_x, k_y, k_z) are related by eqn (10.41) this means we must assign quantum numbers starting from the origin of k_x–k_y–k_z space (excluding $k_x = k_y = k_z = 0$) and moving outwards in spherical fashion. If there are N free electrons to be assigned quantum numbers, the radius k_F at which we stop this procedure will be such that the octant shown in Fig. 10.18 contains $\tfrac{1}{2}N$ points within it. The factor of $\tfrac{1}{2}$ arises, of course, because each point can describe the k-values of two electrons of opposite spin. Since a point in k_x–k_y–k_z space

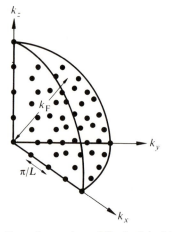

Fig. 10.18. The heavy dots indicate those values of (k_x, k_y, k_z) which have been assigned to two electrons in the metal.

occupies a volume $(\pi/L)^3$ (see Fig. 10.17), and since the octant of radius k_F has a volume $\frac{1}{8}(\frac{4}{3}\pi k_F^3)$, it follows that

$$\frac{1}{2}N = \frac{\frac{1}{8}(\frac{4}{3}\pi k_F^3)}{(\pi/L)^3}$$

$$k_F = \left(\frac{3N\pi^2}{L^3}\right)^{1/3} = (3\pi^2 n)^{1/3}, \tag{10.42}$$

where $n = N/L^3 = N/V$ is the number of free electrons per unit volume of the metal. The energy E_F of those electrons with $k = k_F$ is, from eqns (10.41) and (10.42), given by

$$E_F = \frac{\hbar^2 k_F^2}{2m} = \frac{\hbar^2}{2m}(3\pi^2 n)^{2/3} \tag{10.43}$$

and this energy is referred to as the *Fermi energy*. Values of E_F for a variety of metals will be found in Table 11.2 (p. 375).

The Fermi–Dirac energy distribution function

To find the energy distribution function of the free-electron gas we start by noting that since each point in k_x–k_y–k_z space occupies a volume $(\pi/L)^3$ the number of points lying within an octant-shaped shell of radii k and $k + dk$ (cf. Fig. 10.14(b)) is $\frac{1}{8}(4\pi k^2 \, dk)/(\pi/L)^3 = Vk^2 \, dk/2\pi^2$. Because each of these points can describe the values of (k_x, k_y, k_z) of two electrons (with opposite spins) it follows that the number dN of electrons whose

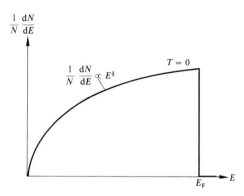

Fig. 10.19. The Fermi–Dirac energy distribution function of a free-electron gas at absolute zero.

circular wavenumbers lie between k and $k+\mathrm{d}k$ is given by

$$\mathrm{d}N = \frac{V}{\pi^2} k^2 \, \mathrm{d}k \tag{10.44}$$

for $k < k_F$ at $T=0$. Now eqn (10.41) tells us that $k^2 = 2mE/\hbar^2$, and thus

$$\mathrm{d}k = \frac{m \, \mathrm{d}E}{\hbar(2mE)^{1/2}}, \tag{10.45}$$

allowing eqn (10.44) to be rewritten in the form

$$\frac{\mathrm{d}N}{\mathrm{d}E} = \frac{V}{2\pi^2} \left(\frac{2m}{\hbar^2}\right)^{3/2} E^{1/2}. \tag{10.46}$$

Dividing eqn (10.46) through by N gives the, so-called, *Fermi–Dirac energy distribution function* of a free-electron gas as

$$\boxed{\frac{1}{N}\frac{\mathrm{d}N}{\mathrm{d}E} = \frac{1}{2\pi^2 n} \left(\frac{2m}{\hbar^2}\right)^{3/2} E^{1/2}} \quad \text{at } T=0. \tag{10.47}$$

This only holds up to $E = E_F$. Since all points with $k > k_F$ are (by definition of k_F) unassigned at $T=0$ it follows that $\mathrm{d}N/\mathrm{d}E = 0$ when $E > E_F$, at $T=0$. Eqn (10.47) is plotted graphically in Fig. 10.19.

The electronic heat capacity

To calculate the heat capacity C_e of the electron gas it is necessary to know its energy distribution function at a finite temperature. If we heat the metal from absolute zero to a temperature T each ion of the solid

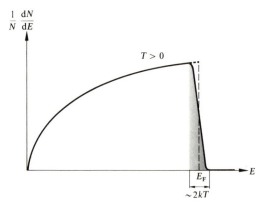

Fig. 10.20. This Fermi–Dirac energy distribution function of a free-electron gas at a finite temperature T. The area shown shaded is equal to the electrons which are thermally excited at temperature T.

will be given a mean energy of order kT. However, the only electrons which can accept this energy as they 'collide' with the ions are those lying within a range of about kT of E_F. Electrons whose energy is lower than about $(E_F - kT)$ cannot accept extra energy, for the states to which they would be excited (meaning the new values of (k_x, k_y, k_z), with due regard to spin, to which they would have to move) are already filled. Electrons whose energy is around E_F can, of course, move up to states corresponding to an energy of about $(E_F + kT)$. Hence we would expect the energy distribution function to take the form shown in Fig. 10.20. The fraction of the electrons with energies in the range $(E_F - kT)$ to $(E_F + kT)$ is equal to the area shown shaded in Fig. 10.20. Remembering that the area of a triangle is $\frac{1}{2}$base × height this fraction is given by

$$\frac{1}{N} \int_{E_F - kT}^{E_F + kT} \frac{\mathrm{d}N}{\mathrm{d}E} \, \mathrm{d}E \approx \frac{1}{2N} (2kT) \left(\frac{\mathrm{d}N}{\mathrm{d}E}\right)_{E_F} = \left(\frac{3}{\pi^4}\right)^{1/3} \frac{Vmn^{1/3}}{N\hbar^2} kT, \quad (10.48)$$

where we have substituted for $\mathrm{d}N/\mathrm{d}E$ from eqn (10.46) and for E_F from eqn (10.43). Each of these electrons—and only these electrons—will have its energy *increased* by $\frac{3}{2}kT$, on average, as it is thermally excited. Hence the total internal energy of the electrons of a metal at temperature T (relative to their energy at $T = 0$) is given by the product of eqn (10.48) and $N(\frac{3}{2}kT)$. Differentiating this product with respect to T gives

$$C_e = \left(\frac{3}{\pi^4}\right)^{1/3} \frac{Vmn^{1/3}}{\hbar^2} \frac{\mathrm{d}}{\mathrm{d}T} (kT \times \tfrac{3}{2}kT).$$

If we assume that there is one mole of metal present and that each atom contributes one free electron then $V = V_m$ and $n = N_A/V_m$. Thus

$$\boxed{C_{e,m} = \gamma T}$$ (10.49)

where

$$\gamma = \left(\frac{3}{\pi}\right)^{4/3} \frac{V_m^{2/3} N_A^{1/3} m k^2}{\hbar^2}.$$ (10.50)

Substituting for the known values of the constants in eqn (10.50) predicts an electronic heat capacity for copper at 300 K of $0 \cdot 15\,\text{J mol}^{-1}\,\text{K}^{-1}$. As this is less than one per cent of $3R$ we see why the molar heat capacity of a metallic and of a non-metallic element are practically equal at high temperatures (Fig. 10.1).

At $T \ll \theta_D$, where the heat capacity of the structure is proportional to T^3 (eqn (10.37)), we may write the molar heat capacity $C_{V,m}$ of the metal as

$$\boxed{\frac{C_{V,m}}{T} = \beta T^2 + \gamma}$$

where β is a constant. A graph of $C_{V,m}/T$ plotted against T^2 should therefore be linear with a slope β and an intercept γ. Fig. 10.21 shows such a plot for copper. The linearity provides striking confirmation of the T^3 dependence of the atomic heat capacity. The intercept gives a value of $\gamma = 7 \cdot 0 \times 10^{-4}\,\text{J mol}^{-1}\,\text{K}^{-2}$, as against the value of $4 \cdot 9 \times 10^{-4}\,\text{J mol}^{-1}\,\text{K}^{-2}$

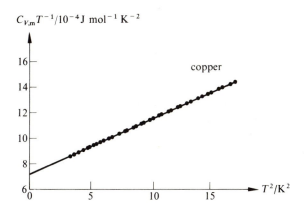

Fig. 10.21. Showing how $C_{V,m}/T$ varies with T^2 for copper over the temperature range $1 \cdot 7$ to 4 K.

predicted by eqn (10.50). The discrepancy arises because the electrons are not as 'free' as we have supposed.

Exercise 10.3

Each atom of solid potassium contributes one electron to the free-electron gas. Calculate (a) the Fermi energy at $T = 0$, (b) the number of states per unit range of energy at E_F in a 1-mol specimen at $T = 0$ (a state being labelled by (k_x, k_y, k_z) and the spin quantum number), (c) the fraction of the free electrons whose energies lie within kT of E_F at 300 K, and (d) the temperature of a classical electron gas with a mean kinetic energy equal to E_F. Take $A_r(K) = 39$ and $\rho = 860$ kg m^{-3}.

Calculations. (a) Before using eqn (10.43) we will need to know the number of electrons per unit volume n. Since each atom contributes one electron to the pool, n will equal the number of atoms per unit volume. Therefore $n = N_A/(39 \times 10^{-3}\,\text{kg}/860\,\text{kg m}^{-3}) = 1 \cdot 33 \times 10^{28}\,\text{m}^{-3}$, giving $E_F = \hbar^2 (3\pi^2 n)^{2/3}/2m = 3 \cdot 28 \times 10^{-19}\,\text{J}$. (b) The number of distinct states per unit range of energy—usually called the *density of states*—is given by $\mathrm{d}N/\mathrm{d}E$ or, substituting $E = E_F = 3 \cdot 28 \times 10^{-19}\,\text{J}$ and $V = V_m\ (= 39 \times 10^{-3}\,\text{kg}/860\,\text{kg m}^{-3})$ into eqn (10.46), by $2 \cdot 76 \times 10^{42}\,\text{J}^{-1}$. (c) Dividing the number of electrons within a range of kT of E_F, as given by $kT(\mathrm{d}N/\mathrm{d}E)_{E_F}$ (see eqn (10.48)), by the total number of electrons, as obtained by integrating eqn (10.46) from $E = 0$ to E_F, gives the fraction of the free electrons within the range kT of E_F at 300 K as $\frac{3}{2}(kT)/E_F = 0 \cdot 019$. (d) Equating $\frac{1}{2}m u^2 = \frac{3}{2}kT$ to $E_F = 3 \cdot 28 \times 10^{-19}\,\text{J}$ gives $T = 1 \cdot 6 \times 10^4\,\text{K}$. It is thus not surprising that attempts to apply classical mechanics to the electron gas fail to work at the temperatures at which metals are solids (below $10^3\,\text{K}$).

10.7. Point defects

If the thermal energy of an atom in a crystal exceeds the energy required to remove an atom from its lattice site then that atom can leave its site. As a result a number of *point defects* will be formed in the structure. The word point is used to denote the localized nature of the defects.

Defect types

Fig. 10.22 shows how point defects may arise. An atom may simply migrate a short distance from the *vacancy* it created on leaving (Fig. 10.22(a)). This localized *interstitial-vacancy* pair is called a *Frenkel defect*. Or it may migrate to the surface leaving an isolated vacancy; a *Schottky defect* (Fig. 10.22(b)). Another possibility is for an atom to diffuse in from the surface of the crystal to form an isolated interstitial (Fig. 10.22(c)). As we would expect, the crystal structure influences the relative number of the different types of defect. In a tightly-packed structure, such as a hcp or fcc metal (Figs. 8.12 and 8.13), the energy required to introduce an interstitial atom of the element is so large as to make their occurrence extremely unlikely. In more open structures, such as that of diamond (Fig. 8.16), interstitials are much more easily accommodated.

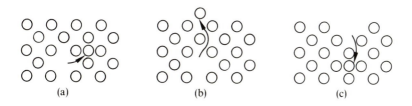

Fig. 10.22. Illustrating the formation of (a) a vacancy-interstitial pair (a Frenkel defect), (b) an isolated vacancy (a Schottky defect), and (c) an isolated interstitial.

Vacancy concentration

We will now consider the temperature dependence of the number of isolated vacancies occurring in a crystal. This type of defect can, and indeed does, occur in all crystals.

The number of vacancies existing in a crystal should be roughly equal to the number of atoms whose vibrational energy E exceeds the energy E_v required to remove an atom from the interior of a crystal to its surface. To find this number we will adopt an Einstein model of the crystal. According to this model (eqn (10.4) with $\beta = 1/kT$) the number of atoms with *one particular energy* E (one which is quantized in integral multiples of $\hbar\omega$) is given by

$$N(E) = A\, e^{-E/kT}. \tag{10.51}$$

This is shown graphically in Fig. 10.23 where the vertical lines indicate that only discrete energies are permitted. If we now look at the energy range dE shown in Fig. 10.23 we see that the number of atoms whose

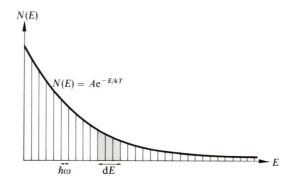

Fig. 10.23. The number of atoms with a particular energy E according to Einstein's model. The vertical lines emphasize that the energy is quantized in multiples of $\hbar\omega$. In the energy interval dE at E there will be $dE/\hbar\omega$ more atoms than there are at energy E. (The number $N(E)$ is effectively constant within the interval dE.)

energies lie within this range is given by $A \exp(-E/kT)$ multiplied by the number of vertical lines (more formally, the number of permitted values of E) within dE, that is $A \exp(-E/kT)(dE/\hbar\omega)$. Integrating from $E = E_v$ to $E = \infty$ gives the number n of atoms with an energy in excess of E_v as

$$n = B \int_{E_v}^{\infty} e^{-E/kT} \, dE \qquad (10.52)$$

where $B = A/\hbar\omega$ is a constant. We can find B by noting that the total number of atoms N is given by

$$N = B \int_{0}^{\infty} e^{-E/kT} \, dE. \qquad (10.53)$$

Substituting for B from eqn (10.53) into eqn (10.52) gives

$$n = N \frac{\displaystyle\int_{E_v}^{\infty} e^{-E/kT} \, dE}{\displaystyle\int_{0}^{\infty} e^{-E/kT} \, dE} = N \frac{[-e^{-E/kT}]_{E_v}^{\infty}}{[-e^{-E/kT}]_{0}^{\infty}}$$

$$\boxed{n = N e^{-E_v/kT}} \qquad (10.54)$$

as the number of vacancies, each of which requires an energy E_v to be formed, in the crystal at a temperature T.

One way to measure the vacancy concentration is to study how the overall volume of a crystal increases with increasing temperature. Most of this increase in volume is due to thermal expansion of the structure. This contribution can be measured by using X-ray diffraction techniques to record how the lattice spacing changes with temperature. The remaining increase must be due to the formation of vacancies with (we assume) the displaced atoms going to the surface of the crystal. By studying the temperature dependence of this excess volume (which is proportional to n)—the results fit eqn (10.54)—one obtains E_v. In solid argon, for example, $E_v = 3 \times 10^{-20}$ J.

We can make a simple theoretical estimate of E_v by noting that it is equal to the energy required to remove an atom from the interior of the crystal to infinity minus the energy released when an atom is condensed on the surface from an infinite distance away. If the atom is surrounded by n nearest neighbours in the interior of the crystal and the dissociation energy of an isolated pair of atoms is ΔE, it requires an energy of approximately $n \, \Delta E$ to remove the atom to infinity. In condensing the atom back onto the surface an energy of about $(n/2) \, \Delta E$ is released—on the surface an atom is surrounded by only half as many neighbours as in the interior—giving $E_v = \frac{1}{2} n \Delta E$. For argon $n = 12$ and $\Delta E = 1 \cdot 7 \times 10^{-21}$ J, yielding $E_v = 1 \cdot 0 \times 10^{-20}$ J, which is reasonably close to the measured value of 3×10^{-20} J.

As a numerical example, eqn (10.54) predicts that the fraction n/N of vacant sites in solid argon at 50 K is $\exp(-3 \times 10^{-20}/1 \cdot 38 \times 10^{-23} \times 50) = 1 \cdot 3 \times 10^{-19}$. Even at the triple-point temperature (83·8 K) the fraction is only 5×10^{-12}.

10.8. Thermal expansion

Up to now we have assumed that the only effect of raising the temperature of a crystal from near $T = 0$ is to increase the amplitude of vibration of the atoms, with the atoms continuing to vibrate about the sites they would (according to classical mechanics) occupy at $T = 0$. If this were really so a solid could never expand as its temperature is raised. In general, the *linear expansivity* α of a solid, defined as the fractional change in length per unit change of temperature, is small but it is finite. For metals α is of order 10^{-5} K^{-1} at room temperature.

To try to discover why solids expand let us consider how an isolated pair of atoms behave. Fig. 10.24 shows such a pair in which atom i is held fixed. In Fig. 10.24(a) it has been assumed that $F(r)$ is linear about the static equilibrium separation r_0 of the pair. To represent a finite temperature, atom j must be given kinetic energy. This we may do by pulling out j to a separation r_2 and letting it go. It will, of course, accelerate inwards, overshoot r_0, and continue on inwards until it comes to a momentary rest at a separation r_1, before returning to separation r_2. It will then continue to vibrate between r_1 and r_2. In pulling atom j out to separation r_2 one increases its potential energy by an amount V equal to the area under the force curve between r_0 and r_2. As j speeds back this energy is exclusively kinetic at r_0, and is again all potential at r_1. Throughout the vibration the *total* energy remains constant—the pair is isolated—but the energy changes from one form to another. The mean separation of the pair is the average of r_1 and r_2. Because the force curve is linear, this mean separation is r_0; in other words a solid will not expand if the force–separation curve is strictly linear. You can, of course, reach the same conclusion by arguing in terms of the potential energy curve. A linear $F(r)$ implies a parabolic $\mathcal{V}(r)$ (the lower curve in Fig. 10.24(a)).

Fig. 10.24(b) shows a more realistic force curve with its rapidly increasing repulsive force as the atoms approach closer than r_0. (This may be simulated by mounting a compression spring inside a weaker extension spring.) Once again we move atom j out to some position r_2', so giving it an energy V equal to the area under the force curve between r_0 and r_2'. On releasing it, atom j will move in to some position r_1' such that the area

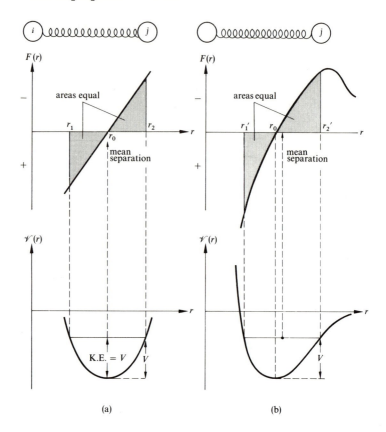

Fig. 10.24. The linear interatomic force characteristic of (a) leads to no change in the mean separation of atoms i and j as the energy of j is increased. In (b) the mean separation of i and j does increase.

under the force curve between r_1' and r_0 is equal to that between r_0 and r_2'. Because of the non-linear nature of the force curve (or, equivalently, because of the asymmetrical potential well) the mean of r_1' and r_2' is greater than r_0. We have thermal expansion.

10.9. Lindemann's theory of fusion

Building on the idea that the amplitude of vibration of the atoms in a solid increases as the temperature is increased, Lindemann in 1910 proposed that when this amplitude of vibration reaches a certain critical fraction of the mean interatomic spacing the vibrations will interfere to such an extent that the structure becomes mechanically unstable—it fuses (melts). He further proposed that this critical fraction should be the same for all crystals with the same structure. We need only expect the model to be applicable to structures with a single atom as basis; molecules have a vibrational complexity which rules out such a simple model. Lindemann adopted the Einstein model of a solid. Here, you will recall, each atom is assumed to vibrate independently of one another with a common circular frequency ω. The mean energy $\bar{E}_x$ of an atom is given by Planck's relation (eqn (10.7) with $\beta = 1/kT$), namely

$$\bar{E}_x = \frac{\hbar\omega}{e^{\hbar\omega/kT} - 1}.$$ (10.55)

Since most solids fuse at a temperature $T \gg \hbar\omega/k$ ($= \theta_E$)—that is, at a temperature at which $C_{V,m}$ has plateaued out at $3R$—we may take

$$\bar{E}_x = kT$$ (10.56)

this being the value of eqn (10.55) when $kT \gg \hbar\omega$. According to eqn (3.52) $\bar{E}_x$ is related to ω and the amplitude A of vibration by

$$\bar{E}_x = \tfrac{1}{2}m\omega^2 A^2.$$ (10.57)

Setting $\bar{E}_x$ as given by eqn (10.56) with $T = T_f$, the fusion temperature, equal to $\bar{E}_x$ as given by eqn (10.57) with $A = cr_0$, where c is the supposedly constant fraction of the mean interatomic spacing r_0 at which A is sufficient for fusion to occur, gives

$$c^2 = \left(\frac{2\hbar^2}{k}\right)\frac{T_f}{\theta_E^2 m r_0^2}$$ (10.58)

where we have written $\hbar\omega/k = \theta_E$, the Einstein temperature (eqn (10.10)). If Lindemann's theory is correct c as calculated from eqn (10.58) should be constant for all elements of the same crystal structure and be less than unity. Table 10.2 sets out data for a variety of elements whose structure is

TABLE 10.2
Values of c as deduced from Lindemann's theory

Element	$r_0/10_m^{-10}$	T_f/K	θ_E/K	$c/10^{-2}$
Ne	3·16	24	50	6·8
Ar	3·76	84	60	6·3
Xe	4·34	161	40	6·3
Pb	3·50	600	58	8·2
Al	2·86	933	326	6·2
Cu	2·56	1356	240	7·4
Pt	2·77	2044	149	7·7

fcc. (Here r_0 is taken to be the nearest-neighbour separation in the crystal structure of the element.) We see that even though T_f changes a hundredfold c remains constant to within 30 per cent.

Despite its seeming success, Lindemann's theory fails in the last analysis for it cannot explain why the structure of a solid breaks down so suddenly within a very narrow range of temperature. It fails because it deals only with the mean energy (eqn (10.56)) and proceeds by assuming that each and every atom has energy kT_f at the fusion temperature. Its limited success—namely that of predicting a constant value of c—arises because of these unwarranted assumptions.

PROBLEMS

10.1. An atom i lies between two atoms j and k as shown in Fig. 3.10(b). Assuming that j and k can be regarded as fixed (their average position is on a lattice site), show that the amplitude A of vibration of i is related to the temperature T of the (one-dimensional) solid by $A = (kT/k_s)^{1/2}$, where k_s is the force constant in an isolated pair of atoms. What percentage of the mean interatomic separation $(2\cdot86\times10^{-10}\,\text{m})$ is A in aluminium at 300 K? Assume $k_s = 20\,\text{N m}^{-1}$.

10.2. Show that in an isolated pair of atoms which interact according to the Mie potential model (eqn (2.18)) the restoring force k_s per unit displacement from equilibrium is $m(n-m)Ar_0^{-(m+2)}$. In an isolated NaCl molecule $r_0 = 2\cdot5\times10^{-10}\,\text{m}$, $A = e^2/4\pi\varepsilon_0$, $m = 1$, and $n = 9$. Calculate k_s and the period of vibration of the molecule (assuming the chlorine ion to be held fixed). $(A_r(\text{Na}) = 23.)$

10.3. Explain how Dulong and Petit's law might be used to determine the relative atomic mass of an element (which is available in solid form).

10.4. Calculate the Einstein frequency ω and the Einstein temperature θ_E in copper, given that the Young's modulus of copper is $1\cdot3\times10^{11}$ Pa and that the interatomic separation $r_0 = 2\cdot6\times10^{-10}\,\text{m}$. Copper has a density of $8\cdot93\times10^3\,\text{kg m}^{-3}$ and a relative atomic mass of $63\cdot5$.

10.5. What, according to Einstein's theory, is the heat capacity of a monatomic solid at the Einstein temperature?

10.6. The molar heat capacity of copper at 1 K is only about 10^{-5} of its value at room temperature (see Fig. 10.1(b)). If a piece of copper at 1 K falls through a distance of 10^{-5} m and the energy is dissipated as heat, what will be the percentage rise in the temperature of the block? (This can be an important consideration in the design of equipment to operate at very low temperatures.) Assume $g = 10\,\text{m s}^{-2}$ and $A_r(\text{Cu}) = 63\cdot5$.

10.7. Crystalline sodium chloride strongly absorbs electromagnetic radiation of frequency 5×10^{12} Hz. Assuming this absorption occurs when the electromagnetic radiation has a frequency equal to the vibrational frequency of a Na^+ ion, calculate the potential energy $V(r-r_0)$ of a Na^+ ion as a function of the displacement $(r-r_0)$ from its equilibrium position. Assume that the ion's vibrations are simple harmonic. Take $A_r(\text{Na}) = 23$.

10.8. Show that the experimental values of $C_{V,m}$ for copper at different temperatures, as plotted in Fig. 10.1(b), are indeed in reasonable agreement with the values predicted by Einstein's relation with $\theta_E = 240$ K. You should also explore the effect of changing θ_E by, say, ten per cent.

10.9. Four gliders of the same mass are connected together by identical springs in a similar fashion to the arrangement shown in Fig. 10.4(a). Draw out all the possible distinct vibrational modes. In deciding if two modes differ disregard the springs and look at how the gliders are behaving. Assuming that each glider has a mass of 0·4 kg and each spring a force constant of 8×10^{-2} N m^{-1}, calculate the circular frequency of each mode on a track of total length 1·8 m. Now plot your results as a graph of ω_q against q.

10.10. By plotting a suitable graph demonstrate that the equation $y = A \cos qx \cos \omega t$ does indeed represent a standing wave of wavelength $2\pi/q$, of period $2\pi/\omega$, and of amplitude A. You will probably find it helpful to examine the waveform at different fixed times, separated by, say, one-eighth of the vibrational period.

10.11. Starting with the dispersion relation for a one-dimensional monatomic structure, eqn (10.16), calculate the velocity at which longitudinal vibrations of large wavelength ($q \approx 0$) travel down the structure. You should compare your answer with the rough estimate for the velocity of sound ($4r_0/T_v$) which we arrived at in §9.2.

10.12. Show that in a one-dimensional monatomic crystal of length L and interatomic separation a the number of standing wave patterns within a range $d\omega$ at a circular frequency ω is $2L \, d\omega/a\pi(\omega_m^2 - \omega^2)^{1/2}$, where ω_m is the maximum circular frequency which the structure can sustain.

10.13. In the Debye approach to calculating the heat capacity of an insulator certain (equivalent) modes are counted twice while other distinct modes are ignored (see Fig. 10.13(a)). What fraction of the distinct modes with values $q_z = 0$ are ignored in Debye's approach? *Clue*: Draw out the q_y–q_z plane of Fig. 10.13(a).

10.14. Calculate the maximum circular frequency of longitudinal waves in a linear one-dimensional crystal composed of atoms of mass 5×10^{-26} kg, when the interatomic force characteristic between an isolated pair of the atoms has the form shown in Fig. 2.7.

10.15. Consider a circularly-shaped chain of atoms of length L containing N identical atoms with an interatomic separation a. Compressional waves may run around the ring in either sense. A wave travelling in one sense is described by the equation $x = A \cos(\omega t - qX)$ and a wave travelling in the other sense by $x = A \cos(\omega t + qX)$. Here X is the distance around the ring. The, so-called, *periodic boundary conditions* demand that the displacement at X and at $X + L$ be equal (X and $X + L$ are the same point). Show that only those waves satisfying the condition

$$q = \pm \left(\frac{2\pi}{L}\right), \pm 2\left(\frac{2\pi}{L}\right), \pm 3\left(\frac{2\pi}{L}\right), + \cdots, \pm \frac{N}{2}\left(\frac{2\pi}{L}\right)$$

are permissible. (All the arguments which we based on standing waves could equally well have been based on travelling waves. The conclusions which we reached would have been unaltered.)

10.16. Adopting a free-electron model, calculate the Fermi energy E_F and the Fermi speed u_F (defined by $\frac{1}{2}mu_F^2 = E_F$) in copper, taking the number density of free electrons to be $8·5 \times 10^{28}$ m^{-3}. At what temperature would a classical gas have a mean thermal energy of E_F?

10.17. Show that the de Broglie wavelength of a free electron at the Fermi level is $(8\pi/3zN)^{1/3}$, where z is the number of free electrons contributed by each atom in the metal, of which there are N per unit volume. What is the value of this wavelength in silver? Take $z = 1$. (Density of silver $= 1 \cdot 0 \times 10^4$ kg m^{-3}. $A_r(\mathrm{Ag}) = 108$.)

10.18. Show that the mean energy of an electron in a free-electron gas at absolute zero is $\frac{3}{5}E_F$ when the gas obeys the Fermi–Dirac energy distribution function.

10.19. What is the energy of an electron with $k_x = 6(\pi/L)$, $k_y = 9(\pi/L)$, and $k_z = 4(\pi/L)$ in a free-electron gas existing within a cube of copper of volume 10^{-9} m^3?

10.20. Prove that dN/dE as given by eqn (10.46)—which was derived assuming the material exists as a cube of volume $V = L^3$—is unchanged if the same material exists as a rectangular block of sides L_x, L_y, and L_z such that $V = L_x L_y L_z$.

10.21. Using the data given in Fig. 10.21 estimate the Debye temperature of copper and the heat capacity at 300 K of the electrons present in 1 mol of copper.

10.22. What fraction of the lattice sites will be unoccupied in a material in which the energy E_v required to form a vacancy is 5×10^{-20} J if the material is at (a) 100 K, and (b) 300 K?

10.23. One method of estimating the linear expansivity of a material proceeds by drawing out the interatomic potential energy curve $V(r)$, as in Fig. 10.24(b), and adding a horizontal line at a height of $\frac{3}{2}kT$ above the minimum. Using this approach estimate the linear expansivity of solid argon. Assume a Lennard-Jones 6–12 potential with $\varepsilon = 1 \cdot 7 \times 10^{-21}$ J and $r_0 = 3 \cdot 7 \times 10^{-10}$ m.

10.24. Summarize the various methods available for estimating the vibrational period of an atom in a solid. You should express the period in terms of macroscopic parameters which can be determined by experiment.

11. Transport processes in solids

As in a gas, diffusion in a solid proceeds by the transport of mass, thermal conduction by the transport of energy, and electrical conduction by the transport of charge. But there the similarity ends. For one thing, a gas atom spends most of its time travelling freely through distances which are large compared with an atomic diameter; an atom in a solid is, for the most part, confined by the attractive forces exerted by its neighbours. For another, the transport properties of a (low-pressure) gas are virtually the same for all substances; the transport properties of a solid can vary drastically from substance to substance. As an example, the electrical conductivity of copper at room temperature is some 10^{22} times greater than that of sulphur. In this chapter we shall look at those aspects of transport processes in solids which can be treated classically or with a minimum of quantum mechanics.

11.1. Bulk diffusion

Unlike diffusion in liquids or gases, diffusion in solids is not part of our everday experience. The layer of silver found on plated-steel cutlery seems to remain stable. The chromium layer on a car bumper does not disappear by diffusing into the underlying steel. However, at high temperatures, 1300 K for example, the silver and chromium will disappear into the steel in a matter of weeks. Another example of a layered structure is the integrated circuit. In manufacturing these devices a silicon chip is held close to its melting point and various impurities are allowed to diffuse into the chip over a period of minutes. Once the device is cooled to room temperature it is essentially stable for all time.

Possible mechanisms

Early theories of diffusion supposed it to proceed by adjacent atoms interchanging places as, for example, in Fig. 11.1(a). To allow such an interchange to occur the surrounding atoms must be pushed far apart. In a close-packed solid this calls for so much energy as to render the process highly unlikely. A much more plausible explanation is that in such solids diffusion proceeds by atoms hopping into adjacent vacant lattice sites (Fig. 11.1(b)); this mechanism operates in fcc metals, in the solidified inert gases (fcc), and in ionic solids such as sodium chloride. The other main mechanism arises from the hopping of interstitials (Fig. 11.1(c)); the carbon and hydrogen atoms present in steel diffuse by this means.

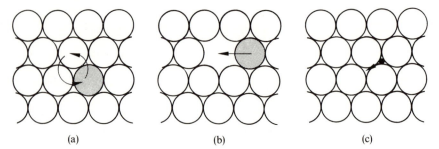

Fig. 11.1. Three possible mechanisms for diffusion in a solid. In (a) two atoms interchange places by rotating about a midway point, forcing neighbouring atoms apart. In (b) diffusion proceeds by atoms hopping into adjacent vacant lattice sites. In (c) interstitial atoms hop to adjacent vacant sites.

An experimental study

For simplicity, we will only look at the diffusion of a monatomic solid into its radioactive equal. In one experiment the diffusion of the radioactive isotope ^{64}Cu through ^{63}Cu was established by depositing electrolytically a layer of ^{64}Cu on to a face of a single crystal of ^{63}Cu. To find out how far the ^{64}Cu had diffused into the crystal, the sample was placed in a

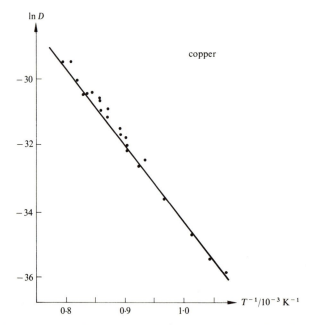

Fig. 11.2. Showing how $\ln D$ varies with T^{-1} for copper. Here D is the coefficient of self-diffusion and is measured in units of $m^2 s^{-1}$.

precision grinding machine and successive layers were ground off the surface. By counting the radioactivity present in each sample—the count rate is proportional to the number of ^{64}Cu atoms present in that sample— the self-diffusion coefficient D was evaluated. (We shall presently see how D is related to the mean diffusion distance.) To find out how the coefficient varied with temperature it was, of course, necessary to repeat the entire operation at a variety of different temperatures. Fig. 11.2 shows the results of such a study, demonstrating that D varies exponentially as T^{-1}. This contrasts strongly with the weak $T^{3/2}$ dependence found in a gas at constant pressure (eqn (7.12)). Some quite different mechanism is evidently at work in a solid.

The self-diffusion coefficient

Fig. 11.3(a) shows the problem facing an atom i as it attempts to diffuse through a close-packed structure. To hop into site j two criteria must be satisfied. First, site j must be vacant. The probability that j is vacant—which is equal to the fraction of vacancies present in the crystal—is given by eqn (10.54), namely $\exp(-E_v/kT)$, where E_v is the

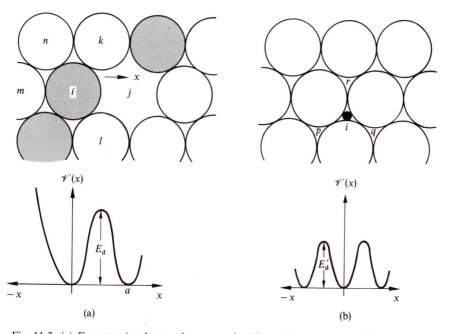

Fig. 11.3. (a) For atom i to hop to the vacant site j it must have an energy E_d or greater. Here $\mathscr{V}(x)$ is the potential energy of i as it is displaced from its lattice site. (b) For an interstitial i to hop to an adjacent site like p it must have an energy E_d' or greater. (Strictly speaking, the $+x$-direction is along line iq and the $-x$-direction along ip.)

energy required to create the vacancy. The second requirement is that i should have the necessary energy to move to site j.

To escape to site j atom i must squeeze atoms k and l apart against their mutual attractive force. It must also escape the backward pull of atoms like m and n. We may represent the net energy required to displace atom i from its lattice site by a potential energy curve as shown in Fig. 11.3(a). In this figure the zero of potential energy $\mathcal{V}(x)$ is taken to be at the undisplaced site of atom i. In calculating the probability that atom i has the energy E_d or greater which is necessary for escape we shall adopt Einstein's model of a solid. Following through exactly similar arguments as those which produced eqn (10.54) leads to the conclusion that a fraction $\exp(-E_d/kT)$ of the atoms of a solid will have an energy $E_x \geq E_d$. We may take it that this fraction is equal to the probability than an *individual* atom will have a vibrational energy $E_x \geq E_d$.

Taking both requirements for diffusion into account we see that the probability that an atom has an adjacent vacant site *and* that it has the necessary energy to hop into this site is given by $\exp(-E_v/kT) \times \exp(-E_d/kT)$, that is, $\exp[-(E_v + E_d)/kT]$. Our next task is to calculate the speed at which an atom hops through the crystal.

According to Einstein's model, the vibrational frequency $\omega_E/2\pi$ of an atom such as i Fig. 11.3(a) remains constant even though it exchanges energy with other atoms of the solid. Thus the number of *attempts* which i makes at escape per unit time as it vibrates back and forth will be constant and equal to the number of occasions it reaches each extremity of its vibration (Fig. 11.3(a)) in unit time. This is just twice the vibrational frequency $\omega_E/2\pi$. For an escape *attempt* to be successful we have seen that an adjacent site must be vacant and that the vibrating atom has the necessary escape energy. Since the probability of both these events occurring simultaneously is $\exp[-(E_v + E_d)/kT]$, it follows that the average number of escape hops per unit time is given by $2(\omega_E/2\pi)\exp[-(E_v + E_d)/kT]$. Now each hop is of length a, the interatomic spacing, so the (average) hopping speed v_d of i is given by

$$v_d = \frac{a\omega_E}{\pi} e^{-(E_v + E_d)/kT}. \tag{11.1}$$

The final task is to calculate the net number of labelled atoms (such as the shaded atoms in Fig. 11.3(a)) which pass through unit area per unit time. To do this we consider a plane A of unit area (Fig. 11.4). Atoms passing through this plane will have hopped there either from a plane P or a plane Q, each a distance a away. If the number density of labelled atoms at $x + \delta x$ is, say, $n + \delta n$ then the actual number of these which cross plane A in unit time is $\frac{1}{6}(n + \delta n)v_d$; the factor of $\frac{1}{6}$ taking care of the fact that, at any instant, only one atom in six of the radioactive species will be

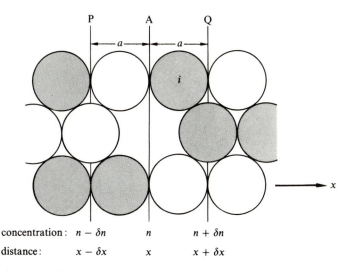

Fig. 11.4. Atoms crossing plane A in a single hop will have originated either from plane P or plane Q, each a distance a away from plane A. The number density n refers to that of the radioactive (shaded) species.

travelling in the $-x$-direction. Likewise, the number of labelled atoms which cross plane A per unit time moving to the right is $\frac{1}{6}(n - \delta n)v_d$, where $n - \delta n$ is the number density of these atoms at $x - \delta x$ (Fig. 11.4). The net number crossing unit area per unit time, the flux j, is therefore

$$j = -[\tfrac{1}{6}(n + \delta n)v_d - \tfrac{1}{6}(n - \delta n)v_d] = -\tfrac{1}{3}v_d\,\delta n, \qquad (11.2)$$

where a minus sign has been inserted to make this equation conform to the sign convention introduced in §7.1 which labels diffusion in the $-x$ direction as negative. Multiplying the right-hand side of eqn (11.2) by a/a, and remembering that the concentration gradient is $\delta n/\delta x = \delta n/a$, leads to

$$j = -D\frac{\delta n}{\delta x},$$

where the coefficient of self-diffusion $D = \frac{1}{3}v_d a$ or, substituting for v_d from eqn (11.1),

$$\boxed{D = D_0\,e^{-(E_v + E_d)/kT}}. \qquad (11.3)$$

Here $D_0 = a^2\omega_E/3\pi$, which is a constant for a given material. Fig. 11.2 confirms the prediction of eqn (11.3) that a graph of $\ln D$ against $1/T$ should be linear with a gradient $-(E_v + E_d)/k$.

TABLE 11.1
Values of D_0 and E for self-diffusion in various materials[a]; $D = D_0 \, e^{-E/kT}$

Material	$D_0/10^{-4} \, m^2 \, s^{-1}$	$E/10^{-19} \, J$
aluminium	1·71	2·37
germanium	7·8	4·77
gold	0·11	2·95
lead	0·89	1·78
potassium	0·16	0.65
silicon	1800	7·67
tin	9·2	1·76
zinc	0·15	1·56

[a] All the materials were in the form of single crystals.

When diffusion proceeds by the movement of interstitials (Fig. 11.3(b)) a very similar analysis applies. Essentially, E_v disappears from eqn. (11.3) and E_d is replaced by E_d'; the activation energy required for the interstitial to hop to an adjacent site. Because both mechanisms lead to the same temperature dependence of D it is not possible to decide which one is involved on the basis of this dependence alone. Other evidence is required, such as that provided by a study of how the volume of a solid changes with temperature (§10.7). Table 11.1 lists the measured values of D_0 and E, as defined by $D = D_0 \exp(-E/kT)$, for self-diffusion in a variety of materials.

Exercise 11.1

Using the data given in Fig. 11.2, calculate the value of D_0 and $E_v + E_d$ in copper. How many successful hops per second will be made by an atom in a copper block (a) at 300 K, and (b) at 1300 K (the melting point is 1358 K)? Assume the Einstein temperature is 240 K.

Solution. It follows from eqn (11.3) that $\ln D_0$ is the value of $\ln D$ at $T^{-1} = 0$. Fig. 11.2, extrapolated back to $T^{-1} = 0$, gives $D_0 = 3 \times 10^{-5} \, m^2 \, s^{-1}$. The slope of the graph gives $(E_v + E_d)/k$ as $2 \cdot 3 \times 10^4 \, K$, or $E_v + E_d = 3 \cdot 17 \times 10^{-19} \, J = 2 \cdot 0 \, eV$ (The *electron volt* (eV) is the energy acquired by an electron when it is accelerated through a potential difference of one volt; it thus has a value of $1 \cdot 6 \times 10^{-19} \, J$.)

The number of successful hops per second made by a copper atom in a block at temperature T is given by

$$\frac{\omega_E}{\pi} e^{-(E_v + E_d)/kT}. \tag{11.4}$$

Adopting eqn (11.4) with $\omega_E/2\pi = k\theta_E/2\pi\hbar = 5 \cdot 0 \times 10^{12} \, s^{-1}$ gives the number of successful hops per second (a) at 300 K as $5 \cdot 0 \times 10^{-21} \, s^{-1}$, and (b) at 1300 K as

$2 \cdot 0 \times 10^5 \, \text{s}^{-1}$. We clearly need not worry about a coating disappearing into the underlying metal at temperatures significantly below the melting point.

Comment. The value of D_0 predicted by eqn (11.3), namely $D_0 = a^2 \omega_E / 3\pi$, $= 2 \cdot 3 \times 10^{-7} \, \text{m}^2 \, \text{s}^{-1}$, differs substantially from the measured value.

11.2. Surface diffusion

Fig. 11.5(a) illustrates some of the features which can exist on the surface of a real crystal. Following custom, this figure represents each atom by a cube. Fig. 11.5(b) shows the potential energy diagram for an adsorbed atom (usually shortened to *adatom*) as it moves across a terrace on the surface. Here E_d is the activation energy for surface diffusion which, for many metal surfaces, lies in the range 1×10^{-19} to 3×10^{-19} J. A curve qualitatively similar to Fig. 11.5(b) could be drawn showing diffusion along a ledge; here, however, we would expect E_d to be greater than

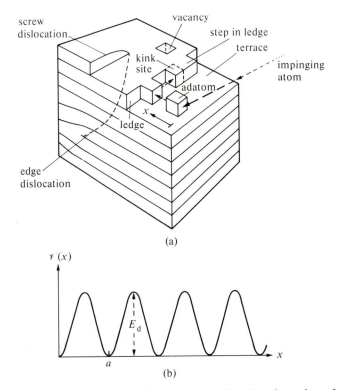

(a)

(b)

Fig. 11.5. (a) Showing some of the main features to be found on the surface of a crystal. (b) The potential energy of an adatom as it moves across the surface. Here a is the separation between equilibrium sites of the adatom on the terrace and E_d is the activation energy required to hop between sites.

on a terrace site since a ledge adatom has the greater number of nearest neighbours.

The self-diffusion coefficient of adatoms can be measured in a variety of ways. One technique, which is analagous to that already described for measuring the diffusion coefficient in the bulk (p. 354), measures the time it takes for, say, a radioactive nickel isotope to diffuse a known distance along the surface of a nickel crystal. The mechanism whereby diffusion occurs is broadly similar to the mechanism which operates within the body of the crystal; an adatom remains at one site until its fluctuating kinetic energy exceeds E_d, whereupon it will hop to an adjacent vacant site. Consequently, we may take over much of the earlier discussion. The only changes we need make are to set $E_v = 0$ in eqn (11.1)—since a surface terrace normally contains but few adatoms every such atom is almost certainly surrounded by unoccupied terrace sites—and to replace the factor of $\frac{1}{6}$ in eqn (11.2) by $\frac{1}{4}$. This immediately leads to

$$D = \frac{a^2}{2}\left(\frac{\omega_E}{\pi}\right)e^{-E_d/kT}, \tag{11.5}$$

where a is the lattice separation in the plane of the surface, and ω_E is the circular vibrational frequency of an adatom.

The random walk

Fig. 11.6 shows the lattice sites on a {100} face of a simple cubic crystal. We may simulate the motion of an adatom i by spinning a four-sided top whose edges are labelled L, R, U, and D. The side on

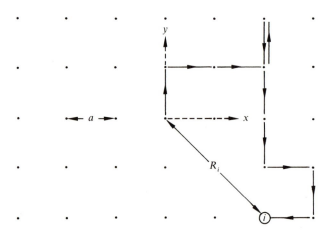

Fig. 11.6. Illustrating the random-walk nature of diffusion. Each dot represents a lattice site on the surface of a crystal. Whether adatom i hops left, right, up, or down by one lattice spacing a is decided by a random process.

which the top comes to rest tells us whether adatom i is to jump left, right, up or down by a distance a from its present lattice site. After each hop the top is spun afresh. The path marked out in Fig. 11.6 was obtained by this means.

If we denote the individual displacements of i as it hops by $x_1, x_2, x_3, \ldots, x_p$ and $y_1, y_2, y_3, \ldots, y_q$ we see from Fig. 11.6 that the overall distance R_i gone in this, so-called, *random walk* is given by

$$R_i^2 = (x_1 + x_2 + x_3 + \cdots + x_p)^2 + (y_1 + y_2 + y_3 + \cdots + y_q)^2$$
$$R_i^2 = (x_1^2 + x_2^2 + x_3^2 + \cdots + x_p^2) + (y_1^2 + y_2^2 + y_3^2 + \cdots + y_q^2)$$
$$+ 2x_1 x_2 + \cdots + 2y_1 y_2 + \cdots . \quad (11.6)$$

A similar expression can be written down for any other adatom, say j, as it too executes a random walk of $n = p + q$ steps. Our interest is the mean square distance $\overline{R^2}$ gone by the average adatom as it takes a random walk of n steps. If there are N adatoms on the lattice we have

$$\overline{R^2} = \frac{\sum\limits_{i=1}^{N} R_i^2}{N}, \quad (11.7)$$

where R_i^2 is given by eqn (11.6). Now each of $x_1, x_2, \ldots, y_1, y_2, \ldots$ are of magnitude a, the lattice spacing, so that $x_1^2 = x_2^2 \cdots = y_1^2 = y_2^2 = \cdots = a^2$. Averaged over the motion of all N adatoms, terms like $2x_1 x_2$ will all equal zero because each of x_1, x_2, etc., has an equal chance of being $+a$ or $-a$. So eqn (11.7) becomes

$$\overline{R^2} = \frac{\sum\limits_{1}^{N} [(pa^2) + (qa^2)]}{N} + 0 = \frac{\sum\limits_{1}^{N} [(p+q)a^2]}{N} = \frac{N(na^2)}{N}, \quad (11.8)$$

giving the root-mean-square distance travelled by an adatom making a random walk of n steps, each of length a, as

$$\boxed{(\overline{R^2})^{1/2} = an^{1/2}} . \quad (11.9)$$

This equation can be rewritten so as to introduce D by noting that the number of jumps n made by an atom in a time t can be expressed as

$$n = \text{jump frequency} \times t = 4t \left(\frac{\omega_E}{\pi} \right) e^{-E_d/kT}, \quad (11.10)$$

where the factor of 4 takes care of the fact that there are four directions in which an adatom can hop in the lattice of Fig. 11.6. Substituting for $(\omega_E/\pi)\exp(-E_d/kT)$ as given by eqn (11.5) and for n as given by eqn

(11.9) leads to

$$\overline{R^2} = 8Dt. \tag{11.11}$$

In many situations we are only interested in the net distance moved in one particular direction, say the x-direction. Since, by symmetry, $\overline{X^2} = \overline{Y^2}$, where $\overline{X^2}$ and $\overline{Y^2}$ are the mean square distance moved in the x- and y-directions, respectively, it follows that $\overline{R^2} = \overline{X^2} + \overline{Y^2} = 2\overline{X^2}$. Hence eqn. (11.11) becomes

$$\boxed{(\overline{X^2})^{1/2} = (4Dt)^{1/2}} \tag{11.12}$$

which shows that D can be deduced from measurements of $\overline{X^2}$ at various times t. The field ion microscope (described in the caption to Fig. 9.21) allows one to observe directly the motion of individual adatoms (which can, of course, be of a different species to the atoms of the solid) and thus to deduce D via eqn (11.12) or eqn (11.11). The measurements would then be repeated at a series of different temperatures and a graph of $\ln D$ plotted against T^{-1}. As predicted by eqn (11.5), such plots are usually linear, allowing E_d to be determined from their gradients. By way of examples, it has been found that E_d for a single tungsten atom diffusing on the (110) surface of a tungsten crystal is $1 \cdot 44 \times 10^{-19}$ J (it is $1 \cdot 47 \times 10^{-19}$ J for a diffusing W_2 cluster) while for a tantalum atom on the (110) tungsten surface $E_d = 3 \times 10^{-19}$ J. In general—and this is only to be expected when we compare Figs. 11.5(a) and 11.3—the values of E_d are markedly smaller for surface self-diffusion than for bulk self-diffusion. Hence surface self-diffusion is a much more rapid process than bulk self-diffusion.

Crystal growth from the gas phase

If a crystal is in equilibrium with its vapour (as it is in the s–g region of Fig. 1.2) the rate at which atoms are adsorbed on to the surface of the crystal must equal the rate at which atoms leave the surface for the gas phase. If a crystal is to grow from the gas phase, a necessary condition is that the rate at which gas atoms are adsorbed must exceed the rate at which solid atoms leave. Since the former is proportional to the vapour pressure (c.f. eqn (5.24)), crystal growth demands that the vapour in contact with the solid is at a pressure greater than its equilibrium value (it is then said to be supersaturated).

It is generally accepted that crystal growth proceeds by the sequence of stages shown in Fig. 11.5(a). Assuming that the incoming gas atom is adsorbed on to a terrace (that is, it does not immediately rebound from the surface; §5.1) it will be able to diffuse across the terrace. Provided the adatom does not evaporate while executing its two-dimensional walk it will eventually reach a ledge. Here it will either be reflected back along the terrace or it will be adsorbed on to the ledge. If the latter happens, the adatom will execute a one-dimensional random walk along the ledge. Again assuming that it does not evaporate on route, it will in time reach a kink site where it will be fairly securely bound (an atom at a kink site has many more nearest neighbours than does an isolated adatom on a terrace). The necessary ledges may be provided by dislocations emerging from the crystal surface (Fig. 9.21). Exactly the same stages occur, but in reverse order, when a solid sublimes.

11.3. Thermal conduction in non-metals

Fig. 11.7 shows how the thermal conductivity of a variety of substances varies with their temperature. Despite a superficial similarity in the form of the curves for metallic and non-metallic materials there is a fundamental difference in the mechanism whereby heat is transported in these two types of material. In metals heat is conducted by electrons; in non-metals it is conducted through coupled vibrations of the atoms. Fig. 11.7 also makes clear that a non-metal like diamond can have a higher thermal conductivity than copper.

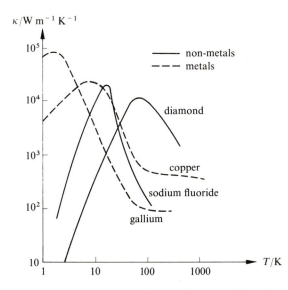

Fig. 11.7. The thermal conductivity κ of metallic and non-metallic solids as a function of temperature. The copper data is for a pure annealed single crystal.

Phonons

In §10.3 we argued that the interatomic coupling present in a solid allows many different vibrational patterns (modes) to be established. We then proved that a mode of circular frequency ω_q and circular wavenumber q has a mean energy given by eqn (10.24). Another way of looking at this equation follows when we rewrite it as

$$\bar{E}_q = n_q \hbar \omega_q \, , \tag{11.13}$$

where

$$n_q = \frac{1}{e^{\hbar \omega_q / kT} - 1} \, . \tag{11.14}$$

Thus, so far as the energy of this particular mode of circular wavenumber q is concerned, it is equal to the energy of $n_q = [\exp(\hbar \omega_q / kT) - 1]^{-1}$ 'particles' each of which has an energy $\hbar \omega_q$. In fact we say that the mode of circular frequency ω_q has n_q *phonons* (each of energy $\hbar \omega_q$) associated with it at temperature T. The number n_q is also referred to as the *phonon occupancy* of the mode of circular frequency ω_q. Fig. 11.8(a) shows the situation graphically for a simple cubic crystal of side L and in which the

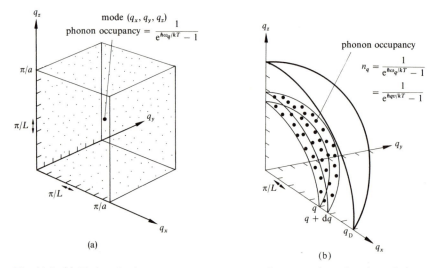

(a)

(b)

Fig. 11.8. (a) Each point in q_x–q_y–q_z space corresponds to a mode and each mode has a circular frequency ω_q determined by the dispersion relation. The phonon occupancy of each mode is given by eqn (11.14). (b) Illustrating Debye's approximation in which the number of states within the octant of radius q_D is equal to the number of states within the cube of side π/a shown in (a).

nearest-neighbour separation is a. Here each mode is designated by a triplet of values q_x, q_y, and q_z. However, as eqn (11.14) shows, we can only evaluate n_q at (q_x, q_y, q_z) if we know the dispersion relation between ω_q and (q_x, q_y, q_z). This relation involves the interatomic force constants and the atomic masses.

The total phonon population

To find the total number N_P of phonons in a solid at a temperature T we add together the phonon occupancies of all the permitted modes. Thus

$$N_P = \sum_q n_q = \sum_q \frac{1}{e^{\hbar\omega_q/kT} - 1}, \tag{11.15}$$

where the summation is from $q_x = \pi/L$ to $q_x = \pi/a$, from $q_y = \pi/L$ to $q_y = \pi/a$, and from $q_z = \pi/L$ to π/a (Fig. 11.8(a)). Adopting arguments similar to those used by Debye in discussing heat capacities (§10.5) we now draw in the octant of radius q_D shown in Fig. 11.8(b) so that it contains the same number of points as does the cube of side π/a shown in Fig. 11.8(a). Again following Debye, we replace the actual dispersion relation by $\omega = qv$ (eqn (10.30)), where v is the velocity of sound in the solid (Fig. 10.13(b)).

We see from Fig. 11.8(b) that the number of points within the octant of radii q and $q + dq$—each of which points occupies a volume $(\pi/L)^3$—is $\frac{1}{8}(4\pi q^2 \, dq)/(\pi/L)^3$. Since each point (that is, each mode) has n_q phonons associated with it, where n_q is given by eqn (11.14) with $\omega = qv$, it therefore follows that

$$N_P = \frac{1}{8}\left(\frac{L}{\pi}\right)^3 \int_0^{q_D} \frac{4\pi q^2 \, dq}{e^{\hbar qv/kT} - 1}.$$

Writing $x = \hbar qv/kT$ so that the upper limit of integration is $x_D = \hbar\omega_D/kT = \theta_D/T$, where θ_D is the Debye temperature, leads to

$$N_P = \frac{4\pi}{8}\left(\frac{L}{\pi}\right)^3 \left(\frac{T}{\theta_D}\right)^3 q_D^3 \int_0^{\theta_D/T} \frac{x^2}{e^x - 1} \, dx. \tag{11.16}$$

Substituting for q_D from eqn (10.29) and multiplying the right-hand side of eqn (11.16) by 3 to allow for the fact there are three types of mode (two transverse and one longitudinal) gives the number of phonons n per unit volume as

$$n = \frac{N_P}{L^3} = \frac{9}{a^3}\left(\frac{T}{\theta_D}\right)^3 \int_0^{\theta_D/T} \frac{x^2}{e^x - 1} \, dx. \tag{11.17}$$

At high temperatures $x \, (= \hbar qv/kT)$ and the upper limit of integration are both small, so that eqn (11.17) approximates to

$$n = \frac{9}{a^3}\left(\frac{T}{\theta_D}\right)^3 \int_0^{\theta_D/T} \frac{x^2}{1+x-1}\,dx$$

$$n = \left(\frac{9}{2a^3\theta_D}\right) T.$$

As a and θ_D are constant,

$$\boxed{n \propto T} \qquad \text{at } T > \theta_D. \tag{11.18}$$

At low temperatures the upper limit of integration in eqn (11.17) may be taken as infinity for all practical purposes. Under these conditions the integral tends to a constant value $(2\cdot77\ldots)$ giving

$$\boxed{n \propto T^3} \qquad \text{at } T < \theta_D. \tag{11.19}$$

The mean phonon energy

The mean energy $\bar{E}$ of a phonon is, by definition, given by

$$\bar{E} = \frac{\sum_q n_q \hbar\omega_q}{\sum_q n_q}, \tag{11.20}$$

where n_q is, of course, the number of phonons of circular frequency ω_q (each of which has an energy $\hbar\omega_q$); see eqn (11.14).

Introducing both of Debye's approximations—his octant and his dispersion relation—transforms eqn (11.20) to

$$\bar{E} = \frac{\int_0^{q_D} (e^{\hbar qv/kT} - 1)^{-1} 4\pi q^2 (\hbar qv)\,dq}{\int_0^{q_D} (e^{\hbar qv/kT} - 1)^{-1} 4\pi q^2 \,dq}$$

or, writing $x = \hbar qv/kT$, to

$$\bar{E} = kT \frac{\int_0^{\theta_D/T} (e^x - 1)^{-1} x^3 \,dx}{\int_0^{\theta_D/T} (e^x - 1)^{-1} x^2 \,dx}. \tag{11.21}$$

At high temperatures x and θ_D/T are both small, so eqn (11.21) approximates to

$$\bar{E} = kT \frac{\displaystyle\int_0^{\theta_D/T} x^2\,\mathrm{d}x}{\displaystyle\int_0^{\theta_D/T} x\,\mathrm{d}x}$$

$$\boxed{\bar{E} = \tfrac{2}{3}k\theta_D} \qquad \text{at} \quad T > \theta_D. \tag{11.22}$$

At low temperatures the upper limit of integration in both the numerator and the denominator of eqn (11.21) may be taken as infinity. Under these conditions the numerator tends to $6\cdot49\ldots$, the denominator† to $2\cdot77\ldots$ giving

$$\boxed{\bar{E} = 2\cdot34kT} \qquad \text{at} \quad T < \theta_D. \tag{11.23}$$

The mechanism of thermal conduction

There is an obvious analogy to be drawn between the phonons in a solid and the atoms in a gas. Thus we can say, for example, that the number and mean energy of the phonons varies as we move from the hot to the cold end of a solid rod. Taking over our gas-kinetic relation (eqn (7.42)), rewritten in a somewhat different form, gives the thermal conductivity of the non-metallic solid as

$$\kappa = \tfrac{1}{3}v\lambda\,(n\tfrac{3}{2}k) = \tfrac{1}{3}v\lambda\,\frac{\mathrm{d}}{\mathrm{d}T}(n\tfrac{3}{2}kT)$$

$$\kappa = \tfrac{1}{3}v\lambda\,\frac{\mathrm{d}}{\mathrm{d}T}(n\bar{E}), \tag{11.24}$$

where v is the speed of sound in the rod, λ is the phonons' mean free

---

† It is not hard to show that this is a plausible value. Ignoring the one in $(e^x - 1)^{-1}$—which we can more or less do at low temperatures when $x(= \hbar qv/kT) \gg 1$—the integral in the denominator becomes

$$\int_0^\infty x^2 e^{-x}\,\mathrm{d}x = -[x^2 e^{-x}]_0^\infty + \int_0^\infty 2x\,e^{-x}\,\mathrm{d}x$$

$$= 0 - 2[x\,e^{-x}]_0^\infty + 2\int_0^\infty e^{-x}\,\mathrm{d}x$$

$$= 2.$$

You can likewise show that when $x \gg 1$ the integral in the numerator of eqn (11.21) has an approximate value of 6. This leads to the result $\bar{E} \approx 3kT$.

path, n is their number density, and $\bar{E}$ their mean energy. Eqn (11.24) can be justified by more rigorous arguments (see problem 11.10). Since v is (essentially) independent of temperature—Debye's dispersion relation $v = \omega/q$ assumes it to be so—the reason for the temperature dependence of the thermal conductivity of non-metals, which is so evident in Fig. 11.7, must be sought in λ, n, and $\bar{E}$.

We will start by considering the thermal conductivity at $T > \theta_D$. In this region $n \propto T$ (eqn (11.18)) and $\bar{E}$ is constant (eqn (11.22)). Also as $\lambda = 1/\pi d^2 n$ (eqn (5.57)) we would expect $\lambda \propto 1/n \propto 1/T$. Making these substitutions into eqn (11.24) gives $\kappa \propto T^{-1}\, dT/dT$, that is

$$\boxed{\kappa \propto \frac{1}{T}} \quad \text{at} \quad T > \theta_D. \tag{11.25}$$

Fig. 11.9(a) shows that, as predicted by eqn (11.25), graphs of log κ for three inert-gas solids plotted against log T are indeed linear at $T > \theta_D$ and have a slope of -1. This figure also shows that as T is reduced below θ_D

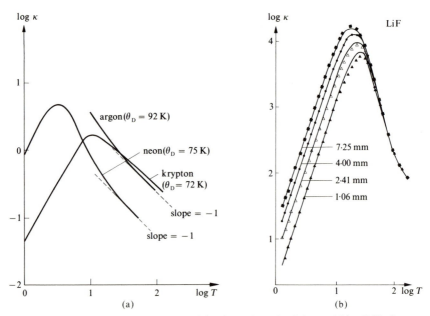

Fig. 11.9. The temperature dependence of the thermal conductivity κ of (a) solidified neon, argon, and krypton, (b) isotopically pure LiF crystals in the form of sandblasted rods of square cross-section (whose width is specified). The thermal conductivity measurements were in units of W m^{-1} K^{-1} and temperatures in units of K. (The LiF data is from Thacher, D. T., (1967). *Phys. Rev.* **156**, 975.)

the thermal conductivity passes through a maximum and decreases towards zero. The reason is not hard to find. As we have just seen, a phonon's mean free path increases with decreasing temperature; eventually it must become comparable to the specimen diameter D. In addition, eqns (11.19) and (11.23) tell us that $n \propto T^3$ and $\bar{E} \propto T$, at $T < \theta_D$. Making these substitutions into eqn (11.24) gives $\kappa \propto D \, dT^4/dT$, that is

$$\boxed{\kappa \propto DT^3} \quad \text{at} \quad T < \theta_D. \tag{11.26}$$

Fig. 11.9(b) shows the results of a detailed study aimed at testing this relation in LiF ($\theta_D = 730$ K). Each of the specimens was in the form of a rod of square cross-section. You can readily verify that $\kappa \propto D$ at any fixed temperature below 10 K and that $\kappa \propto T^3$ for each and every specimen. The dependence of κ on the diameter of the specimen is reminiscent of the low-pressure behaviour of a gas (eqn (7.50)) where the heat flow is limited by the plate separation. Furthermore, one finds that, as in a gas, κ depends on the nature of the material's surface. For example, the low-temperature conductivity of a sapphire crystal with a smooth surface is almost double that of a crystal with a sandblasted surface.

Exercise 11.2

Another way of expressing the thermal conductivity of a non-metallic solid follows on noting that $(C_{V,m}/N_A)n$ in eqn (7.44) is equal to the heat capacity C *per unit volume*. Thus

$$\kappa = \tfrac{1}{3}v\lambda C. \tag{11.27}$$

Using this relation and Debye's T^3 expression for $C_{V,m}$ (eqn (10.37)) deduce the approximate diameter of the diamond specimen used in obtaining the data shown in Fig. 11.7. The Debye temperature of diamond ($A_r(C) = 12$) is 2000 K and its density is 3.5×10^3 kg m^{-3}. Take the speed of sound in diamond to be 5×10^3 m s^{-1}.

Calculation. The maximum in the κ versus T plot for a non-metal occurs when the phonon mean free path is of the order of magnitude of the specimen's diameter. The maximum κ is about 10^4 W m^{-1} K^{-1} and this occurs at $T \approx 80$ K where $C_{V,m} = 234 \, R \, (80/2000)^3 = 0.12$ J mol^{-1} K^{-1}. Dividing the mass of 1 mol of diamond ($A_r(C) = 12$), namely 12×10^{-3} kg, by its density gives the molar volume as 3.4×10^{-6} m^3 mol^{-1}. Thus the heat capacity per unit volume is $C = 0.12/(3.4 \times 10^{-6})$ J m^{-3} K$^{-1} = 3.5 \times 10^4$ J m^{-3} K^{-1}. Substituting for κ, C, and v into eqn (11.27) gives $\lambda = 1.7 \times 10^{-4}$ m. So the specimen probably had a diameter of order of magnitude 0.1 to 1 mm.

11.4. Thermal conduction in metals

The apparent similarities between the temperature dependence of the thermal conductivities of metallic and non-metallic solids to be seen in Fig. 11.7 is entirely coincidental. For one thing, heat is conducted almost exclusively by electrons in a metal (the phonon contribution is usually insignificant) and solely by phonons in a non-metal. For another, the detailed shapes of the $\kappa-T$ curves are quite different in the two cases. In a metal $\kappa \propto T$ at temperatures well below the maximum, while above the maximum $\kappa \propto T^{-2}$. In a non-metal we have seen that the corresponding relations are $\kappa \propto T^3$ and $\kappa \propto T^{-1}$.

The mechanism of thermal conduction

You will recall how we argued in §10.6 that the only electrons which are free to have their energy increased—and thus to participate in thermal conduction—are those within a range of $\pm kT$ of the Fermi energy E_F (Fig. 10.20). We also saw that their number density $n \propto T$ (eqn (10.48)) and their mean thermal energy $\bar{E} = \frac{3}{2}kT$. Furthermore, each of these electrons have a speed close to the *Fermi speed* u_F defined by $\frac{1}{2}mu_F^2 = E_F$, that is

$$u_F = (2E_F/m)^{1/2}, \tag{11.28}$$

which is independent of temperature. Substituting these expressions for n, $\bar{E}$, and u_F into our gas-kinetic relation (eqn (11.24)) shows that the thermal conductivity of a free electron gas $\kappa \propto \lambda \, dT^2/dT$, and so

$$\kappa \propto \lambda T. \tag{11.29}$$

A similar conclusion could have been reached by arguing via eqns (10.49) and (11.27). All that remains is to determine the mean free path of an electron. Since we can expect the free path of an electron to be terminated either through collision with a phonon or with an impurity atom this, in turn, raises the question of how to take two such independent scattering processes into account.

In § 5.12 we showed that the probability of a gas atom colliding with a like atom in travelling a distance dl is dl/λ_g, where λ_g is the mean free path arising from such collisions. If we now suppose that there are impurity atoms present, the probability that a gas atom will collide with one of these impurity atoms in travelling a distance dl is dl/λ_i, where λ_i is the mean value of those free paths which are terminated by collisions with impurity atoms. Overall, the probability that a gas atom collides with *either* a like atom *or* with an impurity atom in travelling a distance dl is $(dl/\lambda_g + dl/\lambda_i) = (1/\lambda_g + 1/\lambda_i) \, dl = \gamma \, dl$, say. (The probability that *either* a 2 *or* a 5 is thrown on a die is $\frac{1}{6} + \frac{1}{6} = \frac{1}{3}$.) To evaluate the mean free path λ with both these scattering processes present we follow the same arguments as those in §5.12 but with α replaced by γ. This gives $\lambda = 1/\gamma$, that is

$$\frac{1}{\lambda} = \frac{1}{\lambda_g} + \frac{1}{\lambda_i}.$$

In the present context we may therefore write the electronic mean free path λ as

$$\frac{1}{\lambda} = \frac{1}{\lambda_T} + \frac{1}{\lambda_i},$$

(11.30)

where λ_T is the mean free path of the electron due to phonon scattering and λ_i is the mean free path due to impurity scattering. Multiplying eqn (11.30) through by $1/AT$, where A is the constant of proportionality which occurs in eqn (11.29), gives

$$\frac{1}{\kappa} = \frac{1}{\kappa_T} + \frac{1}{\kappa_i},$$

(11.31)

where $\kappa_T = A\lambda_T T$, and $\kappa_i = A\lambda_i T$. Now at $T < \theta_D$ the phonon number density $n \propto T^3$ (eqn (11.19)). Since $\lambda_T \propto 1/\pi d^2 n$ it therefore follows that $\kappa_T \propto 1/T^2$. With impurity scattering λ_i should be independent of temperature, implying $\kappa_i \propto T$. Substituting these relations for κ_T and κ_i into eqn (11.31) gives

$$\frac{1}{\kappa} = \alpha T^2 + \frac{\beta}{T} \quad \text{at} \quad T < \theta_D,$$

(11.32)

where α and β are constants. At very low temperatures the second term —arising from impurity scattering—predominates; predicting $\kappa \propto T$. At

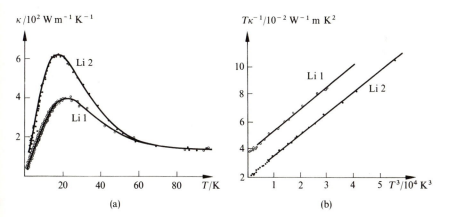

Fig. 11.10. (a) The thermal conductivity κ of two specimens of lithium (Li 2 is purer than Li 1) as a function of temperature T. (b) The same information plotted as T/κ against T^3. The Debye temperature of Li is 344 K. (Data from Rosenberg, H. M. (1956). *Phil. Mag.*, **1**, 738.)

somewhat higher temperatures, but below θ_D, the first term—arising from phonon scattering—predominates, predicting $\kappa \propto 1/T^2$. Fig. 11.7 shows that these predictions are borne out for copper and gallium.

It always makes for easier experimental verification of a theoretical expression relating two quantities, here κ and T, if the expression can be rearranged to predict a linear relationship. Multiplying eqn (11.32) through by T gives

$$\frac{T}{\kappa} = \alpha T^3 + \beta \quad \text{at} \quad T < \theta_D. \tag{11.33}$$

Fig. 11.10(a) shows how the thermal conductivity of two lithium samples of different purity varies with temperature. Fig. 11.10(b) shows that graphs of T/κ potted against T^3 are indeed linear as predicted by eqn (11.33). We also see that the slope α of these lines is, as expected, independent of the impurity level.

11.5. Electrical conduction

Periodic boundary conditions

In discussing the permitted wave functions of a free electron in a metal (§10.6) we only considered functions like those shown in Fig. 10.16(b). There is, however, another way of discussing the permitted wave functions. This approach starts by bending a long block of the metal into a ring (Fig. 11.11(a)) so that what were once opposite parallel faces at each end of the block of length L_x now touch. Under these circumstances waves can travel round the length of the block (Fig. 11.11(b)). Clearly, such waves must have a wavelength λ_x chosen so that an integral number n_x of wavelengths fits around its length L_x (if this condition is not satisfied $\psi(x)$ will be multivalued at each point). Thus with this, so-called, *periodic boundary condition* the allowed wave functions are characterized by $n_x \lambda_x = L_x$ or, in terms of the circular

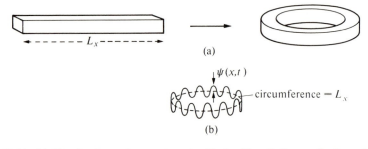

(a)

(b)

Fig. 11.11. (a) Showing how a long rectangular block of length L_x may be formed into a ring. (b) A travelling wave $\psi(x, t)$ must have a wavelength λ_x such that an integral number of wavelengths fit around the ring.

wavenumber k_x $(=2\pi/\lambda_x)$, by

$$k_x = \pm\left(\frac{2\pi}{L_x}\right), \pm2\left(\frac{2\pi}{L_x}\right), \pm3\left(\frac{2\pi}{L_x}\right), \ldots \quad (11.34)$$

Here we have added the possibility of a negative sign to allow for the fact that a wave with a given value of k_x may travel in either sense around the ring. (A wave with equation $y = A\cos(kx + \omega t)$ travels in the opposite sense to one with equation $y = A\cos(-kx + \omega t)$.)

Generalizing the arguments (even though it is topologically impossible to twist a rectangular block so that all three pairs of opposite faces now touch) leads to the following permitted values of k_x, k_y, and k_z in a rectangular block of sides L_x, L_y, and L_z:

$$
\begin{aligned}
k_x &= 0, \pm(2\pi/L_x), \pm2(2\pi/L_x), \pm3(2\pi/L_x), \ldots \\
k_y &= 0, \pm(2\pi/L_y), \pm2(2\pi/L_y), \pm3(2\pi/L_y), \ldots \\
k_z &= 0, \pm(2\pi/L_z), \pm2(2\pi/L_z), \pm3(2\pi/L_z), \ldots
\end{aligned}
\quad (11.35)
$$

The wave function will thus be characterized by (k_x, k_y, k_z) chosen from the above values, and a statement as to its spin direction. One cannot, however, choose $k_x = k_y = k_z = 0$, that is $\lambda_x = \lambda_y = \lambda_z = \infty$, since this implies the electron is not in the metal. Although one may use the set of k-values given *either* by eqn (10.39) *or* by eqn (11.35) to describe the electron wave functions, the latter is clearly the most apposite in discussing the continuous flow of charge which constitutes electrical conduction in a material.

The mechanism of electrical conduction

Adopting the k-values dictated by periodic boundary conditions and assuming that the metal is in the form of a cube of sides $L_x =: L_y = L_z = L$, we see from eqn (11.35) that the permitted points in k_x–k_y–k_z space lie on a simple cubic lattice whose nearest-neighbour separation is $2\pi/L$. Extending the procedures described in §10.6, and illustrated in Fig. 10.18, we may draw in a sphere of radius k_F (Fig. 11.12(a)) such that the number of points within the sphere is equal to $\frac{1}{2}N$, where N is the total number of free electrons within the metal. The factor of $\frac{1}{2}$ arises, yet again, because each point can describe the k-values of two electrons of opposite spin. Since each point in k_x–k_y–k_z space occupies a volume of $(2\pi/L)^3$ and since a sphere of radius k_F has a volume of $\frac{4}{3}\pi k_F^3$ it follows that

$$\frac{1}{2}N = \frac{\frac{4}{3}\pi k_F^3}{(2\pi/L)^3}$$

$$k_F = (3\pi^2 n)^{1/3}, \quad (11.36)$$

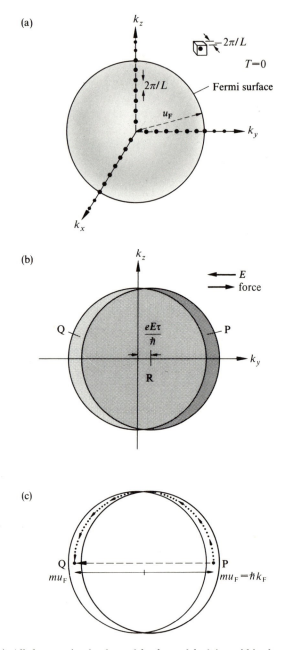

Fig. 11.12. (a) All the permitted values of k_x, k_y, and k_z lying within the sphere—most of which are omitted for clarity—are filled by two electrons of opposite spin. (b) An electric field E shifts the entire sphere by $\delta k_y = eE\tau/\hbar$. Region P has an excess of electrons; region Q a deficiency. (c) Showing how electrons are scattered around the Fermi surface from P to Q.

where $n = N/L^3$ is the number of free electrons per unit volume of the metal. This is identical to the result obtained with the k-values determined by fixed boundary conditions (eqn (10.43)). Since $E = \hbar^2 k^2/2m$ (eqn (10.41)) it follows that all those electrons lying on the surface $k = k_F$ will have a constant energy; the Fermi energy $E_F = \hbar^2 k_F^2/2m$. The name *Fermi surface* is used to denote a surface drawn in k_x–k_y–k_z space on which the electrons have a constant energy; in the case of our free electron gas the Fermi surface is, of course, a sphere (Fig. 11.12(a)).

Because for every occupied point at (k_x, k_y, k_z) in Fig. 11.12(a) there is another occupied point at $(-k_x, -k_y, -k_z)$ there will be no net flow of electrons along the conductor in the absence of an electric field. If, however, a field E is now applied, as in Fig. 11.12(b), its effect will be to shift the entire sphere of Fig. 11.12(a) in the direction of the applied force. Taking over eqn (7.61), which was originally deduced to describe the behaviour of gaseous ions, and applying it to the free-electron gas, we see that each and every electron will have its y-component of momentum increased by $mv_d = (-e)(-E)\tau$, where τ is the (mean) time between collisions, $-e$ is the charge on an electron ($e = +1.6 \times 10^{-19}$ C), and where E is the electric field (the minus sign being required since E is the force per unit *positive* charge). Thus the value of k_y of each and every electron will increase by

$$\delta k_y = \frac{1}{\hbar}(mv_d) = \frac{eE\tau}{\hbar}. \tag{11.37}$$

The resulting shift of the entire sphere and its contents is shown schematically in Fig. 11.12(b) (drawn looking in along the k_x-axis). Now when an entire squad of *identical* men take one step to the right the only changes which are evident are those occurring at the ends of the squad. Likewise,

TABLE 11.2
Electronic properties of metals at 295 K

Metal	Electrical conductivity $\sigma/10^7\,\Omega^{-1}\,m^{-1}$	Electron density[a] $n/10^{28}\,m^{-3}$	Fermi energy[b] $E_F/10^{-19}\,J$	Fermi speed[c] $u_F/10^6\,m\,s^{-1}$	Electron mean free path[d] $\lambda/10^{-8}\,m$
lithium	1·05	4·5	7·4	1·3	1·0
sodium	2·10	2·5	5·0	1·1	3·1
potassium	1·40	1·3	3·2	0·85	3·2
rubidium	0·80	1·1	2·9	0·80	2·1
caesium	0·50	0·85	2·4	0·75	1·5
copper	5·90	8·5	11·3	1·6	3·9
silver	6·20	5·8	8·8	1·4	5·3
gold	4·55	5·9	8·9	1·4	3·8

[a] Assumes one free electron per atom. [b] Calculated using eqn (10.43); $E_F = \hbar^2(3\pi^2 n)^{2/3}/2m$. [c] Calculated from $\frac{1}{2}mu_F^2 = E_F$. [d] Calculated using eqn (11.38); $\lambda = mu_F\sigma/ne^2$.

the only real changes to occur to the electron gas are those which take place in the vicinity of the Fermi surface; all interior points remaining filled with their quota of two electrons per point.

Looking at Fig. 11.12(b) we see that, compared with the situation shown in Fig. 11.12(a), there is an excess of electrons in region P and a deficiency of electrons in region Q. We must now consider what mechanisms can take excess electrons in region P back into the deficient region Q without sending them through the filled region R. Clearly, the only route which is available is around the Fermi surface, as shown in Fig. 11.12(c). Without such a scattering mechanism the Fermi surface would move off to even larger and larger values of k_y; in other words, τ must be finite in eqn (11.37). Before looking at possible scattering mechanisms it is worth defining the electron mean free path as $\lambda = u_F\tau$, where u_F is the effectively constant *speed* of those electrons which are scattered around the Fermi surface. This allows eqn (7.65), with $q = -e$, to be written as

$$\sigma = \frac{ne^2\tau}{m} = \frac{ne^2\lambda}{mu_F}. \tag{11.38}$$

Instead of talking of the conductivity of a metal we may talk instead of its *electrical resistivity* ρ, defined as $1/\sigma$. Thus

$$\rho = \frac{m}{ne^2\tau} = \frac{mu_F}{ne^2\lambda}. \tag{11.39}$$

The advantage of introducing the resistivity is that the mean free path λ appears in the denominator. In view of eqn (11.30)—or its equivalent form expressed in terms of relaxation times (see problem 11.18)—the resistivities attributable to different scattering processes may be added. Table 11.2 lists values of λ as deduced for a variety of metals from σ and u_F via eqn (11.38); the value of u_F being obtained from $\frac{1}{2}mu_F^2 = E_F$, with E_F given by eqn (10.43). This data demonstrates that in these metals the electrons' mean free path at 295 K is about a hundred atomic diameters.

Exercise 11.3

Copper at 295 K has an electrical conductivity of $5\cdot9\times10^7\ \Omega^{-1}\,\mathrm{m}^{-1}$ and (with a pure enough sample) a conductivity of some 10^5 times greater at 4 K. On the assumption that there are as many free electrons as there are copper atoms, calculate the electronic mean free path at these two temperatures. Compare these values with the mean free path as calculated classically assuming the copper atoms to be hard spheres and the electron to have zero diameter. For simplicity, suppose copper to have a simple cubic lattice (it is fcc). The density of copper is $8\cdot9\times10^3\ \mathrm{kg\ m}^{-3}$ and $A_r(\mathrm{Cu}) = 63\cdot5$. Take $u_F = 1\cdot6\times10^6\ \mathrm{m\ s}^{-1}$ (you calculated it in problem 10.16).

Calculation. We know from the density and relative atomic mass of copper that there are $N_A = 6\times10^{23}$ atoms in $(63\cdot5\times10^{-3}/8\cdot9\times10^3)\ \mathrm{m}^3 = 7\cdot1\times10^{-6}\ \mathrm{m}^3$ of copper. Therefore each copper atom occupies a volume of roughly $(7\cdot1\times10^{-6}/6\times10^{23})\,\text{m}^3 = 12\times10^{-30}\,\text{m}^3$, and so has a diameter of about $(12\times10^{-30})^{1/3}\,\text{m} = 2\cdot3\times10^{-10}\,\text{m}$. We also deduce that there are $6\times10^{23}/7\cdot1\times10^{-6} = 8\cdot5\times10^{28}$ atoms m^{-3}. We are to assume that the number of free electrons per unit volume is the same as the number of copper atoms per unit volume. So $n = 8\cdot5\times10^{28}\,\text{m}^{-3}$. Substituting for n, e, m, σ, and u_F into eqn (11.38) gives an electronic mean free path of $4\times10^{-8}\,\text{m}$ at 295 K and one of $4\times10^{-3}\,\text{m}$ at 4 K.

We argued in Exercise 7.3 that the classical mean free path of a particle of effectively zero size as it moves through an array of hard spheres, each of radius r, is $\lambda = 1/\pi r^2 n$, where n is the number density of the spheres. It follows therefore that the classical mean free path of a free electron in copper is $[\pi\times(1\cdot15\times10^{-10})^2\times8\cdot5\times10^{28}]^{-1}\,\text{m} = 2\cdot8\times10^{-10}\,\text{m}$. In other words, the classical mean free path is of the same order as the lattice spacing.

Comment. As this exercise has demonstrated, the mean free path of an electron in a metal is considerably greater than can be accounted for by classical arguments; an electron in a metal is seemingly not scattered by a regular atomic structure. We must therefore look elsewhere for scattering processes which can bring about the desired change in the velocity of an electron (Fig. 11.12(c)).

Impurity scattering

Just as a ping-pong ball rebounds with little loss of speed when it strikes a billiard ball so we can expect an electron in a metal to behave similarly when it strikes a massive imperfection in the lattice, and thus to produce the type of scattering we are seeking. The associated mean free path λ_i should be independent of temperature when the scattering is caused by impurity atoms, and be almost independent of temperature when the scattering is caused by dislocations (except close to the melting temperature, where the dislocation density will depend on T). There is thus a temperature-independent contribution to the resistivity given by eqn (11.39) with $\lambda = \lambda_i$, namely

$$\rho_i = \frac{mu_F}{ne^2\lambda_i}. \tag{11.40}$$

Phonon scattering

It is an experimental fact that the electrical resistivity of all pure metals increases with increasing temperature. Were the resistivity of tungsten to decrease with increasing temperature the familiar electric light bulb would quickly burn out! This temperature dependence must clearly be sought in the temperature dependence of the lattice vibrations, that is, in the interaction of electrons with phonons.

Since we are looking for a mechanism which will change the momentum of an electron from $+mu_F\,(=\hbar k_F)$ to $-mu_F\,(=-\hbar k_F)$, as shown in Fig. 11.12(c), it is natural to begin by calculating the mean momentum $\bar{p}$

of a phonon. Now, by definition,

$$\bar{p} = \frac{\sum\limits_{q} n_q p}{\sum\limits_{q} n_q}, \tag{11.41}$$

where n_q is the number of phonons associated with mode q. Assuming that we can apply de Broglie's relation $\lambda = h/p$ to a phonon, we see that

$$p = \frac{h}{\lambda} = \hbar q, \tag{11.42}$$

where $q = 2\pi/\lambda$ is the circular wavenumber of the mode. Appealing once more to Debye's dispersion relation, $\omega_q = qv$, allows us to rewrite eqn (11.42) as $p = \hbar\omega_q/v$. Substituting this result into eqn (11.41) gives

$$\bar{p} = \frac{1}{v} \frac{\sum\limits_{q} n_q \hbar\omega_q}{\sum\limits_{q} n_q}.$$

Apart from the constant multiplier $1/v$ this is identical to eqn (11.20). We may therefore take over the results which were obtained on evaluating eqn (11.20); in particular eqns (11.22) and (11.23). Multiplying the right-hand sides of these equations by $1/v$ and approximating the numerical factors of $2/3$ and $2\cdot34$ by unity tells us that

$$\boxed{\bar{p} = \frac{k\theta_D}{v}} \quad \text{at} \quad T > \theta_D \tag{11.43}$$

and that

$$\boxed{\bar{p} = \frac{kT}{v}} \quad \text{at} \quad T < \theta_D. \tag{11.44}$$

At low temperatures the phonon momentum kT/v is insufficient to change the momentum of an electron from mu_F to $-mu_F$ in one go. Thus the only way that a phonon in a metal at a low temperature can reverse the momentum of an electron is to 'nudge' it around the Fermi surface. Fig. 11.13(a) shows a single electron–phonon interaction; Fig. 11.13(b) the required sequence of interactions which is required to reverse the momentum of the electron. However, there is no guarantee that the

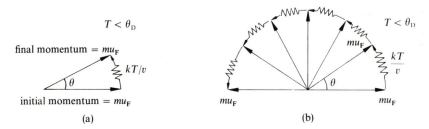

Fig. 11.13. (a) In a single encounter with a phonon the momentum of an electron is but little changed. (b) It takes many such encounters to change the electron's momentum from mu_F to $-mu_F$.

progression will occur in this ordered fashion. An electron–phonon collision is equally likely to remove momentum from the electron (this would be illustrated by swapping the 'initial' and 'final' labels in Fig. 11.13(a) and reversing the momentum of the phonon). So in going round the surface it may be two steps forward, then one back. This is essentially the random walk problem of §11.2 in a new guise.

Taking over eqn (11.9), it tells us that the number of steps, each of length a, which are required to cover a r.m.s. distance of l is given by l^2/a^2. In the present context the total distance to be travelled by the tip of the vector representing the electron's momentum is one-half of the circumference of a circle of radius mu_F, that is πmu_F (see Fig. 11.13(b)). Since each step is of length kT/v it follows that

$$\text{number of steps required} = \frac{(\pi mu_F)^2}{(kT/v)^2} \propto \frac{1}{T^2} \quad \text{at} \quad T < \theta_D, \quad (11.45)$$

where we have used the fact that u_F and v are independent of T. As it takes this number of encounters with phonons for the momentum of a particular electron to be reversed it follows that, at any moment, the number of phonons which are effective in turning electrons around is the total number of phonons present divided by the number of steps required to effect a reversal. (By way of analogy, it takes an average of 6 throws of a die to secure a desired result, say a 4. Thus if 100 dice are thrown together the number producing the desired result (4) is about 100/6.) Since, at $T < \theta_D$, the total number of phonons present is proportional to T^3 (eqn (11.19)) and since the number of steps required is proportional to $1/T^2$ (eqn (11.45)), we deduce that

$$\text{effective number of phonons} \propto \frac{T^3}{1/T^2} \propto T^5. \quad (11.46)$$

Because the mean free path λ_T of an electron is inversely proportional to the number of phonons capable of causing the momentum reversal we conclude, from eqn (11.46), that $\lambda_T \propto 1/T^5$. Inserting this relation into eqn (11.39) gives

$$\boxed{\rho_T \propto T^5} \quad \text{at} \quad T < \theta_D. \tag{11.47}$$

At $T > \theta_D$ the mean phonon momentum $k\theta_D/v$ (eqn (11.43)) is sufficient to reverse the momentum of an electron in a single encounter. Therefore *all* phonons are effective in scattering electrons through 180° at these temperatures. Since the number of phonons is proportional to T at $T > \theta_D$ (eqn (11.18)) this tells us that $\lambda_T \propto 1/T$ at such temperatures. Inserting this result into eqn (11.39) gives

$$\boxed{\rho_T \propto T} \quad \text{at} \quad T > \theta_D. \tag{11.48}$$

Experimental results

   Multiplying eqn (11.30) through by mu_F/ne^2 (see eqn (11.39)) predicts that the overall resistivity of a metal $\rho(T)$ at temperature T should be the sum of the resistivity ρ_T due to phonon scattering of the conduction electrons and the resistivity ρ_i due to impurity scattering of the

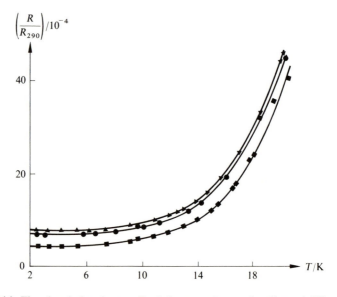

Fig. 11.14. The electrical resistance R of three specimens of sodium of different purity, expressed as a fraction of their resistance at $T = 290$ K. (Data from MacDonald, D. K. C. and Mendelssohn, K. (1950). *Proc. R. Soc.*, **A202**, 103.)

electrons:

$$\rho(T) = \rho_i + \rho_T. \tag{11.49}$$

Substituting for ρ_i from eqn (11.40) and for ρ_T from eqn (11.47) gives

$$\boxed{\rho(T) = \rho_0 + AT^5} \quad \text{at} \quad T < \theta_D, \tag{11.50}$$

where ρ_0 and A are constants.

Fig. 11.14 shows how the resistance of three samples of sodium, each of different purity, vary with temperature. Clearly, as the temperature falls towards $T = 0$ the resistance of each sample does tend to a constant value, in accordance with eqn (11.50). Eqn (11.50) also predicts that $\rho(T) - \rho_0$ plotted against T should be the same for all specimens. This is so; all three curves in Fig. 11.14 do superimpose if we subtract from each curve the constant resistance close to $T = 0$. In addition, the temperature dependence at $T < \theta_D$ does conform to that predicted by eqn (11.50). Finally, at $T > \theta_D$ (for sodium $\theta_D = 158$ K), $\rho(T) - \rho_0$ is proportional to T, as is predicted when eqn (11.48) is substituted into eqn (11.49).

The Wiedemann–Franz law

As long ago as 1853 Wiedemann and Franz showed that the ratio of the thermal conductivity κ to the electrical conductivity σ is very nearly the same for all metals at room temperature. This law was generalized by Lorenz in 1881 who demonstrated that the so-called *Lorenz number L*, defined by

$$L = \frac{\kappa}{\sigma T}, \tag{11.51}$$

is very nearly constant over a wide range of temperatures.

The constancy of L follows from the value of κ as given by eqn (11.27), with $v = u_F$ and the electronic heat capacity per unit volume $C = C_{e,m}/V_m = (3/\pi)^{4/3} n^{1/3} m k^2 T/\hbar^2$ (see eqns (10.49) and (10.50)), and from the value of σ as given by eqn (11.38). Substituting these values of κ and σ into eqn (11.51) and using the relation $\frac{1}{2} m u_F^2 = E_F$, with E given by eqn (10.43), leads to

$$\boxed{L = 3 \frac{k^2}{e^2} = 2 \cdot 23 \times 10^{-8} \, \text{W} \, \Omega \, \text{K}^{-2}}. \tag{11.52}$$

Experimental values of L lie between about $2 \cdot 2 \times 10^{-8} \, \text{W} \, \Omega \, \text{K}^{-2}$ and $3 \cdot 0 \times 10^{-8} \, \text{W} \, \Omega \, \text{K}^{-2}$. As an example, $L = 2 \cdot 28 \times 10^{-8} \, \text{W} \, \Omega \, \text{K}^{-2}$ for copper at 290 K, and remains constant to within about twenty per cent over the range $T = 100$–1000 K. At low temperatures, meaning $T \ll \theta_D$, the measured values of L start to decrease rapidly with decreasing temperature.

Long before the birth of wave mechanics Drude had deduced the value of L on the basis of the classical free-electron gas model outlined in exercise 7.3. This model supposes that each electron has a heat capacity $\frac{3}{2}k$, giving $C = \frac{3}{2}nk$, and a mean speed $\bar{u}$ of $(8kT/\pi m)^{1/2}$. Thus eqn (11.27) predicts $\kappa = n\lambda (2k^3 T/\pi m)^{1/2}$. Eqn (11.38) with u_F replaced by $\bar{u}$, predicts $\sigma = ne^2 \lambda (\pi/8kTm)^{1/2}$. Hence Drude's model predicts

$$L = \frac{4}{\pi} \frac{k^2}{e^2} = 0.95 \times 10^{-8} \text{ W } \Omega \text{ K}^{-2}. \tag{11.53}$$

In view of the untenable nature of its assumptions and the consequent wrong predictions as to the temperature dependence of κ and σ it is remarkable how close Drude's model comes to predicting correct values of L. Finally, it should be pointed out that other authors may arrive at slightly different forms of eqns (11.52) and (11.53). These discrepancies usually arise from different approximations being made in the evaluation of κ and σ. Thus, for example, one may take the electron mean free path as $2\tau\bar{u}$ when eqn (11.38) is applied to Drude's model; this gives $\sigma = ne^2\lambda/2m\bar{u}$ and $L = 1.9 \times 10^{-8} \text{ W } \Omega \text{ K}^{-2}$. The main point, however, is that both models predict that L is of order k^2/e^2.

11.6. Metals, semiconductors, and insulators

Despite its undoubted success in accounting for many of the properties of a metal, the free-electron Fermi-gas model fails to explain why some materials are conductors, others are insulators, and yet others are semiconductors. The range of electrical conductivities encompassed by these classes of materials can be truly enormous; as an example, the room temperature conductivity of copper is some 10^{19} times that of diamond. One's first thoughts might be that an insulator arises when none of the material's electrons stray from their parent atom. Such a model cannot, however, explain why a covalently-bonded material like diamond—a diamond is effectively one giant molecule—is an insulator and not, as would seem more reasonable, a conductor. In fact we shall shortly discover that diamond is an insulator because there is no way that its 'roving' electrons can acquire energy from an applied electric field. It is *not* an insulator because of an absence of roving electrons.

We begin by reviewing the main arguments employed in applying the free-electron theory to a one-dimensional solid. The first step is to smear out the positive ions into a uniform jelly. This means an electron will have a constant potential energy throughout the length of the solid (Fig. 11.15(a)). In the next step we imagine the two ends of the solid to be

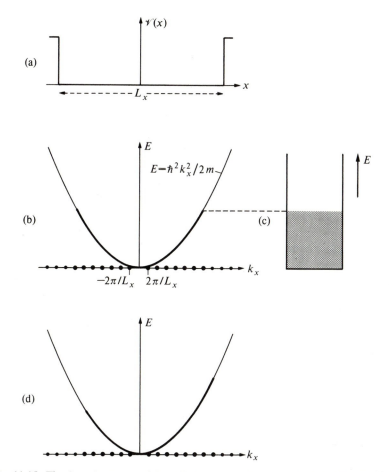

Fig. 11.15. The free-electron model applied to a one-dimensional solid at $T = 0$. (a) The potential energy $V(x)$ of an electron. (b) The permitted values of k_x are shown by dots on the abscissa (drawn large if assigned). The corresponding values of the energy E of the electron (which is wholly kinetic; $V(x) = 0$, by definition). (c) A block diagram in which filled states are indicated by shading. (d) Showing the effect of an electric field when applied around the ring.

joined together to form a ring, around which the electrons can travel in either direction. Introducing the boundary condition that $\psi(x) = \psi(x + L_x)$, where L_x is the length of the solid, leads to the conclusion that the allowed wave functions will have circular wavenumbers given by $k_x = \pm(2\pi/L_x)$, $\pm2(2\pi/L_x)$, $\pm3(2\pi/L_x)$, etc. (eqn (11.34)). These permitted wavenumbers are set out along the abscissa of Fig. 11.15(b); the ordinate giving the corresponding energies $E = \hbar^2 k_x^2/2m$ of the electron (eqn (10.41)). The final step in the argument is to assign each k_x-value to two electrons of opposite spin, starting from $k_x = \pm(2\pi/L_x)$ and working outwards to ever higher values of k_x (that is, to higher energies) until all the free electrons have been assigned 'homes'. There are several ways of indicating this filling-up process. One is to thicken the line on the $E - k_x$ plot up to the maximum occupied value of k_x (Fig. 11.15(b)). Another is to use the rather uninformative block diagram shown in Fig. 11.15(c). Here shading is used to indicate that states up to a particular value of E are full; blank paper indicates empty states. If an electric field is now applied around the ring (this could be done by placing the ring in an increasing magnetic field directed normal to the plane of the ring), the distribution will shift to something like that shown in Fig. 11.15(d). Two key facts emerge from these arguments. First, any number of free electrons can be accommodated in the metal. Secondly, free electrons can always have their energy increased by an applied electric field. In short, this model says that any material which has free electrons will be a conductor.

We must now face up to the fact that smearing out the ions of a solid into a uniform jelly really is a bit bogus. So we shall introduce the ions into the discussion along with the associated potential energy curve of an electron shown in Fig. 11.16(a). (You can see that this curve makes sense by recalling that as an electron approaches a positive charge of, say, e at $x = 0$ its potential energy will fall off as $-e^2/4\pi\varepsilon_0 x$.) If we now go through the business of assigning electrons to ever-increasing values of k_x something important happens when we reach $k_x = \pm\frac{1}{2}N(2\pi/L_x)$, that is when we give electrons travelling (in either sense) around the ring a wavelength $\lambda = 2\pi/k_x = 2L_x/N = 2a$, where N is the total number of *ions* present and a $(= L_x/N)$ is the nearest-neighbour separation. This value of λ happens to satisfy Bragg's law (eqn (8.4) with $d_{hkl} = a$, $\theta = 90°$, and $n = 1$; that is $2a = \lambda$). Thus electrons with $k_x = \pm\frac{1}{2}N(2\pi/L_x)$ will be reflected back on themselves as they attempt to travel around the ring. As with transverse waves on a rope, this leads to the formation of standing waves (but see footnote on p. 337) of wavelength $\frac{1}{2}(2a) = a$. Such waves can either have $|\psi(x)|^2 \propto \cos^2(\pi x/a)$ or $|\psi(x)|^2 \propto \sin^2(\pi x/a)$; see Fig. 11.16(b). Recalling the meaning of $|\psi(x)|^2$ we see that the $\cos^2(\pi x/a)$ function gives a high probability of finding the electron close to the ion cores, whereas the $\sin^2(\pi x/a)$ function gives a high probability of finding the electron between the ion cores. A moment's thought will show that the potential energy of an electron with $|\psi(x)|^2 \propto \cos^2(\pi x/a)$ is lower than the potential energy of a travelling electron (where $|\psi(x)|^2$ is constant), and that the potential energy of an electron with $|\psi(x)|^2 \propto \sin^2(\pi x/a)$ is higher than the potential energy of a travelling electron. Thus an *energy gap*, or *band gap*, of width E_g, say, appears at $k_x = \pm\frac{1}{2}N(2\pi/L_x)$ as shown in Fig. 11.16(c). It is worth noting that since the electron is not travelling at these values of

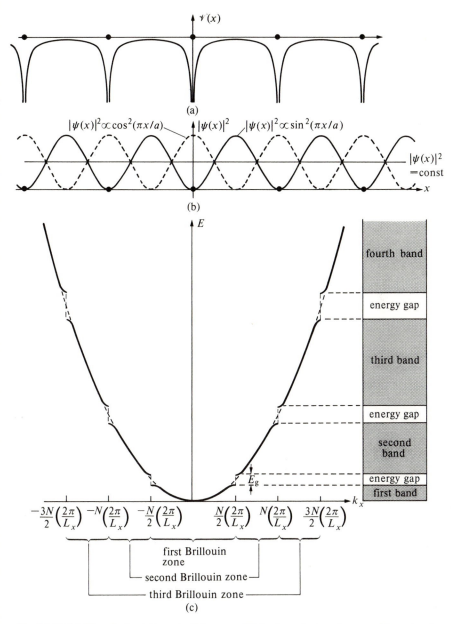

(a)

(b)

(c)

Fig. 11.16 (a) The electrostatic potential energy $\mathscr{V}(x)$ of an electron in a one-dimensional solid. (b) Showing the probability density $|\psi(x)|^2$ of the two permitted wave functions at $k_x = \pm\frac{1}{2}N(2\pi/L_x)$ and of a travelling wave function. (c) Illustrating the energy gaps which appear at $k_x = \pm\frac{1}{2}N(2\pi/L_x)$, $\pm N(2\pi/L_x)$, $\pm\frac{3}{2}N(2\pi/L_x)$, etc. Here E is the total energy (electrostatic plus potential) of an electron. The dotted line shows only the kinetic energy $\hbar^2 k_x^2/2m$.

k_x its kinetic energy is zero here and all its energy exists as electrostatic potential energy. Conversely, near $k_x = 0$, where λ is very large, the lattice exerts little influence on the electron; here its energy may be taken as wholly kinetic. Although we have only considered first-order Bragg reflection, similar effects occur at the second-order reflection, where $2a = 2\lambda$, that is $k_x = \pm N(2\pi/L_x)$; at the third-order reflection (where $k_x = \pm\frac{3}{2}N(2\pi/L_x)$), and so on (see Fig. 11.16(c)). The region between $k_x = -\frac{1}{2}N(2\pi/L_x)$ and $k_x = +\frac{1}{2}N(2\pi/L_x)$ is usually referred to as the first *Brillouin zone*: This and the second, third, etc. Brillouin zones are shown in Fig. 11.16(c). We may likewise speak of the first, second, third, etc. allowed energy bands; these are also indicated in Fig. 11.16(c). Fig. 11.16(c) shows that the first Brillouin zone contains $\frac{1}{2}N + \frac{1}{2}N = N$ allowed values of k_x. Since each such value of k_x can describe the translational momenta of two electrons of opposite spins it follows that the first Brillouin zone (and hence the first allowed energy band) can accommodate $2N$ electrons, where N is equal to the number of *ions* in the solid. The same is also true of all other Brillouin zones, as can readily be verified from Fig. 11.16(c).

The final step is to see the effect of having different numbers of 'nearly free' electrons present in the solid. Sodium, for example, has one conduction electron per atom. Thus (picturing it to be in the form of a one-dimensional crystal) we see that its electrons will only fill the states from $k_x = -\frac{1}{4}N(2\pi/L_x)$ to $k_x = +\frac{1}{4}N(2\pi/L_x)$, as shown in Fig. 11.17(a). We also see that its highest-energy electrons are well removed from a band edge, where the electrostatic potential energy of an electron becomes significant. This explains our earlier success in applying the free-electron model to sodium. Extending the arguments further, we see that elements with an odd number of valence electrons per atom should be conductors, as is generally the case. Fig. 11.17(b) illustrates the situation of three outer electrons per atom. Clearly if there are an even number of valence electrons per atom we will have an insulator. Diamond has four such electrons (Fig. 11.17(c)). In the case of diamond the energy gap E_g which the top electrons have to cross before they can enter the next allowed energy band is around 7 eV ($= 7 \times 1 \cdot 6 \times 10^{-19}$ J). This is much larger than the thermal energy $\frac{3}{2}kT$ ($= 0 \cdot 04$ eV at 300 K) possessed by these electrons. If, however, the material has a small energy gap, say around 1 eV, an appreciable number of electrons can be thermally excited across the gap. Under these circumstances we have what is called an *intrinsic semiconductor* (Fig. 11.17(d)). Examples are silicon and germanium. The distinction between an intrinsic semiconductor and an insulator lies only in their values of E_g and how these values compare with $\frac{3}{2}kT$. In fact, all intrinsic semiconductors are insulators when at $T = 0$.

Finally, we ought to remind ourselves that our arguments were

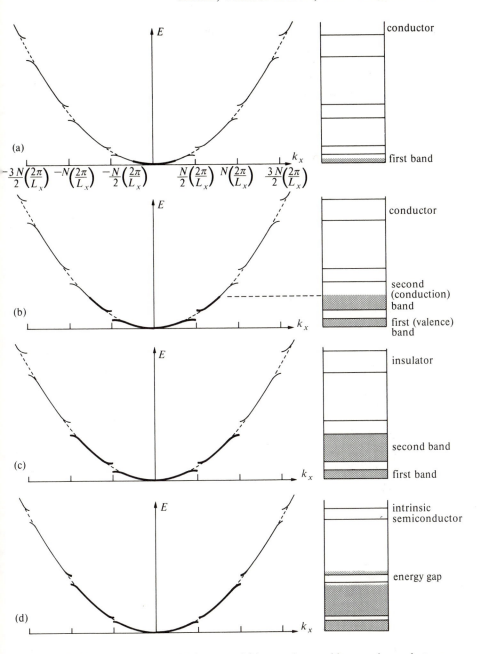

Fig. 11.17. Illustrating the essential features of (a) a conductor with one valence electron per atom, (b) a conductor with three valence electrons per atom, (c) an insulator with four valence electrons per atom, and (d) an intrinsic semiconductor with four valence electrons per atom.

centred around a hypothetical one-dimensional crystal. Although the arguments are broadly similar in a three-dimensional structure, they are a good deal more difficult to illustrate. For example, the form of Fig. 11.16(c) will often be different in different crystal directions. This may ensure that an allowed energy band in one direction can 'bridge' an energy gap in another direction. In fact, this explains why divalent metals like magnesium are electrical conductors rather than insulators, as our one-dimensional theory would predict.

PROBLEMS

11.1. Show that in a one-dimensional random walk the r.m.s. distance travelled after n hops, each of length a, is $an^{1/2}$. Also show that the mean square distance travelled in a time t is $2Dt$, where D is the self-diffusion coefficient. Surface diffusion along the length of a crack, or along the line of contact of two crystals, are examples of one-dimensional diffusion.

11.2. In a process called *carburizing*, low-carbon steels are given a thin skin of high-carbon steel. Many parts of a car engine, for example, need to have a hard surface without being brittle (which low carbon steel is, in bulk). This is achieved by packing carbon (and other substances) around the low-carbon steel part and heating them both to around 1170 K. The carbon diffuses into the steel forming the high-carbon layer. If it takes 25 hours to produce a 'case' $\frac{1}{8}$ in (3·2 mm) thick at 1140 K, how long will it take to produce the same thickness case at 1230 K? Assume that the activation energy required for the interstitial carbon to hop from site to site is $1 \cdot 4 \times 10^{-19}$ J.

11.3. Construct a four-sided top out of a piece of card with a matchstick passing through its centre. Label the sides, 'left', 'right', 'up', and 'down'. Spin the top and note on which side it comes to rest. This will tell you in which direction an adatom lying on a {100} face of a simple cubic crystal is to make its first hop. Repeat the operation of spinning the top to determine the direction of the next hop, tracing out the path followed (as in Fig. 11.6). Check out whether the r.m.s. distance gone is given by $(\overline{R^2})^{1/2} = an^{1/2}$, where a is the lattice spacing, and n is the number of hops. You will need to include a considerable number of hops. You may find it convenient to use the 5 mm squares on a piece of graph paper as your grid.

11.4. A thin layer of N atoms is deposited on a surface of area A on a thick solid. After a certain time t some of the surface atoms will have diffused into the solid. Show (by substitution into the expression derived in problem 7.5) that the concentration $n(x)$ of these atoms as a function of distance x into the solid is, at time t, given by

$$n(x) = \frac{N}{A(\pi Dt)^{1/2}} \exp(-x^2/4Dt).$$

Make a rough sketch of this relation at time $t = 1/4D$, $t = 1/2D$, and $t = 1/D$. Find the value of x at the point of inflexion, and compare your value with the r.m.s. distance travelled as given by eqn (11.12).

11.5. When visible light strikes a silver bromide crystal (which may be in a photographic emulsion) free electrons are produced. (This happens because the photon energy is sufficient to excite electrons from a full valence band to an empty conduction band.) The electrons so released can then be trapped at various impurities. The next stage involves the diffusion of interstitial Ag^+ ions to these trapped electrons, leading to the formation of Ag atoms and rendering the crystal developable.

Considering the case of an electron trapped on the surface of the silver bromide crystal, how long will it take an Ag^+ adion to diffuse a distance of 10^{-6} m (of order half the circumference of a typical crystal incorporated in an emulsion) at temperatures of 100 and 300 K if the diffusion coefficients of silver ions on the surface of silver bromide is 10^{-13} m^2 s^{-1} at 100 K and 10^{-17} m^2 s^{-1} at 300 K? The exponential temperature dependence of D explains why silver bromide photography is not possible at low temperatures.

11.6. This problem gives a flavour of the sort of calculations which must be performed in estimating the theoretical growth rate of a crystal from the vapour phase. By calculating the time t that it takes for an adatom to diffuse through a r.m.s. distance l (this might be half the width of a terrace; see Fig. 11.5(a)) and deducing the probability that an adatom will actually escape back into the gas phase during time t, show that the probability that an adatom will reach a ledge without evaporating *en route* is given by

$$1 - \frac{l^2}{4a^2} e^{-(E_s - E_d)/kT}.$$

Here E_s is the energy required for an adatom to evaporate from the surface of the crystal, E_d is the activation energy required for surface diffusion (as in eqn (11.5), and a is the nearest-neighbour separation between surface sites (assuming them to be on a simple square lattice). Taking values of $l = 10^{-8}$ m, $a = 3 \times 10^{-10}$ m, $E_d = 1 \times 10^{-19}$ J, and $E_s = 2 \times 10^{-19}$ J (it is often between two and three times E_d), deduce the probability that an adatom will *fail* to reach the ledge at 300 and at 600 K. *Note*: The expression you derive may differ slightly from that quoted above; it all depends on the precise arguments you employ.

11.7. Using the data given in Fig. 11.2 calculate the time taken for ^{64}Cu atoms, deposited on the surface of a ^{63}Cu block, to diffuse in through a r.m.s. distance of 10^{-4} m if the block is at 1300 K. It will be sufficient to assume that $\overline{X^2} \approx Dt$.

11.8. What is the total number of phonons present in a 1 cm × 1 cm × 1 cm block of copper at 500 K? Assume that the structure of copper is simple cubic with a nearest-neighbour separation of $2 \cdot 6 \times 10^{-10}$ m. The Debye temperature of copper is 240 K.

11.9. What is the mean energy and momentum of a phonon in a block of aluminium ($\theta_D = 430$ K) held at a temperature of (a) 200 K, and (b) 600 K? Take the speed of sound to be 5×10^3 m s^{-1} in aluminium (it is in fact different for longitudinal and transverse waves).

11.10. The derivation of eqn (11.24) as given on p. 367 is a bit of a cheat since it involves taking n—which is not a constant in the case of phonons—inside $d(n\frac{3}{2}kT)/dT$. Fig. 11.18 shows the actual situation in a phonon gas. Using a 'one-sixth' model, show that the net heat current Φ flowing to the right through plane A is given by

$$\Phi = \frac{v}{3}(\bar{E}\, \delta n + n\, \delta \bar{E}) = \frac{v}{3}\, \delta(n\bar{E}), \qquad (11.54)$$

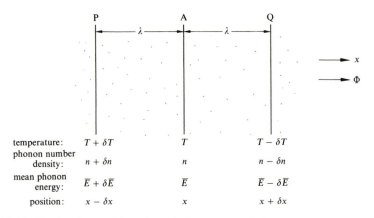

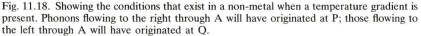

Fig. 11.18. Showing the conditions that exist in a non-metal when a temperature gradient is present. Phonons flowing to the right through A will have originated at P; those flowing to the left through A will have originated at Q.

where v is the speed of sound in the material. Also show that eqn (11.24) follows as a consequence of multiplying the right-hand side of eqn (11.54) by $\delta x/\delta x$, where $\delta x = \lambda$ is the distance within which the phonon density changes by δn and the mean phonon energy changes by $\delta \bar{E}$ (Fig. 11.18).

11.11. Sodium chloride at 83 K—which is above the temperature of the maximum in κ—has a thermal conductivity κ of 27 W m^{-1} K^{-1} and a heat capacity of $1\cdot0\times10^6$ J m^{-3} K^{-1}. Assuming the speed of sound is 5×10^3 m s^{-1} in NaCl, what is the phonon mean free path at 83 K?

11.12. Using the information contained in Fig. 11.10, calculate the ratio of the impurity concentration in Li 1 to that in Li 2. Explain your reasoning.

11.13. Show that the thermal conductivity κ of a metal (in which heat is transported only by electrons) should become constant, independent of temperature, at temperatures significantly greater than the Debye temperature of the material. In the case of copper ($\theta_D = 240$ K), for example, κ is constant to within about ten per cent over the temperature range 300–950 K.

11.14. Using the data shown in Fig. 11.9(b), confirm that the low-temperature thermal conductivity κ of LiF is proportional to DT^3, where D is the diameter of the specimen.

11.15. Investigate whether the mean momentum of a phonon in copper ($\theta_D = 240$ K) at room temperature is sufficient to change the momentum of an electron from mu_F to $-mu_F$. Take $u_F = 1\cdot6\times10^6$ m s^{-1} (as calculated in problem 10.16) and assume that the speed of sound is 4×10^3 m s^{-1} in copper.

11.16. Aluminium at 300 K has a resistivity of $2\cdot7\times10^{-8}$ Ω m. Calculate the drift velocity of an electron in a $0\cdot1$ m length of aluminium which is connected across a 250 V direct current supply. If the aluminium is connected instead to a 250 V, 50 Hz supply, roughly how far will the electron travel before the current reverses? Assume $n = 1\cdot8\times10^{29}$ m^{-3}.

11.17. In large samples the electrical resistivity of a certain material is ρ_b and the electron mean free path is λ_b. Show that the resistivity ρ_D of the material when prepared as a cylinder of diameter D is given by

$$\frac{\rho_D}{\rho_b} = 1 + \frac{\lambda_b}{D}.$$

Assume that the momentum of an electron may be changed by $2mu_F$ in each and every collision which it makes with the surface of the cylinder. (In other words, disregard any effects which may, and indeed do, arise from incomplete momentum accommodation at the surface.) The resistivity of a metal prepared in a very thin sample, of thickness less than λ_b, is greater than the resistivity of the metal in a large sample. Assume that the temperature of the material is close to 0 K.

11.18. Show that the overall 'mean free time' τ between collisions of an electron with either a phonon or an impurity atom is given by

$$\frac{1}{\tau} = \frac{1}{\tau_T} + \frac{1}{\tau_i},$$

where τ_T is the mean free time between collisions with phonons, and τ_i is the mean free time between collisions with impurity ions. *Clue:* Study the arguments leading up to eqn (11.30).

11.19. Using the data given in Fig. 11.14, confirm that the temperature-dependent part of the low-temperature resistance of solid sodium is proportional to T^5.

11.20. As we saw in §11.5, an electron may be taken around a Fermi surface by means of a random walk, in which each step has a length of kT/v in momentum space, at $T < \theta_D$ (eqn (11.44)). How many such randomly-directed steps are required to change the speed of a free electron in copper at 100 K from $+u_F$ to $-u_F$? Take the speed of sound to be $4 \times 10^3 \text{ m s}^{-1}$ and u_F to be $1 \cdot 6 \times 10^6 \text{ m s}^{-1}$.

11.21. Potassium at 78 K has an electrical resistivity of $1 \cdot 38 \times 10^{-8} \, \Omega \text{ m}$. On the supposition that there are as many free electrons as there are potassium atoms, calculate the electronic mean free path at 78 K. For simplicity, assume potassium to have a simple cubic lattice (it is bcc). Its density is $8 \cdot 62 \times 10^3 \text{ kg m}^{-3}$ and $A_r(K) = 39$.

11.22. Sketch the permitted forms of $|\psi(x)|^2$ at $k_x = \pm N(2\pi/L_x) = \pm 2\pi/a$, and at $k_x = \pm\frac{3}{2}N(2\pi/L_x)$ in a one-dimensional crystal of length L_x and containing N ions. Describe in words how the energy gap E_g at $k_x = \pm 2\pi/a$ might be calculated numerically via $|\psi(x)|^2$ and the supposedly-known form of the electron's potential energy $V(x)$.

11.23. The closest analogue to viscous flow in a solid is the process of creep. A familiar example of creep is the motion of lead on an old church roof; over the centuries it flows down under the stress provided by gravity. Experiments show that, in many materials, the strain-time curve has the form shown in Fig. 11.19. At temperatures up to about one-third the melting temperature the strain (ε) versus time (t) curve is linear over an appreciable period (that is $d\varepsilon/dt$ is constant) at a fixed temperature. Increasing the temperature T increases the steady-state creep rate. Under conditions of constant stress it is found that

$$\frac{d\varepsilon}{dt} = \text{constant } e^{-E/kT}, \tag{11.55}$$

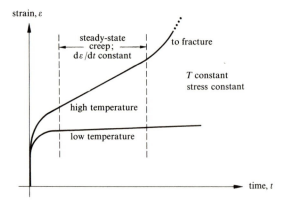

Fig. 11.19. When a material is subjected to a constant stress the strain does not remain constant, although at low temperatures the creep rate dε/dt is often negligible. At high temperatures the steady-state region (where dε/dt is constant) may eventually give way to a region in which the creep rate accelerates, leading to fracture.

where E is a constant activation energy (often the energy required to unpin dislocations from imperfections which temporarily impede their free movement). The activation energy involved in creep of a pure aluminium single crystal is $1 \cdot 95 \times 10^{-19}$ J (within the temperature range of 600 to 800 K). What temperature change is necessary to double the steady-state creep rate in such a crystal if its present temperature is 700 K?

12. The liquid phase

IN this chapter we shall attempt to discover what a liquid is like at the atomic level. We shall find, however, that, unlike the situation in a gas or solid, it requires mathematics to fully describe the structure of even a simple liquid. No neat picture of a liquid will emerge, one that can be described so succinctly as that for a perfect gas or a perfect solid. However, simulation studies will give us some feel for what a liquid may be like at the atomic level.

Of course, a liquid may also be studied at the macroscopic level. One might, for example, measure its heat capacity at constant pressure or its compressibility at constant temperature. The natural next step would be to try to account for these measurements, starting each time from a microscopic picture of the liquid. In fact, one does not need to keep returning to a microscopic picture: The laws of thermodynamics assure us that provided we know a substance's equation of state and its internal energy, $U_m(T, V_m)$, *all* the equilibrium properties of that substance can be categorized. So we shall devote some time to the derivation of these functions. We did not adopt this tactic in discussing perfect solids and perfect gases because their structure could be simply described in microscopic terms. Such a direct approach is never easy in a liquid, where the atoms are part bound and part free.

Throughout this chapter we will focus on the properties of simple liquids—which we may define loosely as liquids whose molecules are roughly spherical and which interact according to a Mie or Lennard-Jones type of potential model. This definition encompasses liquids like carbon tetrachloride and the liquefied inert gases. However, we shall briefly examine the properties of liquid crystals which, as the name suggests, exhibit properties reminiscent of both the solid and liquid phases.

12.1. Liquids *vis-à-vis* gases and solids

In §3.1 we argued that the liquid phase might be characterized by the rough equality

$$\tfrac{1}{2}m\overline{u^2} \approx \Delta E,$$

where $\tfrac{1}{2}m\overline{u^2}$ is, of course, an atom's mean kinetic energy, and ΔE is the pair dissociation energy.

Perhaps we should begin our study of the liquid phase by asking whether it is really necessary to roughly equate $\tfrac{1}{2}m\overline{u^2}$ and ΔE. Could one not simply regard a liquid as a very imperfect gas; one in which ΔE is a significant fraction (say, $\frac{2}{3}$) of $\frac{1}{2}\overline{mu^2}$? Or perhaps a truer model might be of a thermally 'imperfect' solid; one in which $\frac{1}{2}\overline{mu^2}$ is a significant fraction of ΔE. We shall seek to answer these questions by contrasting the structural, thermal, and mechanical properties of liquids with the corresponding properties of gases and solids.

The $p-V_m-T$ surface

You may remember the confusion over whether to label the region shown shaded in Fig. 1.2 a 'liquid' or a 'gas'. The confusion arose because, starting from a point where liquid and gas are undoubtedly both present (point i in Fig. 1.9) we could proceed to a point f within the region in question by two distinct routes. Following one route (that in full line) we would see the volume of liquid expand as the temperature is increased until the substance appeared to be wholly 'liquid'. Following another route (that in dashed line) we would see the volume of liquid diminish until the substance appeared to be wholly 'gaseous'. We resolved the problem of nomenclature by agreeing to use the word liquid to denote the single phase present when its state is represented by points lying within the shaded region of Fig. 1.2. Thus, in terms of the $p-T$ projection (see Fig. 1.11), the liquid phase extends from the triple point temperature T_{tr} to the critical temperature T_c. In terms of the $p-V_m$ projection it extends from $V_{tr,m}$ to $V_{c,m}$. Both of the ratios T_c/T_{tr} and $V_{c,m}/V_{tr,m}$ are usually in the range 1–5 (see Table 1.1), so there really is very little latitude possible in assigning values to the independent variables T and V_m if the substance is to be in the liquid phase. Bear in mind that the scales of p, V_m, and T extend to near infinity.

The evidence gleaned from the $p-V_m-T$ surface clearly suggests that a liquid more closely resembles a gas than a solid; particularly so in the vicinity of the critical point where the two phases become indistinguishable.

Thermal properties

(a) *Triple and critical temperatures.* If the arguments based on competing binding and kinetic energies are correct the relation $\frac{1}{2}\overline{mu^2}$ ($= \frac{3}{2}kT$) $\approx \Delta E$ must hold true *somewhere* in the liquid range. Table 12.1 records the values of the ratio $(\Delta E/k)/T$ computed at $T = T_{tr}$ and $T = T_c$ in the case of the inert gases. We see that the ratio is indeed constant as expected and that it most nearly has a value of 1·5 at T_{tr}. This conclusion also holds for other substances; for example, $\Delta E/kT_{tr} = 1·5$ for liquid nitrogen and 1·2 for liquid sodium. (There are exceptions; see p. 71.)

(b) *Molar enthalpies of fusion and evaporation.* To melt (fuse) one mole of a solid at a constant pressure p requires energy $H_{m,f}$—the molar enthalpy of fusion (at pressure p). To evaporate one mole of a liquid at

TABLE 12.1
Triple and critical temperatures of the inert gases

Substance	$(\Delta E/k)/K$	T_{tr}/K	T_c/K	$(\Delta E/k)/T_{tr}$	$(\Delta E/k)/T_c$
Ne	35·6	24·5	44·4	1·45	0·80
Ar	120	84	151	1·43	0·80
Kr	171	116	209	1·48	0·82
Xe	221	161	290	1·38	0·76

constant pressure p requires energy $H_{m,e}$—the molar enthalpy of evaporation (at pressure p).

Experiments show that at pressures close to p_{tr} the value of $H_{m,e}$
always exceeds $H_{m,f}$. As examples, $H_{m,e}/H_{m,f}$ is 5·5 for argon, 3·0 for
carbon dioxide, 7·0 for water, and 25 for mercury. However, on repeating
the experiments at higher pressures one finds that $H_{m,e}$ falls towards zero
as p approaches p_c but that $H_{m,f}$ remains (virtually) constant. These
findings suggest that, close to the triple point, the molar internal energy of
a liquid is nearer to that of a solid than to that of a gas but that the
situation is reversed close to the critical point. The same conclusions are
presented more formally in Fig. 1.15.

(c) *The molar heat capacity.* Another way of probing the internal
energy is to study how $C_{V,m} = (\partial U_m/\partial T)_{V_m}$ (eqn (1.12)) changes when a
substance undergoes a phase transition. You will recall (eqns (5.37) and
(5.38)) that in a perfect monatomic gas $U_m = \frac{3}{2}RT$, and therefore $C_{V,m} = \frac{3}{2}R$. You will also recall (eqns (10.1) and (10.2)) that in a classical
monatomic solid (or in a real solid at $T > \theta_D$) $U_m = 3RT$, and therefore
$C_{V,m} = 3R$. So if a solid melts to form a liquid which is essentially
solid-like there should be no change in $C_{V,m}$. If, on the other hand, it
melts to form a perfect gas-like liquid there should be a substantial
reduction in $C_{V,m}$. In practice the value of $C_{V,m}$ changes but little when
most solids melt. As an example, it decreases by about 15 per cent when
solid argon melts.

Taken overall, the thermal evidence suggests that, near the triple
point, the liquid phase is closer to the solid than the gaseous phase, but
that the opposite is true near the critical point.

Mechanical properties

(a) *Tensile strength.* Because of the way we can splash about in water
and thereby separate one mass of water from another you might infer that
a liquid is incapable of withstanding tensile stresses (often called
negative pressures). That liquids can, in fact, withstand negative pressures
may be demonstrated by taking a thick-walled tube, fitted with a valve,

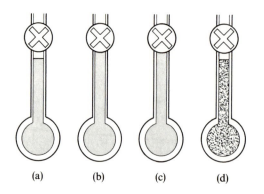

Fig. 12.1. In (a) the sealed tube contains the liquid and a small amount of vapour (all other gases are rigorously excluded). (b) The tube is heated, the liquid expands, and the vapour bubble disappears. (c) When the liquid is cooled it contracts but the bubble fails to reappear at the expected temperature. (d) With further cooling the liquid suddenly ruptures.

and filling it almost full of liquid (Fig. 12.1(a)) leaving only a small bubble of vapour. (All other gases, such as air, must be excluded.) If the tube is heated slightly (Fig. 12.1(b)) the liquid will expand to fill the entire tube and the bubble will disappear. On cooling the tube the vapour bubble does not reappear at the expected temperature—the liquid is now in a state of tension, being 'stretched' to fill the tube (Fig. 12.1(c)). On further cooling, the liquid suddenly ruptures with the appearance of many tiny vapour bubbles (Fig. 12.1(d)). These then rise to reform the vapour bubble. Negative pressures corresponding to stresses of, for example, about 10^6 Pa for liquid argon and about 10^7 Pa for mercury are observed by this method. By comparison, steel has a tensile strength of around 10^9 Pa.

(b) *Compressibility.* The rapidly rising shaded portion of the $p-V_m-T$ surface shown in Fig. 1.11 tells us that it takes a large increase in pressure to bring about a small reduction in the volume of a liquid near the triple point. But the still more rapidly rising solid region of the surface tells us that an even larger increase in pressure is required to bring about the same fractional change in the volume of a solid. Of course, the compressibility of a liquid increases as the critical point is approached.

We know from our discussions in §9.1 that there is relatively little free volume in a solid and that the low compressibility reflects the repulsive forces between closely-packed atoms. Now when a solid melts to form a liquid there is a small change in the volume of the substance; typically around 10–15 per cent. Perhaps then a liquid is more compressible than a solid because this *free volume*, so-called, can be relatively easily squeezed out. One might be tempted to conclude that a liquid near the triple point is only a thermally very-imperfect solid with 10–15 per cent of its lattice sites vacant. Taken overall, the mechanical evidence (like the thermal evidence) suggests that, near the triple point, the liquid phase is closer to the solid than the gaseous phase, but that the opposite is true near the critical point.

12.2. The radial density

We must now consider how we might describe whatever structure does exist in a liquid. Fig. 12.2 shows a hypothetical snapshot of what a liquid might look like; its exact nature is of no immediate concern. If the mean number of atoms per unit volume at radius r is $\rho(r)$—this implies averaging over a great number of such snapshots—then the total number of atoms whose centres lie within a shell of radii r and $r + \delta r$ is given by

number in the shell = (volume of shell)(atomic number-density in shell)

$$= 4\pi r^2 \, \delta r \, \rho(r). \tag{12.1}$$

This total number of atoms may also be written as $g(r) \, \delta r$, so that

$$\boxed{g(r) = 4\pi r^2 \rho(r)} . \tag{12.2}$$

We shall call $\rho(r)$—the mean number of atoms per unit volume at a distance r from our reference atom—the *radial density*.†

X-ray diffraction studies give the value of $g(r)$ and therefore the value of $\rho(r)$. Since a diffraction photograph requires a time exposure which is very much greater than the vibrational period of the atoms and since it collects information from all over a macroscopic sample the values it gives for $g(r)$ are both time-averaged and space-averaged. Fig. 12.3(a) shows $g(r)$ for argon at five points along the saturated liquid line (ac of Fig. 1.3). We see that as the temperature of the liquid is raised the peaks become ever less pronounced and the curves approximate ever more closely to the value to be expected for a perfect gas, where $\rho(r)$ is constant at all r. This causes $g(r)$ to be proportional to r^2 (indicated by dotted line in Fig. 12.3(a)).

The number of nearest neighbours to the atom at the origin can be obtained by estimating the area under the first peak in the $g(r)$ curve as

---

† Many and varied are the names given to $\rho(r)$ and $g(r)$. The name *radial distribution function* is commonly given to $\rho(r)$. We use the words radial density since $\rho(r)$ is indeed a number density (number per unit volume) and the description is shorter! Other authors call our $g(r)$ (eqn (12.2)) the radial distribution function. Yet other authors interchange the symbols $\rho(r)$ and $g(r)$ in eqn (12.3). Our usuage of $\rho(r)$ and $g(r)$ is the same as that to be found, for example, in *The liquid state*, by J. A. Pryde (Hutchinson, 1966).

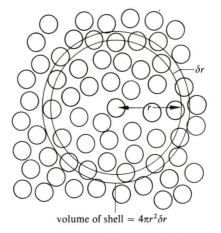

volume of shell $= 4\pi r^2 \delta r$

Fig. 12.2. A hypothetical 'snapshot' of conditions in a liquid, showing the disposition of neighbours about an arbitrarily-chosen central atom. Although illustrated in two dimensions we consider the liquid to be three-dimensional.

shown in Fig. 12.3(b). This result follows because

$$\int_{r_1}^{r_2} g(r)\,\mathrm{d}r = \int_{r_1}^{r_2} 4\pi r^2 \rho(r)\,\mathrm{d}r$$

$$= \text{number of atoms within shell of radii } r_1 \text{ and } r_2. \quad (12.3)$$

This number is called the *first coordination number*. Fig. 12.3(c) records the first, second, etc. coordination numbers for solid argon (here at a temperature approaching absolute zero). The value of $N(r)$ indicates the number of atoms which are to be found at any specified distance r from an arbitrarily chosen central atom. If you like, this is another way of presenting the information that solid argon has an fcc structure. Comparing Figs. 12.3(c) and 12.3(a) we see that even at 0·6 K above the triple point (83·8 K) the long-range order which is characteristic of the solid phase has all but disappeared although short-range order persists.

Applying the technique illustrated in Fig. 12.3(b) to the data of Fig. 12.3(a) shows that the first coordination number decreases from 12 in the solid (Fig. 12.3(c)) to about 10 in the liquid near the triple point to about 4 near the critical point. It is worth noting that the nearest-neighbour separation, given by the value of r at the maximum of the first peak, does not change significantly in going from the solid phase right through to the liquid phase close to the critical point ($T_c = 150\cdot7$ K).

We may eliminate the 'gas' effect ($4\pi r^2$) inherent in plots of $g(r)$ by plotting instead $g(r)/4\pi r^2$, that is $\rho(r)$ (eqn (12.2)). In the limit as $r \to \infty$ the value of $\rho(r)$ will tend to $\rho_0\,(= N_A/V_m)$, the average number density

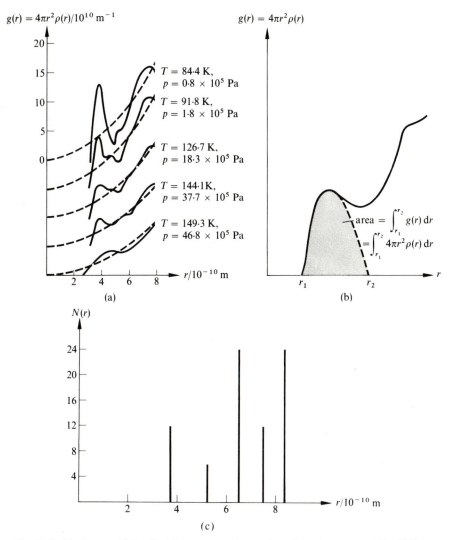

Fig. 12.3. (a) Shows $g(r)$ for liquid argon at various points along the saturated liquid line. For clarity the ordinate values of successive curves are displaced by $5 \times 10^{10} \, \text{m}^{-1}$. (b) Shows how the first coordination number may be estimated. (c) The number of atoms $N(r)$ at a distance r from any one atom in solid argon. (Liquid argon data from Eisenstein, A. and Ginrich, N. S. (1942). *Phys. Rev.*, **62**, 261.)

of the liquid as a whole. Fig. 12.4 shows $\rho(r)/\rho_0$ for liquid argon. The peaks now show up a little more clearly than they did in Fig. 12.3(a). It is also perhaps more evident that their position does not alter significantly as we move along the saturated liquid line from near the triple point to near the critical point.

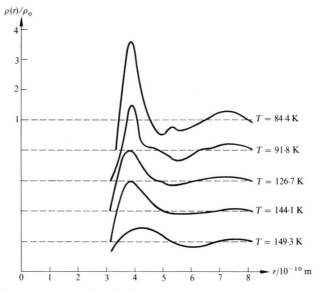

Fig. 12.4. The radial density $\rho(r)$ of liquid argon, expressed in units of ρ_0, the mean number density in the liquid, at various temperatures along the saturated liquid line; the same temperatures as in Fig. 12.3(a).

12.3. Clausius virial theorem: the equation of state

We can highlight the problems likely to be encountered in developing the equation of state of a liquid by asking how it would behave were it to be nothing more than a very dense perfect gas. Taking water as our example, perfect-gas theory predicts (see problem 12.9) that an externally applied pressure of 1351×10^5 Pa is required to contain it at 293 K. Since we know that the pressure actually required is only around 1×10^5 Pa it follows that there must be an internal (negative) pressure of 1350×10^5 Pa. Because of the large value demanded of this 'correction term' it would clearly be foolish to adopt an 'imperfect gas' equation of state, such as the van der Waals equation, at regions of the p–V_m–T surface near the triple point. These equations are, however, remarkably successful near the critical point where the correction terms to p and V_m are usually of the same order as p and V_m. What we need to do is to derive the equation of state afresh, this time with the interatomic force characteristic built in *ab initio* rather than put in as an afterthought.

How a single atom behaves

We will consider a vessel in the form of a cube of side l (Fig. 12.5) containing N_A atoms of a monatomic fluid, each atom having a mass m. As the atoms move about in the fluid each will be subjected to a rapidly

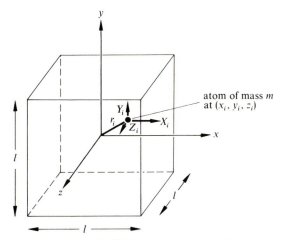

Fig. 12.5. The atom of mass m of the fluid located at (x_i, y_i, z_i) is acted on by the forces X_i, Y_i, Z_i.

fluctuating force. At any one instant the force on atom i located at point (x_i, y_i, z_i)—we take the coordinate origin to be at the centre of the cube—may be resolved into its cartesian components X_i, Y_i, Z_i. The response of the atom to this force is, of course, determined by Newton's second law:

$$X_i = m \frac{d^2 x_i}{dt^2} \tag{12.4}$$

or, multiplying both sides by x_i,

$$X_i x_i = m x_i \frac{d^2 x_i}{dt^2}. \tag{12.5}$$

Similar equations apply for the other two force components along the y- and z-directions. We now make use of the following identity (the identity is nothing more than the rule for differentiating a product):

$$\frac{d}{dt}\left(m x_i \frac{dx_i}{dt}\right) = m x_i \frac{d^2 x_i}{dt^2} + m \left(\frac{dx_i}{dt}\right)^2. \tag{12.6}$$

Substituting for $m x_i \, d^2 x_i / dt^2$, as given by eqn (12.6), into eqn (12.5) yields

$$X_i x_i = -m \left(\frac{dx_i}{dt}\right)^2 + m \frac{d}{dt}\left(x_i \frac{dx_i}{dt}\right). \tag{12.7}$$

In view of the identity

$$\frac{1}{2}\frac{d}{dt}\left[\frac{d(x_i^2)}{dt}\right] = \frac{d}{dt}\left(x_i \frac{dx_i}{dt}\right),$$

eqn (12.7) can be rewritten as

$$X_i x_i + m\left(\frac{dx_i}{dt}\right)^2 = \frac{m}{2}\frac{d}{dt}\left[\frac{d(x_i^2)}{dt}\right]. \tag{12.8}$$

Expressions similar to eqn (12.8) apply to the y- and z-components. Adding together all three such expressions gives

$$\frac{m}{2}\frac{d}{dt}\left(\frac{d}{dt}r_i^2\right) = X_i x_i + Y_i y_i + Z_i z_i + mu_i^2, \tag{12.9}$$

where the distance r_i of atom i from the centre of the box is given by $r_i^2 = x_i^2 + y_i^2 + z_i^2$, and where the speed u_i of atom i is related to the component speeds $u_{xi}\,(=dx_i/dt)$, u_{yi}, and u_{zi} by $u_i^2 = u_{xi}^2 + u_{yi}^2 + u_{zi}^2$. Although we have considered only a single atom of the fluid, exactly the same analysis applies to each and every atom of the fluid. So writing down similar expressions to eqn (12.9) for all N_A atoms present we obtain, on adding all these expressions together,

$$\sum_i \frac{m}{2}\frac{d}{dt}\left(\frac{d}{dt}r_i^2\right) = \sum_i (X_i x_i + Y_i y_i + Z_i z_i) + \sum_i mu_i^2. \tag{12.10}$$

This represents what is happening at one instant of time in the fluid. As time goes by the distances r_i will vary as an atom changes position in the box; as it executes a random walk, in fact, However r_i cannot increase without limit for each atom is contained within the box. When a time-average is taken the left-hand side of eqn (12.10) vanishes,† giving

$$0 = \left\langle \sum_i (X_i x_i + Y_i y_i + Z_i z_i)\right\rangle + \left\langle \sum_i mu_i^2\right\rangle \tag{12.11}$$

where we use $\langle\ \rangle$ to indicate averaging.

---

† It is not difficult to prove this result. We rewrite each term inside the summation on the left-hand side of eqn (12.10) (ignoring the constant term m) as follows

$$\frac{1}{2}\frac{d}{dt}\left(\frac{d}{dt}r_i^2\right) = \frac{d}{dt}\left(r_i\frac{dr_i}{dt}\right).$$

We now sum this over all atoms and then take the time-average of our summed expression. We have, by definition of time-average (over some time τ),

$$\left\langle \sum_i \frac{d}{dt}\left(r_i\frac{dr_i}{dt}\right)\right\rangle = \frac{1}{\tau}\int_0^\tau \sum_i \frac{d}{dt}\left(r_i\frac{dr_i}{dt}\right)dt$$

$$= \frac{1}{\tau}\int_0^\tau \sum_i d\left(r_i\frac{dr_i}{dt}\right)$$

$$= \frac{1}{\tau}\sum_i \left[\left(r_i\frac{dr_i}{dt}\right)_{t=\tau} - \left(r_i\frac{dr_i}{dt}\right)_{t=0}\right], \tag{12.12}$$

The forces acting on an atom

So far we have said nothing of the origin of the force on atom i. This force is made up of two contributions. First, there will be the forces acting on it due to all other atoms in the *fluid*. Secondly, there will be forces acting on it due to all the atoms of the *solid* which make up the walls of the box. We may therefore write the component force X_i as

$$X_i = X'_i + X''_i, \tag{12.13}$$

where X'_i is the x-component of the resultant force acting on atom i attributable to all other atoms of the fluid, and X''_i is the x-component of the resultant force attributable to all the atoms of the box. Substituting eqn (12.13) and the two equivalent expressions for Y_i and Z_i into eqn (12.11) gives

$$\left\langle \sum_i (x_i X'_i + x_i X''_i + y_i Y'_i + y_i Y''_i + z_i Z'_i + z_i Z''_i) \right\rangle = -\left\langle \sum_i mu_i^2 \right\rangle, \tag{12.14}$$

where the summation is made over all N_A atoms in the box, and then this sum is time-averaged. We would expect X''_i, Y''_i, and Z''_i only to be significant when atoms come to within a few atomic diameters of the wall. Therefore terms like $\sum x_i X''_i$ will make an essentially zero contribution except when $x = -\frac{1}{2}l$ and when $x = +\frac{1}{2}l$, and similarly for the y and z terms. Writing X''_i, for example, at $x = \frac{1}{2}l$ as $X''_i(\frac{1}{2}l)$ allows eqn (12.14) to be expressed as

$$\left\langle \left[\sum_i (x_i X'_i + y_i Y'_i + z_i Z'_i) + \tfrac{1}{2}l \sum_i X''_i(\tfrac{1}{2}l) + \tfrac{1}{2}l \sum_i Y''_i(\tfrac{1}{2}l) + \tfrac{1}{2}l \sum_i Z''_i(\tfrac{1}{2}l) \right.\right.$$
$$\left.\left. - \tfrac{1}{2}l \sum_i X''_i(-\tfrac{1}{2}l) - \tfrac{1}{2}l \sum_i Y''_i(-\tfrac{1}{2}l) - \tfrac{1}{2}l \sum_i Z''_i(-\tfrac{1}{2}l) \right] \right\rangle = -\left\langle \sum_i mu_i^2 \right\rangle. \tag{12.15}$$

The summations in the last six terms on the left-hand side of eqn (12.15) represent the *total* force between each face (such as that at $x = \frac{1}{2}l$) and the adjacent layer of atoms (those within a few atomic diameters of the face).

---

assuming $\sum$ and $\int$ commute (i.e., that their order can be 'swapped'). It is not difficult to accept that provided τ is sufficiently large there will be no correlation between what happens at time 0 and at time τ, that is

$$\sum_i \left(r_i \frac{dr_i}{dt} \right)_{t=0} = \sum_i \left(r_i \frac{dr_i}{dt} \right)_{t=\tau} \quad \text{as} \quad \tau \to \infty.$$

Even if the term in square brackets in eqn (12.12) is not identically zero for a particular atom i, it is equally likely to be positive or negative for any atom; the sum over all atoms will therefore more nearly vanish. Even if this suspicion is unjustified the presence of $1/\tau$ multiplying the right-hand side of eqn (12.12) should cause it to vanish as $\tau \to \infty$.

Evaluating $\sum X_i''(\frac{1}{2}l)$, etc., might seem a daunting proposition. Fortunately we may appeal to Newton's third law, from which we deduce that the force exerted by the wall on the fluid will be equal in magnitude but opposite in direction to the force exerted by the fluid on the wall. We know that force exerted by the fluid on, say, the yz face of area l^2 at $x = \frac{1}{2}l$, is $+pl^2$, where p is the pressure in the liquid. It therefore follows that $\sum X_i''(\frac{1}{2}l) = -pl^2$. Writing down similar expressions, with due regard to sign, at each of the other five faces and substituting these expressions into eqn (12.15) gives

$$\left\langle \sum_i (x_i X_i' + y_i Y_i' + z_i Z_i') \right\rangle + 3[\frac{1}{2}l(-pl^2) - \frac{1}{2}l(pl^2)] = -\left\langle \sum_i mu_i^2 \right\rangle,$$

where we have dispensed with $\langle\ \rangle$ around the terms involving p since p is already time-averaged (as measured experimentally). Remembering that V_m, the volume of the liquid, is l^3 gives

$$\left\langle \sum_i (x_i X_i' + y_i Y_i' + z_i Z_i') \right\rangle - 3pV_m = -\left\langle \sum_i mu_i^2 \right\rangle. \tag{12.16}$$

Introducing the interatomic force

You might anticipate that the first three terms on the left-hand side of eqn (12.16) will vanish in the averaging process. In fact, they fail to vanish because the atoms of the liquid are not independent.

Consider a pair of atoms i and j separated by distance r_{ji} (Fig. 12.6). Let X_{ij}' denote the *part* of the net force X_i' acting on i (the first suffix in X_{ij}') which is due to atom j (the second suffix). We may likewise write X_{ji}' to denote that *part* of the net force X_j' acting on j which is due to i. By Newton's third law, $X_{ji}' = -X_{ij}'$. The contribution which X_{ij}' and X_{ji}' make to the first term in eqn (12.11) is

$$x_i X_{ij}' + x_j X_{ji}' = X_{ji}'(x_j - x_i). \tag{12.17}$$

In the liquid inert gases the force between two atoms is directed along the line joining their centres. With such a *central* force present we can write

$$X_{ji}' = F(r_{ji})\cos\theta,$$

where θ is the angle r_{ji} makes with the x-axis, and $F(r_{ji})$ is the central force at interatomic separation r_{ji}. But $\cos\theta = (x_j - x_i)/r_{ji}$, allowing eqn (12.17) to be written as

$$x_i X_{ij}' + x_j X_{ji}' = F(r_{ji}) \frac{(x_j - x_i)^2}{r_{ji}}$$

and likewise for the y- and z-contributions. Feeding all three such

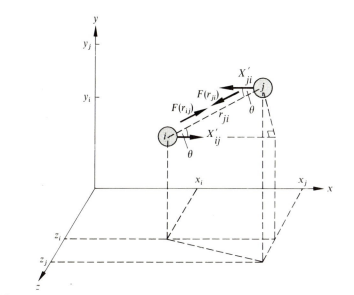

Fig. 12.6. Showing the interaction of two atoms i and j in the liquid. The interatomic force is directed along the line joining their centres.

relations into eqn (12.16) gives

$$\left\langle \sum_{\text{pairs}} \frac{F(r_{ji})}{r_{ji}} [(x_j - x_i)^2 + (y_j - y_i)^2 + (z_j - z_i)^2] \right\rangle - 3pV_m = - \left\langle \sum_i mu_i^2 \right\rangle.$$

(12.18)

Since the expression in square brackets is equal to r_{ji}^2 (see Fig. 12.6) and since $\langle \frac{1}{2}mu_i^2 \rangle = \frac{3}{2}kT$—this implies that the right-hand side of eqn (12.18) is $3N_A kT = 3RT$—we finally arrive at

$$\boxed{pV_m = RT + \tfrac{1}{3} \overline{\sum_{\text{pairs}} F(r_{ji})r_{ji}}},$$

(12.19)

where we have reverted to bar notation for averages. Eqn (12.19) is known as the *Clausius virial theorem* for atoms with central forces. It says that to find the equation of state of a fluid all one needs to do is multiply the central force $F(r_{ji})$ between a pair of atoms by their distance r_{ji} apart between centres, sum these products over *all* pairs in the fluid, take the average of this result as measured at different times, and then substitute the result into eqn (12.19). It makes no difference whether the fluid be a liquid or a gas for no specific atomic model was assumed. In a fluid composed of atoms of effectively zero size in which $F(r) = 0$ for all r, eqn

(12.19) reduces to the equation of state of a perfect gas, $pV_m = RT$, as it must if the analysis is correct.

Introducing the radial density

Although eqn (12.19) can be used as it stands, it is much more useful to have the radial density $\rho(r)$ in the result as this is the quantity that is determined experimentally from X-ray studies. If we choose i as the reference atom as shown in Fig. 12.7 and j as one of the atoms in the spherical shell of radius $r = r_{ji}$, then $F(r_{ji}) = -dV(r)/dr$, where $V(r)$ is the interatomic pair potential. The atoms whose centres lie within the shell of radius r and thickness δr make a contribution of

$$\overline{\sum_{\text{shell}} r_{ji} F(r_{ji})} = r\left(-\frac{dV(r)}{dr}\right)(4\pi r^2 \, dr \, \rho(r)) \qquad (12.20)$$

for the particular atom i. The term $4\pi r^2 \, dr\rho(r)$ represents the number of atoms whose centres lie within the shell, each of which is at a distance r from i (eqn (12.1)). To find the contribution to atom i of all other atoms eqn (12.20) must be integrated from $r = 0$ to $r = \infty$:

$$\overline{\sum_{j} r_{ji} F(r_{ji})} = -\int_0^\infty \frac{dV(r)}{dr} \rho(r) 4\pi r^3 \, dr. \qquad (12.21)$$

However this only includes some of the pairs; the interaction between i and the other atoms. Eqn (12.19) affirms that *all* pairs must be included: We must include the interaction of atom s, say, with all other atoms (Fig. 12.7). This may be achieved by multiplying the right-hand side of eqn

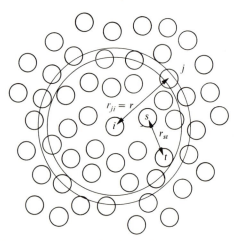

Fig. 12.7. Showing how the atoms of a liquid are distributed about a particular atom i.

(12.21) by $\frac{1}{2}N_A$; the factor of $\frac{1}{2}$ being required to ensure that each pair is only counted once. Substituting this modified version of eqn (12.21) into eqn (12.19) leads to

$$pV_m = RT - \frac{N_A}{6} \int_0^\infty \frac{d\mathcal{V}(r)}{dr} \rho(r) 4\pi r^3 \, dr \qquad (12.22)$$

as the more immediately useful form of the Clausius virial theorem. It is worth reminding ourselves that $\rho(r)$ depends on the state of the liquid, that is on, say, p and T (Fig. 12.3(a)). We should therefore have written $\rho(r, p, T)$ rather than $\rho(r)$ in eqn (12.22); for economy we shall continue to use $\rho(r)$. If $\mathcal{V}(r)$ is given the form shown in Fig. 6.1(c) (in which $\mathcal{V}(r)$ is such that the attractive force falls off as $1/r^5$ or faster) eqn (12.22) reduces to the van der Waals equation when the density of the fluid is such that there is a negligible chance of three or more atoms lying within a volume over which the attractive forces are significant.

Exercise 12.1

This exercise will allow you to predict the pressure of a liquid at a stated volume and temperature. By repeating the procedure at other volumes and temperatures you could arrive at the (empirical) form of the equation of state.

Using the Clausius virial theorem predict the pressure of liquid argon at a (mass) density of 0.98×10^3 kg m^{-3} at 143 K. Fig. 12.8(a) shows the form of $\rho(r)$ at this density and temperature. The radial density $\rho(r)$ is expressed in units of ρ_0, the mean number density of the liquid as a whole. Also shown in Fig. 12.8(a) is a plot of r^3 and the derivative with respect to r of the Lennard-Jones 6–12 potential $\mathcal{V}(r)$ for the interaction between two argon atoms (eqn (2.20)). In argon $\varepsilon = 1.66 \times 10^{-21}$ J and $\sigma = 3.4 \times 10^{-10}$ m. To help you, the product $d\mathcal{V}(r)/dr[\rho(r)/\rho_0]r^3$ is shown in Fig. 12.8(b). In evaluating the integral term in eqn (12.22) you can do so graphically ('counting squares') or by using Simpson's rule. This rule asserts that if $f_0, f_1, f_2, \ldots, f_{n-1}, f_n$ are the values of $f(x)$ at $x = a, a+h, a+2h, \ldots, a+(n-1)h, a+nh\ (=b)$, where n is an even integer, then

$$\int_a^b f(x) \, dx \approx \frac{h}{3}[f_0 + f_n + 4(f_1 + f_3 + \cdots + f_{n-1}) + 2(f_2 + f_4 + \cdots + f_{n-2})].$$

Compare your predicted pressure with the experimental value of 6.7×10^6 Pa at the density of 0.98×10^3 kg m^{-3} at 143 K and try to track down any discrepancy. Take $A_r(\text{Ar}) = 40$.

Calculation. The Clausius virial theorem (eqn (12.22)) may be written as

$$pV_m = RT - \frac{4\pi\rho_0 N_A}{6} \int_0^\infty \frac{d\mathcal{V}(r)}{dr} \frac{\rho(r)}{\rho_0} r^3 \, dr. \qquad (12.23)$$

The integral is, of course, equal to the area under the plot of $d\mathcal{V}(r)/dr[\rho(r)/\rho_0]r^3$ against r between $r = 0$ and $r = \infty$. When this area is measured in Fig. 12.8(b) it comes out at 3.55×10^{-50} N m^4. (To find the net area under the curve one must subtract the area under the abscissa from that above the abscissa.)

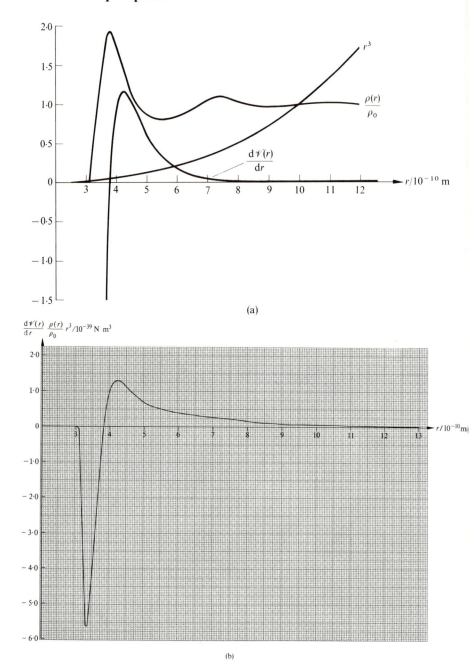

Fig. 12.8. (a) Shows the dependence of $d\mathcal{V}(r)/dr$, $\rho(r)/\rho_0$, and r^3 on r for liquid argon at a temperature of 143 K and a density of $0\cdot98 \times 10^3$ kg m^{-3}. Each scale division of the ordinate corresponds to $d\mathcal{V}(r)/dr = 1 \times 10^{-11}$ N, $\rho(r)/\rho_0 = 1$, and $r^3 = 1\cdot0 \times 10^{-27}$ m^3. (b) The dependence of $d\mathcal{V}(r)/dr[\rho(r)/\rho_0]r^3$ on r. (The radial density data comes from X-ray studies by Mikolaj, P. G. and Pings, C. J. (1967). *J. Chem. Phys.* **46**, 1401.)

The mean number density ρ_0 follows on dividing the mass density of $0.98 \times 10^3 \, kg \, m^{-3}$ by the mass of an argon atom. Since $A_r(Ar) = 40$ it follows that $40 \times 10^{-3} \, kg$ contain N_A atoms, giving the mass of an atom of argon as $(40 \times 10^{-3}/6.02 \times 10^{23}) \, kg = 6.64 \times 10^{-26} \, kg$. Therefore $\rho_0 = 1.48 \times 10^{28} \, m^{-3}$, giving $4\pi\rho_0 N_A/6 = 1.87 \times 10^{52} \, m^{-3} \, mol^{-1}$. Hence the second term on the right-hand side of eqn (12.23) has a value of $1.87 \times 10^{52} \times 3.55 \times 10^{-50} \, J \, mol^{-1} = 664 \, J \, mol^{-1}$. Substracting this from $RT = 1188 \, J \, mol^{-1}$ at 143 K tells us that the right-hand side of eqn (12.23) has a net value of $524 \, J \, mol^{-1}$. To find the value of V_m (and hence p) we divide the mass of 1 mol of argon, namely $40 \times 10^{-3} \, kg \, mol^{-1}$, by the density, namely $0.98 \times 10^3 \, kg \, m^{-3}$, to give $V_m = 4.08 \times 10^{-5} \, m^3 \, mol^{-1}$. Therefore $p = (524 \, J \, mol^{-1}/4.08 \times 10^{-5} \, m^3 \, mol^{-1}) = 1.28 \times 10^7 \, Pa$.

The experimental value of p is $6.7 \times 10^6 \, Pa$. To 'force' the theoretical and the experimental values of p to agree we would have to assign the area under the graph in Fig. 12.8(b) a value of $4.9 \times 10^{-50} \, N \, m^4$. This only requires a slight change in the value of either the positive or negative area under the curve of Fig. 12.8(b). A slight change in the form of $\rho(r)/\rho_0$, particularly at low r, can have a profound change in the form of this curve. When the experimental errors inherent in the X-ray results are considered, a net area of $4.9 = 10^{-50} \, N \, m^4$ is quite on the cards!

Comment. This exercise will have brought home to you the importance of having accurately known values of $\rho(r)$. The sort of accuracy which is required if worthwhile p (or V_m or T) values are to be predicted exceeds the accuracy that is realizable in X-ray studies.

12.4. The internal energy

In calculating the molar internal energy U_m of a liquid we shall adopt the strategy which we followed in calculating U_m for an imperfect gas (§6.6). Starting with the atoms an infinite distance apart and at rest (this defines $U_m = 0$) we first allow them to come together and assume the positions they would have in a time-averaged 'snapshot' of the liquid. We then set the atoms moving so that the mean kinetic energy of an atom is $\frac{3}{2}kT$, where T is the temperature of the liquid.

Supposing Fig. 12.7 to show the time-averaged distribution of atoms about a particular atom i, we note that the potential energy (stored in the field) between atom i and each atom, such as j, lying within the shell of radius r and thickness dr is $\mathcal{V}(r)$. Since there are $4\pi r^2 \, dr \, \rho(r)$ atoms within the shell the potential energy of i due to all these atoms is $4\pi r^2 \rho(r) \mathcal{V}(r) \, dr$. Integrating from $r = 0$ to $r = \infty$ gives

$$\text{potential energy of } i = \int_0^\infty \mathcal{V}(r)\rho(r)4\pi r^2 \, dr. \tag{12.24}$$

To find the total potential energy of all N_A atoms present in one mole of liquid we multiply eqn (12.24) by $\frac{1}{2}N_A$; the factor of $\frac{1}{2}$ being necessary to prevent each pair of atoms from being counted twice. In imagination, we now set the atoms moving, so that their total kinetic energy is $N_A(\frac{3}{2}kT) = \frac{3}{2}RT$. Hence the total energy of the N_A atoms is given by

$$U_m = \frac{3}{2}RT + \frac{N_A}{2}\int_0^\infty \mathcal{V}(r)\rho(r)4\pi r^2\,dr \quad . \tag{12.25}$$

If $\rho(r)$ is specified at different molar volumes and temperatures we should, more correctly, write $U_m(V_m, T)$ in place of U_m. If, as in Fig. 12.3(a), $\rho(r)$ is specified at different pressures and temperatures, we should write $U_m(p, T)$. For certain purposes it may be more convenient to have eqn (12.25) expressed in the form

$$U_m = \frac{3}{2}RT + \frac{1}{2}\sum_{\text{pairs}}\mathcal{V}(r_{ji}) \quad . \tag{12.26}$$

Now that we have established expressions for the equation of state and internal energy of a liquid we could, if we wished, apply the techniques of thermodynamics to elucidate the equilibrium properties of the liquid. As examples, eqns (1.12) and (1.14) would allow $C_{V,m}$ and $C_{p,m}$ to be determined.

Exercise 12.2

Predict the molar internal energy of liquid argon at a mass density of 0.98×10^3 kg m^{-3} at a temperature of 143 K. Fig. 12.9(a) shows the variation of $\rho(r)/\rho_0$ with r at this density and temperature, where ρ_0 is the average number density of the liquid as a whole. Also shown is a plot of r^2 and the Lennard-Jones potential $\mathcal{V}(r)$ for the interaction between two argon atoms, namely eqn (2.20) with $\varepsilon = 1.66\times10^{-21}$ J and $\sigma = 3.4\times10^{-10}$ m. To help you, the product $\mathcal{V}(r)\times[\rho(r)/\rho_0]r^2$ is plotted in Fig. 12.9(b). Compare your predicted value for U_m with the experimental value of -1.7×10^3 J mol^{-1}.

Calculation. The molar internal energy (eqn (12.25)) may be written as

$$U_m = \frac{3}{2}RT + 2\pi\rho_0 N_A\int_0^\infty \mathcal{V}(r)\frac{\rho(r)}{\rho_0}r^2\,dr. \tag{12.27}$$

Evaluating the integral by direct measurement of the area under the plot of $\mathcal{V}(r)[\rho(r)/\rho_0]r^2$ against r (Fig. 12.9(b)) gives a value of -6.53×10^{-50} J m^3. Following the same arguments as in exercise 12.1 gives $2\pi\rho_0 N_A = 5.6\times10^{52}$ m^{-3} mol^{-1}, so the second term on the right-hand side of eqn (12.27) has a value of -3655 J mol^{-1}. At 143 K the value of $\frac{3}{2}RT$ is 1782 J mol^{-1}. Therefore $U_m = -1.9\times10^3$ J mol^{-1}. Bearing in mind the uncertainty in the precise form of $\rho(r)/\rho_0$ this is in satisfactory agreement with the experimental value of -1.7×10^3 J mol^{-1}.

Comment. In working through this exercise you will have discovered that the value calculated for U_m depends crucially on the form of $\rho(r)$. X-ray studies are incapable of providing sufficiently accurate information on $\rho(r)$ to make the predicted values of U_m worth adopting in practice. A major objective of the structural theory of liquids must be to obtain accurate values of $\rho(r)$.

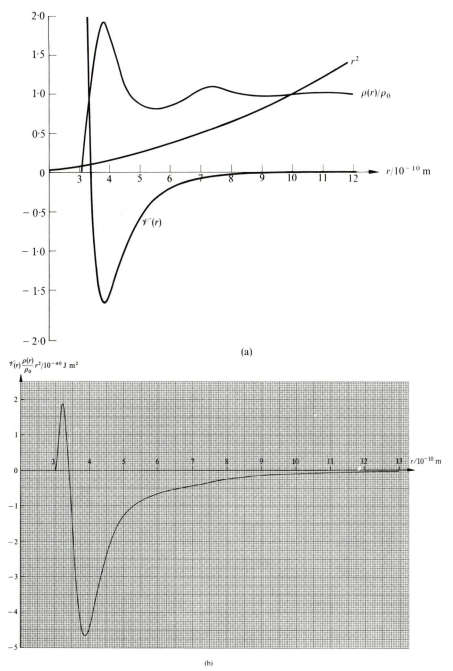

Fig. 12.9. (a) Shows the dependence of $\mathcal{V}(r)$, $\rho(r)/\rho_0$, and r^2 on r for liquid argon at a density of 0.98×10^3 kg m^{-3} and a temperature of 143 K. Each scale division of the ordinate corresponds to $\mathcal{V}(r) = 1 \times 10^{-21}$ J, $\rho(r)/\rho_0 = 1$, and $r^2 = 1 \times 10^{-18}$ m^2. (b) The dependence of $\mathcal{V}(r)[\rho(r)/\rho_0]r^2$ on r. (The radial density data comes from X-ray studies by Mikolaj, P. G. and Pings, C. J. (1967). *J. Chem. Phys.* **46**, 1401.)

12.5. The structure of simple liquids

Theoretical approach

So far we have treated $\rho(r)$ and $\mathscr{V}(r)$ as if they are independent of one another, yet a moment's thought will show that this is very unlikely to be the case. Consider the two hypothetical forms of $\mathscr{V}(r)$ shown in Fig. 12.10. If $\mathscr{V}(r)$ has the form shown in Fig. 12.10(a), there is a hard-sphere repulsive force at $r \leq r_0$ and a constant attractive force at $r_0 < r < r_1$. If it has the form shown in Fig. 12.10(b), the only forces are the hard-sphere repulsive force at $r \leq r_0$ and a very strong, but short-ranged, attractive force at $r = r_1$. It would be surprising if the radial densities arising out of these interactions were the same in both cases. We might expect more closely-knit groups of atoms to exist in a liquid in which $\mathscr{V}(r)$ has the form of Fig. 12.10(a) than in a liquid in which $\mathscr{V}(r)$ has the form of Fig. 12.10(b).

The connection between $\rho(r)$ and $\mathscr{V}(r)$ is, however, not as simple as these arguments might suggest. You will remember that $\rho(r)$ is the number density of atoms which are present *on average* at a radial distance r from a particular (reference) atom. To find $\rho(r)$ we would therefore have to elucidate all possible atomic configurations consistent with the external constraints (p, T) and, from this, calculate the mean configuration. Now the probability that a collection of N_A interacting atoms has a particular configuration—and by configuration we mean that the position

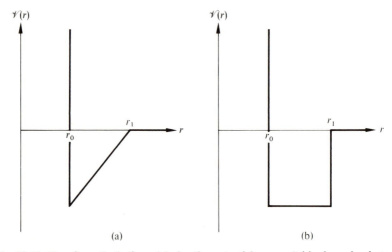

(a)   (b)

Fig. 12.10. Two (hypothetical) models for the potential energy $\mathscr{V}(r)$ of a pair of atoms a distance r apart. In (a) there is a constant attractive force over the range $r_0 < r < r_1$ and a hard-sphere repulsive force at $r \leq r_0$. In (b) there is very strong, but short-ranged, attractive force at $r = r_1$ and a hard-sphere repulsive force at $r \leq r_0$.

and momentum of all N_A atoms are specified—depends on the total energy of that configuration. This energy is the sum of the kinetic and potential energies of all N_A atoms. The mutual potential energy of all the atoms depends, of course, on how the atoms are arranged in that configuration! There are several techniques available for breaking this 'radial density depends on the potential energy which depends on how the atoms are arranged' loop, but none of these techniques can be easily described in words. Each one of these methods involves making approximations which necessarily limit the range of application of the results. No single analytic theory can describe how the properties of a liquid change as one moves throughout the liquid region, from the triple point to the critical point.

Mechanical simulations

Faced with a knotty theoretical problem we once again turn to simulation techniques. If we can make visible the motion of the 'atoms', we will be able to determine the empirical forms of the equation of state via eqn (12.19) and of the internal energy function via eqn (12.26). As an added bonus we will be able to form an overall impression of what the liquid is like at any instant of time and at how this picture changes with time. Analytic theories, by contrast, only aim at finding $\rho(r)$; a quantity which is spatially-averaged (throughout the body of the fluid) and time-averaged.

The first serious attempts at simulating a monatomic liquid were those made by J. D. Bernal in the 1950s. He argued that a liquid consists of a *heap* of atoms, unlike a solid which consists of a *pile* of atoms. A pile is produced when spheres, such as marbles, are carefully poured into a box; the 'atoms' crystallize into a regular structure. A heap results when the spheres are tossed into the box. By studying how the spheres were arranged, Bernal deduced that the first coordination number in his simulated liquid was $9 \cdot 3 \pm 0 \cdot 8$. This is consistent with the experimental value of about 10 in liquid argon near its triple point.

In an independent study Scott poured four thousand steel balls into a cylinder whose inner surface was dimpled (to prevent 'crystallization'). After shaking them down, molten wax was added and allowed to set. A travelling microscope was used to determine the x-, y-, and z-coordinates of each ball in the resulting caviar-like cluster. (These was achieved by measuring the coordinates of a lamp filament imaged by the polished surface of a ball. After the measurements were complete the surrounding wax was scraped away and the ball removed.) In all, the coordinates of about a thousand balls were measured. An arbitrarily selected ball near the middle of the cluster was used as origin and the distances to all other balls were calculated from the coordinate data. The radial density was

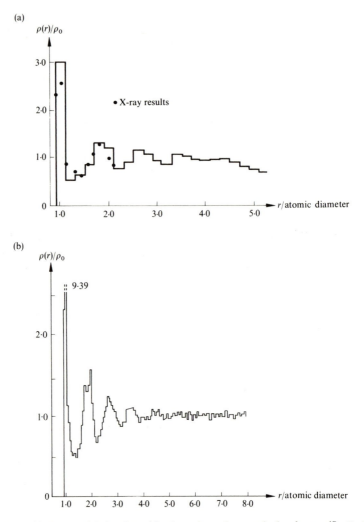

Fig. 12.11. (a) The radial density $\rho(r)$ of random close-packed spheres. (Scott, G. D. (1962). *Nature* **194,** 956.) Here ρ_0 is the mean number density in the heap of balls. Also shown is the radial density in liquid argon close to its triple point (determined from X-ray diffraction data). (b) Scott's data reanalysed to take account of the distribution about 268 balls. Here the intervals in r are one-twentieth of a ball diameter (compared to one-fifth in (a)). (Data from Mason, G. (1968). *Nature* **217,** 734.)

then deduced for intervals of one-fifth of the ball diameter (via eqn (12.1)). The entire calculation was repeated using 25 different balls as origin. The average radial density was then found from these 25 radial density calculations. Fig. 12.11(a) shows the resulting average radial density $\rho(r)$, expressed in units of the mean number density ρ_0. Fig.

12.11(a) also includes the experimental values of $\rho(r)/\rho_0$ for liquid argon close to its triple point as determined by X-ray diffraction studies. Fig. 12.11(b) shows the result of a more recent analysis of Scott's 1962 data. Here, instead of determining the radial density about a mere 25 balls as centre, the calculations were made about 268 balls. Naturally this reduces the uncertainty in the distribution. Even so, the data is still too inaccurate to be used in predicting either the equation of state or the internal energy of a monatomic liquid.

There are a number of criticisms which we might make of Bernal's simulator. First, it but poorly reproduces the interatomic forces present between real atoms; the force between two balls is only hard-sphere repulsive. (Because the balls have weight there is, however, an effective downward vertical force between a ball and the one immediately above it.) Secondly, it is static—and we know that the atoms' kinetic energy cannot be ignored in the liquid phase. Thirdly, it only simulates conditions at a fixed density near the triple point. In §3.2 we described a two-dimensional simulator which is free of these particular defects.

The molecular dynamics approach

With the advent of large-capacity electronic computers the need for mechanical simulators has come to an end. In one technique, that of *molecular dynamics,* the computer is used to solve the equations of motion of several hundred atoms by step-wise numerical integration. This technique currently affords the most powerful method of deducing the microscopic and macroscopic properties of large assemblies of classical particles which interact according to a specified interparticle potential. Such is the technique's success that it is really only limited by the accuracy to which the interatomic potential is known.

Although the details vary from worker to worker, the following procedures are involved in performing a molecular dynamics simulation:

1. Allocate randomly-chosen coordinates to all N atoms—the value of N is determined by the computer's capacity, but is typically several hundred—so as to locate them at random somewhere inside a fixed volume V, usually in the form of a cube. These coordinates are, of course, stored in the computer.

2. Using the known interatomic separations, calculate the forces $\boldsymbol{F}(r_1)$, $\boldsymbol{F}(r_2)$, $\boldsymbol{F}(r_3)$, etc. which an atom will experience due to the remaining $N-1$ atoms at distances r_1, r_2, r_3, etc. away. These forces are determined from the interatomic potential $\mathscr{V}(r)$ via $F(r) = -\mathrm{d}\mathscr{V}(r)/\mathrm{d}r$ (Fig. 12.12(a)). Now calculate the resultant force $\boldsymbol{F} = \boldsymbol{F}(r_1) + \boldsymbol{F}(r_2) + \boldsymbol{F}(r_3) + \cdots$ acting on the atom. Follow through exactly the same procedures on the other $N-1$ atoms. The resultant forces acting on all N atoms are therefore known (shown by heavy line in Fig. 12.12(b)).

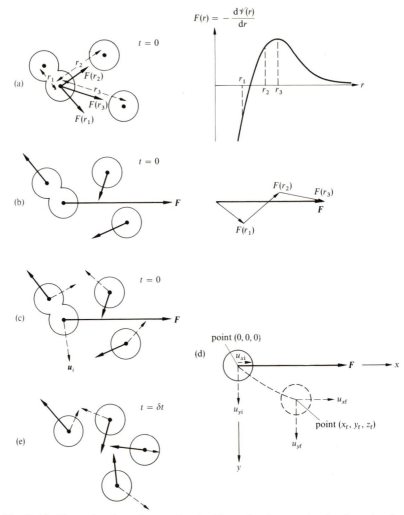

Fig. 12.12. Illustrating the main steps involved in performing a molecular dynamics simulation. (a) The forces experienced by an atom due to the other atoms are calculated using the interatomic force characteristic. (b) Showing how the resultant force acting on an atom is found by vector addition. (c) The atoms are given random velocities (shown dashed). (d) Showing how an individual atom with velocity components u_{xi} and u_{yi} reacts to a constant force F. (e) The situation at a time δt later than that shown in (c).

3. Allocate a random velocity to each atom (shown by dotted line in Fig. 12.12(c)). In practice one often gives each atom the same speed but in a randomly-chosen direction. This obvious bias in the speeds will quickly be eliminated as the 'experiment' proceeds. Fig. 12.12(c) thus gives the initial configuration at time $t = 0$, in which the starting positions and starting velocities of all N atoms are known.

4. Calculate the new positions and velocities of the atoms at the end of a short time interval δt, chosen so that the forces acting on each and every atom may be taken as constant throughout δt. One might take $\delta t = 10^{-14}$ s when simulating liquid argon at 100 K; during $\delta t = 10^{-14}$ s an argon atom will, on average, move by $2 \cdot 3 \times 10^{-12}$ m, which is only $0 \cdot 6$ per cent of an atomic diameter. It is a simple problem in Newtonian mechanics to calculate where an atom of mass m with a velocity u_i at time $t = 0$ will be at a later time δt, and what its new velocity will be. First resolve the velocity vector u_i of an atom at $t = 0$ three components; u_{xi} along the x-axis (taken for convenience, as being in the direction in which F acts), u_{yi} and u_{zi} along axes perpendicular to each other and to F (Fig. 12.12(d)). If we denote the final position of the atom at time δt by x_f, y_f, z_f (Fig. 12.12(d)) then, since it started off at time $t = 0$ from the origin of our coordinate system,

$$x_f = u_{xi}\, \delta t + \tfrac{1}{2}(F/m)\, \delta t^2; \qquad y_f = u_{yi}\, \delta t; \qquad z_f = u_{zi}\, \delta t.$$

Likewise, if we denote the component speeds of our atom at time δt by u_{xf}, u_{yf}, and u_{zf} (Fig. 12.12(d)) then clearly

$$u_{xf} = u_{xi} + (F/m)\, \delta t; \qquad u_{yf} = u_{yi}; \qquad u_{zf} = u_{zi}.$$

5. Using the position data for each atom at time δt, calculate the *new* resultant forces acting on all N atoms (via the procedures outlined in step 2). Regarding these new forces (shown in heavy line in Fig. 12.12(e)) as constant during the interval $t = \delta t$ to $t = 2\,\delta t$, calculate the effect of these forces on the atoms whose initial velocities are those existing at time δt (shown dashed in Fig. 12.12(e)) and which were calculated in step 4.

6. Repeat the entire sequence of steps at times $3\,\delta t$, $4\,\delta t$, $5\,\delta t$, etc. until the velocity distribution of the atoms is Maxwell–Boltzmann in character. When this is achieved we may take it that the original contrived configuration has given way to one which truly represents conditions in a system of atoms interacting according to the chosen potential $\mathscr{V}(r)$. The time taken to reach equilibrium depends on the form of $\mathscr{V}(r)$ but it may be of order $100\,\delta t$. The average kinetic energy of the atoms, $\tfrac{1}{2}m\overline{u^2}$, is then used to deduce the temperature of the system (from $\tfrac{1}{2}m\overline{u^2} = \tfrac{3}{2}kT$).

7. Once the data on each equilibrium configuration—a configuration tells us the location and momenta of all N atoms at a particular time t—have been stored in the computer the final step is to use this data to calculate p via eqn (12.19), with V_m determined by the volume V occupied by the N atoms, and U_m via eqn (12.26). In averaging over all configurations we are, of course, taking a time-average, which is what the theory requires.

As described so far, p and U_m have only been calculated at one particular value of T and V_m. To explore the behaviour of the system at, say, a higher temperature the entire 'experiment' must be repeated using increased initial speeds for the atoms. To change V_m the volume V occupied by the N atoms is changed, and the entire 'experiment' is repeated. Since the equation of state and the internal energy are, in effect, known, the computer can be programmed to predict any other equilibrium properties of the system. The time taken to perform a single 'experiment' obviously depends on the computer being used but, to put round figures to it, a large machine may evaluate 1000 configurations every hour in a system of 1000 atoms.

Results of computer studies

Fig. 12.13 shows the excellent level of agreement which exists between the T^{-1} dependence of the compression factor (pV_m/RT) of liquid argon as determined by a molecular dynamics simulation, and as measured experimentally. This level of agreement is normally realized in substances whose interatomic potential can, as in argon, be represented by a Lennard-Jones 6–12 potential. Indeed, such is the success of molecular dynamics calculations that small discrepancies between theory and experiment can often be traced to slight departures of the interatomic potential from the assumed relation. Furthermore, the calculations can be used to predict, with some confidence, the properties of a substance in technically awkward ranges of p, V_m, and T. This is perhaps part of the reason why these simulations are referred to as 'experiments'.

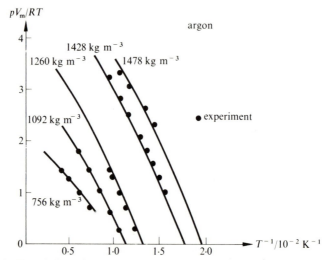

Fig. 12.13. The full lines show the dependence of pV_m/RT on T^{-1} at various densities (and hence at various V_m) as predicted by a molecular dynamics simulation of argon. (Data from Verlet, L. (1967). *Phys. Rev.*, **159**, 98.) A Lennard-Jones 6–12 potential was used with $\varepsilon/k = 119.8$ K and $\sigma = 3.405 \times 10$ m. The experimental results are those of many workers.

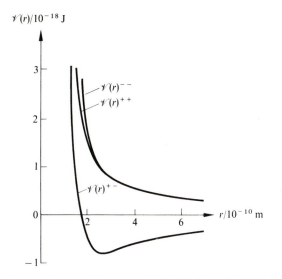

$\mathscr{V}(r)/10^{-18}$ J

Fig. 12.14. The interatomic potential energy of a K^{+}–Cl^{-} ion pair ($\mathscr{V}(r)^{+-}$), a Cl^{-}–Cl^{-} pair ($\mathscr{V}(r)^{--}$), and a K^{+}–K^{+} pair ($\mathscr{V}(r)^{++}$) as assumed by L. V. Woodcock and K. Singer (1971) (*Trans. Faraday Soc.*, **67**, 12) in computing the data presented in Fig. 12.15.

As another example of the application of computer simulation techniques we will consider a liquid alkali halide, such as liquid potassium chloride. Here there is not just one potential $\mathscr{V}(r)$ but three; $\mathscr{V}(r)^{+-}$ between a K^{+} and a Cl^{-} ion; $\mathscr{V}(r)^{++}$ between two K^{+} ions, and $\mathscr{V}(r)^{--}$ between two Cl^{-} ions. These potentials are shown graphically in Fig. 12.14. You should be able to work out for yourself why $\mathscr{V}(r)^{--}$ and $\mathscr{V}(r)^{++}$ are always positive and are proportional to r^{-1} at large r. (The reason why $\mathscr{V}(r)^{--}$ and $\mathscr{V}(r)^{++}$ differ at small interatomic separations is that the Cl^{-} and the K^{+} ions have slightly different sizes so that repulsive (overlap) forces come into play at different r in these two interactions.) Fig. 12.15(a) shows how closely the predictions made by computer simulation—these predictions were actually made by, so-called, Monte Carlo techniques, rather than by molecular dynamics studies—agree with the experimental values of the internal energy. Because there are three different potential energy functions there are three different radial density curves; namely the radial density $\rho_{1}(r)$ of ions of like sign to that of the reference ion, the radial density $\rho_{u}(r)$ of ions of unlike sign, and the mean radial density $\rho_{m}(r)$ ignoring sign. All of these can, of course, be computed from the data stored in the computer. Fig. 12.15(b) shows the values of the three radial densities in liquid potassium chloride at a temperature just above the fusion point (1045 K) and at $V_{m} = 4\cdot88 \times 10^{-5}$ m^{3} mol^{-1}. You should be able to explain why each peak in the

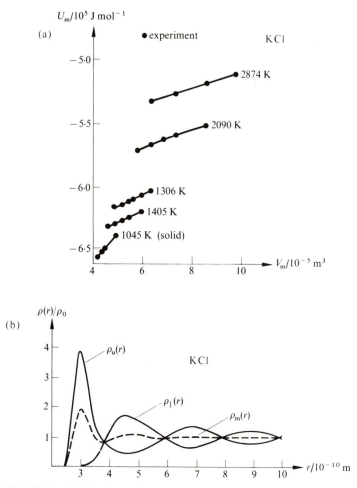

Fig. 12.15. (a) The full lines show the dependence of U_m on V_m at various temperatures as predicted by a Monte Carlo simulation of liquid KCl. (b) The radial density predicted for ions of like sign, $\rho_1(r)$, of unlike sign, $\rho_u(r)$, and the mean radial density $\rho_m(r)$ ignoring sign in liquid KCl at $T = 1045$ K, $V_m = 4 \cdot 88 \times 10^{-5}$ m^3 mol^{-1}. (Data from Woodcock, L. V. and Singer, K. (1971). *Trans. Faraday Soc.* **67**, 12.)

$\rho_1(r)$ curve occurs at a larger r than does the corresponding peak in the $\rho_u(r)$ curve.

Although of no fundamental importance from a quantitative viewpoint, the positions of the atoms as calculated for each configuration can be shown on a visual display unit. The computer can also be programmed to draw in circles about these points to give the appearance of atoms and to add shading to create the illusion of depth. If each configuration is

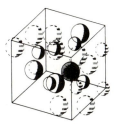
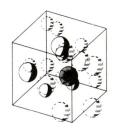
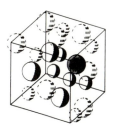
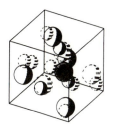
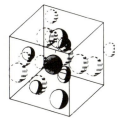
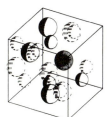
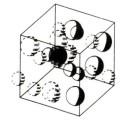
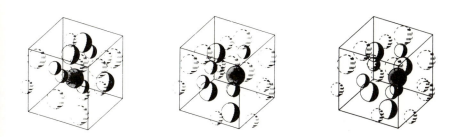

Fig. 12.16. A series of stills (every fifteenth frame) from a film of a molecular dynamics simulation of liquid KCl. The large spheres are Cl^- ions; the small spheres K^+ ions. The box has been drawn in to aid visualization of the three-dimensional nature of the liquid. When outside the box the ions are given dotted outlines. From a molecular dynamics study by J. W. E. Lewis. (Reprinted with permission of the Open University.)

photographed with a cine camera (one frame exposure per configuration) the resulting film gives a remarkably vivid impression of what conditions are like in a liquid at a particular (p, V_m, T). In view of the close agreement between the computer-predicted and the experimentally-determined properties of the liquid we can be confident that such a film gives a highly accurate impression of conditions in the liquid. You may be able to get some feel for the ions' behaviour in liquid KCl by studying the sequence of photographs taken from such a film and shown in Fig. 12.16. Here the large spheres show the Cl^- ions; the small spheres the K^+ ions. For clarity, only a few of the 'atoms' stored in the computer have been displayed. To appreciate the kinetic nature of the simulation concentrate on one of the ions (such as the Cl^- ion which has been coloured black) and move your eyes from frame to frame in comic-strip fashion. Of all the various techniques for demonstrating conditions in a liquid a film of a molecular dynamics simulation undoubtedly provides the most accurate information. As an inexpensive alternative, a randomly-shaken tray of oil-covered balls is not without merit.

12.6. Liquid crystals

When heated, certain organic crystals (usually composed of long rod-like molecules) fail to transform abruptly from a solid into a liquid. Instead they pass through a series of transitions involving new phases, whose symmetry properties are intermediate between those of a solid and of an isotropic liquid. For this reason such materials are known as *liquid crystals*. This behaviour was first observed by Reinitzer in 1888 when he noted that solid cholesteryl benzoate changed to a turbid liquid at 418 K and into a clear liquid when heated further to 452 K. It is now known that liquid crystals fall into three basic types, known as *nematic, cholesteric,* and *smectic* liquid crystals (or 'mesophases'). As a given material is heated through its transition region it may pass, in turn, through several different mesophases. By way of example, p-(p'-ethoxybenzyl-ideneamino)-ethylcinnamate passes through the following sequences of phases (the terms will be explained presently); solid $\xrightarrow{354\,K}$ smectic B $\xrightarrow{392\,K}$ smectic A $\xrightarrow{430\,K}$ nematic $\xrightarrow{432\,K}$ isotropic liquid.

Nematic liquid crystals

Fig. 12.17 shows how the molecules are arranged in the nematic mesophase. The key features to note are that the centres of mass of the individual molecules are distributed as in an ordinary liquid and that, when averaged over space or time, the long molecular axes are aligned parallel to a preferred direction (indicated by the vector n of unit length

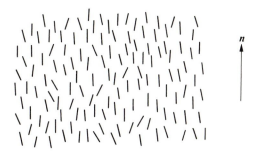

Fig. 12.17. Illustrating the arrangement of molecules in a nematic liquid crystal.

shown in Fig. 12.17). The changes in the direction of the *local* alignment throughout the body of the liquid crystal is the cause of the turbid appearance of the mesophase. In the absence of external forces, such as those provided by the container walls, the direction of *n* is entirely arbitrary. Furthermore, the directions *n* and −*n* are indistinguishable. This means that even if the individual molecules have a permanent electric dipole moment as many molecules will point in one direction as in the opposite direction.

Nematic liquid crystals sandwiched between two plates of glass, each coated with a transparent conducting layer, are at the heart of the ubiquitous 'liquid crystal display' (LCD). In the absence of an applied voltage the sandwiched mesophase appears transparent. The application of a local electric field (via segments photoetched on one of the plates) produces optical scattering-centres in the mesophase, which leads to the familiar local darkening of the display.

Cholesteric liquid crystals

The cholesteric mesophase is really a special case of the nematic mesophase. If sandwiched between two parallel plates the molecules of such a liquid crystal will arrange themselves as shown in Fig. 12.18. Within any sheet (meaning at a particular value of z) the distribution is 'nematic'; that is, the centres of masses of the molecules lack long-range order, although the molecules themselves are aligned about a fixed direction *n*. The direction of *n* varies in helical fashion around the z-axis. Thus the structure is periodic along the z-axis and with a period of half the pitch (the factor of $\frac{1}{2}$ arising because layers *n* and −*n* are equivalent). The pitch is normally very sensitive to temperature. Now changes in pitch lead to changes in the wavelength of the light reflected by the mesophase. Thus by noting the colour of the liquid crystal one can get a good indication of its temperature. Cholesteric liquid crystals applied to a

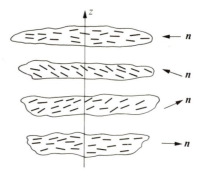

Fig. 12.18. Illustrating the arrangement of molecules in a cholesteric liquid crystal. Although shown as distinct sheets (for clarity) the distribution is a continuous function of z. If the pitch is infinite the structure is nematic.

patient's skin have been used to detect regions of slightly-raised temperature (as may be caused by a tumour). In fact, liquid crystals have been designed to respond to a temperature variation of about 4 K; the coldest regions being blue and the warmest red.

Smectic liquid crystals

All smectic liquid crystals have a layered structure. By this we mean that the centre of masses of the molecules lie on planes with a well-defined interplane separation d (which can be measured by X-ray diffraction techniques). However, within the broad classification of the smectic mesophase there are different sub-classes, the most important of which are smectic A, smectic B, and smectic C. These differ in the way that the centres of mass of the molecules and their orientation vary across a layer.

In smectic A (shown looking sideways-on in Fig. 12.19(a)) the centres of mass of the molecules are arranged as they would be in a two-dimensional liquid. The directional vector n lies normal to each layer, with the result that the interplane separation d is approximately equal to the length l of the constituent molecules.

In smectic C (Fig. 12.19(b)) the centres of mass are still arranged as in a two-dimensional liquid but now n is tilted at an angle ω with respect to the z-axis. This model is confirmed by X-ray diffraction measurements, which show that $l \cos \omega = d$.

Smectic B differs from smectic A and C in that the centres of mass of the molecules of a layer appear to form a two-dimensional solid. Consequently, this is the most ordered of the three smectic mesophases. If a single material can display A, B, and C mesophases then smectic B will always appear first on raising the temperature of the material through the transition region.

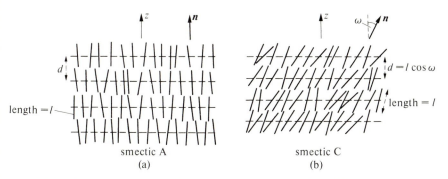

Fig. 12.19. Illustrating the main characteristics of (a) smectic A, and (b) smectic C liquid crystals. In both cases the centres of mass of the molecules lie on well-defined layers and are arranged like atoms in a two-dimensional liquid. In smectic A the unit vector n lies along the normal to the layers; in smectic C it lies at an angle ω to the normal.

PROBLEMS

12.1. Using the information contained in Fig. 3.2(b), plot a radial density curve for the two-dimensional simulated liquid composed of weakly-attracting hard spheres. You will find it helpful to use a piece of tracing paper on which you have drawn a series of concentric circles of radii d, $1\cdot2d$, $1\cdot4d$, etc., where d is the ball diameter in Fig. 3.2(b). Place the ring system over each ball in turn and count the number of centres lying within each annulus.

12.2. Using the data given in Fig. 12.3(a), deduce how the first coordination number in liquid argon changes in going from near the triple point ($T_{tr} = 83\cdot8$ K) to near the critical point ($T_c = 150\cdot7$ K) along the saturated liquid line.

12.3. What does the area under the $g(r)$ plot from $r = 0$ to $r = \infty$ represent in a monatomic liquid?

12.4. Using the plots of $\rho(r)/\rho_0$ shown in Fig. 12.15(b), estimate the first coordination number between ions of (a) unlike sign, and (b) like sign in liquid potassium chloride at 1045 K and at $V_m = 4\cdot88 \times 10^{-5}$ m^3 mol^{-1}. Compare your answer with the corresponding numbers for like and unlike ions in solid potassium chloride.

12.5. One way we might attempt to estimate the free space which exists in a liquid is to see how much solid is required to form a saturated solution (one in which no more solute can be dissolved). Try this experiment using, say, salt and water or sugar and water. You may be surprised by the result!

12.6. In imagination, dissolve a solid (or *solute*) in a liquid (or *solvent*). If N solute atoms are all to be widely dispersed in the solution so formed—and by widely dispersed we mean several atomic diameters or more—then energy of roughly $\frac{1}{2}Nn\,\Delta E$ is required (eqn (2.26)), where ΔE is the dissociation energy of a pair of solute atoms, and n is the number of nearest neighbours to an atom of the solid. If the liquid is contained within adiabatic walls this energy can *only* come from the kinetic energy of the solution's atoms. According to these arguments the temperature of any liquid should drop when a solid is dissolved in it. Try dissolving commonly-available solutes (such as sugar or salt) in water contained in, say, a foam-polystyrene cup (this provides a rather good adiabatic enclosure), noting whether the temperature of the liquid falls as expected. If you only have a clinical thermometer available, start with lukewarm water. Should the temperature of the water not change in the predicted fashion, attempt to account for the fact.

12.7.   A possible model of a liquid might be one in which a fraction x of the atoms are assumed to be perfect gas-like and the remainder are assumed to behave like atoms in a classical solid. On the basis of this speculative model, relate x to the change $\Delta C_{V,m}$ which occurs in the heat capacity of a solid on melting.

12.8.   It has been pointed out (Temperley, 1947) that the tensile strength of a liquid can be predicted assuming the van der Waals equation applies. Fig. 6.4 shows that at temperatures well below T_c part of the isotherm corresponds to the liquid being at negative pressure, that is under tension. Therefore the pressure at point F must correspond to the tensile strength of the liquid at this temperature. Show that the pressure p at the minimum is given by

$$p = \frac{a(V_m - 2b)}{V_m^3}.$$

Now the van der Waals equation (eqn (6.6)) tells us that, at $T = 0$, $V_m = b$. We may therefore write the (limiting) tensile strength of a liquid as

$$p = -\frac{a}{b^2}.$$

Drawing on a result from § 6.4 show that this predicts a tensile strength of

$$p = 27 p_c.$$

Since $p_c = 4\cdot9 \times 10^6$ Pa in argon, for example, this predicts a tensile strength of around $1\cdot3 \times 10^8$ Pa. The experimental value is around 10^6 Pa. Before being too harsh on the theory it is worth remembering how predictions as to the tensile strengths of solids were several orders of magnitude too large in most instances. Indeed, as with solids, the experimentally-measured tensile strengths of a liquid depend on the surface conditions—here the nature of the container walls in contact with the liquid. A water–glass system, for example, can withstand tensions of the order 3×10^6 to 5×10^6 Pa, whereas a water–steel system can only withstand tensions of order 1×10^6 to 3×10^6 Pa.

12.9.   What pressure would be required to contain water at 293 K and at a density of 998 kg m^{-3} were the water molecules to behave as a perfect gas? Take $M_r(H_2O) = 18$.

12.10.   It is an experimental fact (Trouton's rule) that for many liquids $H_{m,e}$ divided by T_b, the boiling temperature (at a fixed temperature—usually atmospheric) is approximately constant. As examples, $H_{m,e}/T_b$ is 75 J mol^{-1} K^{-1} for Ar, 85 J mol^{-1} K^{-1} for Na, and 93 J mol^{-1} K^{-1} for Hg. Show that this rule is consistent with our first primitive picture of the conditions existing at the atomic level when a substance is in the liquid phase.

12.11. The following table gives the values of $\rho(r)/\rho_0$ for argon at 148·1 K and at a density of $0\cdot78\times10^3$ kg m^{-3}. Using this data, and the value of $V(r)$ given by the Lennard-Jones model (eqn (2.20)) with $\varepsilon/k = 120$ K and $\sigma = 3\cdot4\times10^{-10}$ m, predict the pressure of argon at $T = 148\cdot1$ K and at a density of $0\cdot78\times10^3$ kg m^{-3}. The experimentally measured value is $4\cdot5\times10^6$ Pa. Take $A_r(\text{Ar}) = 40$.

$r/10^{-10}$ m	$\rho(r)/\rho_0$	$r/10^{-10}$ m	$\rho(r)/\rho_0$	$r/10^{-10}$ m	$\rho(r)/\rho_0$
3·0	0	5·6	0·8684	9·4	0·9939
3·1	0·0900	5·7	0·8751	9·6	0·9982
3·2	0·4587	5·8	0·8814	9·8	1·0021
3·3	0·8882	5·9	0·8921	10·0	1·0058
3·4	1·3014	6·0	0·9032	10·2	1·0091
3·5	1·6378	6·1	0·9173	10·4	1·0119
3·6	1·8482	6·2	0·9305	10·6	1·0144
3·7	1·9463	6·3	0·9472	10·8	1·0151
3·8	1·9534	6·4	0·9637	11·0	1·0145
3·9	1·9095	6·5	0·9800	11·2	1·0132
4·0	1·8498	6·6	0·9961	11·4	1·0109
4·1	1·7365	6·7	1·0136	11·6	1·0090
4·2	1·6078	6·8	1·0293	11·8	1·0068
4·3	1·4866	6·9	1·0455	12·0	1·0049
4·4	1·3686	7·0	1·0587	12·4	1·0033
4·5	1·2606	7·2	1·0829	12·8	1·0033
4·6	1·1599	7·4	1·0908	13·2	1·0039
4·7	1·0398	7·6	1·0808	13·6	1·0040
4·8	0·9971	7·8	1·0595	14·0	1·0045
4·9	0·9479	8·0	1·0391		
5·0	0·9148	8·2	1·0201		
5·1	0·8921	8·4	1·0067		
5·2	0·8812	8·6	0·9968		
5·3	0·8724	8·8	0·9913		
5·4	0·8666	9·0	0·9891		
5·5	0·8658	9·2	0·9908		

Data from Mikolaj, P. G. and Pings, C. J. (1967), *J. Chem. Phys.*, **46**, 1401.

12.12. Making use of the data given in problem 12.11, calculate the molar internal energy U_m of argon at 148·1 K and at a density of $0\cdot78\times10^3$ kg m^{-3}. The experimental value is $U_m = -1\cdot8\times10^3$ J mol^{-1}.

12.13. During the course of performing a molecular dynamics simulation of liquid argon it is observed that, at a particular instant, eight argon atoms occupy the corners of a cube of side 4×10^{-10} m and that one argon atom lies at the centre of the cube. What is the total potential energy of this group of atoms? Assume that all other atoms are remote from this group. Adopt a Lennard-Jones 6–12 potential with $\varepsilon/k = 120$ K and $\sigma = 3\cdot4\times10^{-10}$ m.

12.14. (a) What is the resultant force acting on each of the argon atoms of problem 12.13? (b) Assuming that all these atoms are stationary at the instant described in problem 12.13, what will be their new velocities and positions at a time of $\delta t = 1\times10^{-14}$ s later? In keeping with the assumptions made in molecular dynamics simulations, assume that the resultant force acting on each atom is constant during the time interval δt. (c) By what percentage will the resultant force acting on each atom have actually changed during time δt? Take $A_r(Ar) = 40$.

12.15. A neon atom with a speed of 150 m s^{-1} at a particular instant of time t is acted on by a resultant force of $F = 3 \times 10^{-10} \text{ N}$ which makes an angle of $30°$ with the velocity vector. Calculate the new position and velocity of the atom at a time $\delta t = 2 \times 10^{-14} \text{ s}$ later, assuming that F remains constant during δt. Take $A_r(Ne) = 20$. You may wish to assume that at time t the atom is at $x = 0$, $y = 0$, $z = 0$ and is moving along the x-axis.

12.16. How far will an atom of liquid argon move, on average, during a time interval of 10^{-14} s if the liquid is at a temperature of 90 K? Take $A_r(Ar) = 40$.

12.17. You might care to have a go at trying to apply the techniques of molecular dynamics to a one-dimensional system of, say, five to ten atoms. To aid computation—and so remove the necessity for a computer—assume that the interatomic force characteristic has the form shown in Fig. 12.20. Decide on appropriate values for F_m, r_0, and r_R. (They should more or less match the maximum interatomic force, the equilibrium separation, and the 'range' of the force in whatever substance you have chosen.) Disregard the first few configurations which you calculate—to remove any bias in the starting conditions—and calculate $\rho(r)$ from the next few configurations. To eliminate boundary effects caused by having but few atoms, surround the 'cell' which contains the atoms by replicas of itself. An atom near the cell boundary will therefore be subjected to forces both from real and from 'ghost' atoms. (This technique is always employed in molecular dynamics calculations employing fewer than several hundred atoms.)

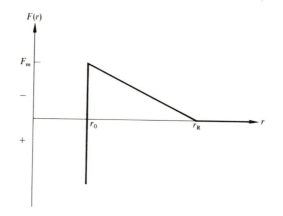

Fig. 12.20. Modelling the interatomic force characteristic.

13. Transport processes in liquids

IN this chapter we will consider the macroscopic transport properties of a liquid and will attempt to explain these properties in terms of what is happening at the atomic level. We will also look at how the surface tension of a liquid is determined by the nature of the gas–liquid interface. Throughout our discussions we will contrast the underlying mechanisms which are responsible for these transport processes with the corresponding mechanisms which apply in the gaseous and solid phases.

This chapter does not aim at reviewing all the models which have been put forward to explain transport phenomena. Indeed it will concentrate almost exclusively on just one model; the cell model. Although other models can be employed—often to better effect—they usually require a level of mathematics well beyond that assumed in this book.

13.1. The cell model

In the last chapter we saw how a liquid can be simulated by 'directing' the component particles to interact in the same way as do the atoms in the substance under investigation. Although it takes a cine-film to represent the dynamic nature of the simulated liquid, individual frames from such films show that in the case of a three-dimensional monatomic liquid each atom has around ten nearest neighbours (as opposed to twelve in a close-packed solid) and that in the case of a two-dimensional monatomic liquid (Fig. 3.2(b)) each atom has around five nearest neighbours (as opposed to six in a close-packed solid; Fig. 3.2(a)). Thus each atom has more free space in which to move in a liquid than it has in a solid. An examination of a film of a molecular dynamics simulation shows that an atom is, for the most part, prevented from wandering too far by its surrounding neighbours. These *seem* to form a *cage* or *cell* about the atom in question. If you examine the sequence of frames shown in Fig. 12.16 you will see that any one atom (such as the shaded atom) does indeed move around but that, in the short term, it does not make any large-scale excursions.

Drawing on these impressions we may therefore picture an atom as being confined to a cell, whose walls are its nearest neighbours. Fig. 13.1(a) shows how the cell walls will appear, on a time-averaged basis, to an individual atom i; if you like, we have averaged the locations of the wall atoms over many frames of a cine-film. The area shown shaded in Fig. 13.1(a) indicates the region in space over which the centre of atom i is more or less free to roam. In a solid (Fig. 13.1(b)) the wall atoms p, q, etc.

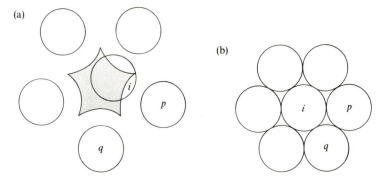

Fig. 13.1. (a) The time-averaged arrangement of atoms in a (two-dimensional) liquid about the 'wanderer' i. The centre of i may move within the region shown shaded. Atoms p, q, etc., comprise the cell walls. (b) The corresponding situation which prevails in a close-packed solid where atom i has little room for movement.

would be close-packed, allowing i but little room for movement. All cell models of a liquid assume that the atoms constituting the wall remain fixed as the 'wanderer' moves about inside the cell. It is paradoxical that, depending on one's viewpoint, each individual atom may be regarded either as a 'wanderer' or as a 'wall' atom, but this facet of the cell model must be tolerated if progress is to be made.

By supposing the wanderer to move freely within its cell we are endowing the liquid with some gas-like attributes. By supposing the bounding cell wall to be fixed—or, rather, by supposing that an activation energy is required for the wanderer to escape—we are endowing the liquid with solid-like attributes. Since the liquid phase is in many ways intermediate in character between that of the gaseous and solid phases, a model combining features of both phases contains, at least, the seeds of future success. As we shall see presently, the model does indeed bear fruit in discussions of most transport processes. However, the cell model is of only limited value in predicting the equation of state and the internal energy of a liquid. In the absence of a single model which can cope with both the equilibrium and the non-equilibrium (transport) properties of a liquid the best we can do with present knowledge is to employ a small number of models which, between them, can accomplish all that is required of them.

13.2. Diffusion

The self-diffusion coefficient.

We will limit our discussion—as we did in the gaseous and solid phases—to self-diffusion in a liquid. Fig. 13.2(a) shows the time-averaged environment of an atom i contained within a cell whose wall atoms are p, q, r, s, and t. For most of the time atom i is confined within its present cell. Occasionally, as a result of fortuitous collisions with the wall atoms, it may acquire enough energy to push the wall atoms apart and so to escape to an adjoining cell. We will suppose that the act of escaping eliminates the cell which formerly surrounded atom i. A new cell will, however, form around atom i in its new location. By definition, the walls of this cell will be the time-averaged nearest neighbours to i.

The energy required to displace i from an arbitrarily chosen starting position—say, the centre of the cell—is given by the potential energy diagram shown in Fig. 13.2(b). Perhaps the easiest way to see why $V(x)$ has this form is, in imagination, to 'freeze' everything and then move i out in, say, the $+x$-direction. In so doing energy is required to pull i away from the attractive force of atoms r, s, and t, and also to separate atoms p and q sufficiently to allow i through the wall of the cell. Once through the 'hole in the wall' atoms p and q can close up, restoring energy to us; the remaining energy being restored as a new cell forms about i. The potential energy curve will therefore be periodic with a period equal to the cell

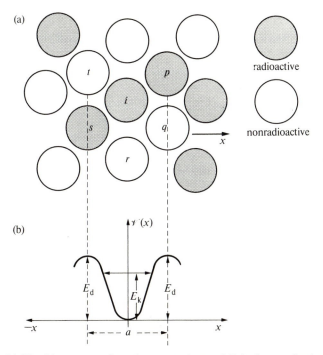

Fig. 13.2. (a) The (time-averaged) environment of atom i. It is thus confined within a cell whose wall atoms are p, q, r, s, and t. (b) The potential energy $V(x)$ of atom i as it is displaced along the x-direction from the centre of the cell.

diameter a (Fig. 13.2(b)). On average—averaged over many cells—$\mathscr{V}(x)$ will be the same along all diameters.

If atom i is to escape from its cell it must acquire a kinetic energy of at least E_d, the activation energy for diffusion. Treating the problem as a two-dimensional one, we may take over the discussion of §5.13 where we showed (eqn (5.79)) that the probability of a gas atom having a kinetic energy of at least E_d is given by $\exp(-E_d/kT)$. You will recall from §4.10 that the Maxwell–Boltzmann energy distribution function (which was assumed in deriving eqn (5.79)) does indeed describe the kinetic energy distribution in a liquid. Since the cell has a diameter of a, and the 'wanderer' a mean speed of $\bar{u} = (\pi kT/2m)^{1/2}$ (see problem 4.9), it follows that it will collide with the cell wall at intervals of about $a/\bar{u} = a/(\pi kT/2m)^{1/2}$, where m is the mass of a liquid atom. Hence the number of escape attempts per unit time is $\bar{u}/a = (\pi kT/2m)^{1/2}/a$. Multiplying this by the fraction of the escape attempts which are successful, namely $\exp(-E_d/kT)$, gives the number of escape hops per unit time, f_d, as

$$f_d = \frac{1}{a}\left(\frac{\pi kT}{2m}\right)^{1/2} e^{-E_d/kT}. \tag{13.1}$$

When the wanderer hops from its present cell into its new cell it moves a distance a. Thus the average speed v_d with which an atom hops is given by

$$v_d = f_d a = \left(\frac{\pi kT}{2m}\right)^{1/2} e^{-E_d/kT}. \tag{13.2}$$

As in a solid and a gas, net diffusion of a species in a liquid depends on there being a concentration gradient of that species. In §11.1 we saw that the self-diffusion coefficient D was related to the hopping speed v_d and the hop distance a by $D = \frac{1}{3}v_d a$. The corresponding result in two dimensions is $D = \frac{1}{2}v_d a$. (This assumes that one-quarter of the atoms are, at any instant, moving along the $+x$-, $-x$-, $+y$-, and $-y$-directions.) Substituting for v_d from eqn (13.2) gives

$$D = \frac{a}{2}\left(\frac{\pi kT}{2m}\right)^{1/2} e^{-E_d/kT}. \tag{13.3}$$

Over the fairly small range of T in which a substance exists as a liquid the $T^{1/2}$ term changes much more slowly than does $\exp(-E_d/kT)$. We may therefore write, to a good approximation, that

$$\boxed{D = D_0\, e^{-E_d/kT}}, \tag{13.4}$$

where D_0 is a constant. Although our treatment assumed a two-dimensional liquid, the same dominant exponential factor reappears in a full three-dimensional discussion.

Experimental techniques

Fig. 13.3 is a (somewhat simplified) diagram of an apparatus which has been used by Naghizadeh and Rice (1961) to study self-diffusion in the liquid inert gases over a fairly wide range of pressure and temperature. The key item is capillary C which can be opened and closed from both ends by valves V_1 and V_2. The capillary is formed from a stainless steel block and has a bore of 0·5 mm and a length of 18·5 mm. The experiment involves four basic steps:

(1) Keeping the lower valve V_1 closed, the bath B is filled through F with a suitable mixture of pure and radioactive fluid. The capillary is then filled with pure liquid.
(2) The pressure in the capillary and the bath are equalized.
(3) After closing the upper valve V_2, the lower valve V_1 is opened so that diffusion from the isotope-enriched bath into the capillary may proceed.

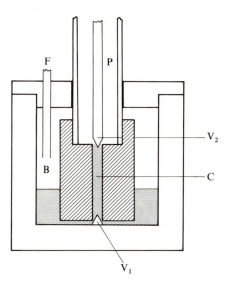

Fig. 13.3. An apparatus for determining the self-diffusion coefficient in the liquid inert gases. On opening valve V_1 radioactive tracers diffuse into the pure liquid contained in the capillary tube C. (Naghizadeh, J. and Rice, S. A. (1961). *J. Chem. Phys.*, **36**, 2710.)

(4) Once a suitable diffusion time t has elapsed the lower valve is closed. The upper valve is then opened and by pumping through tube P the contents of the capillary are transferred to a sample tube where the level of radioactivity is counted.

Knowing the (mean) level of radioactivity in the capillary and the diffusion time t, the value of D can be readily calculated. A high radioactive assay, for example, implies that the atoms have diffused far into the tube. It follows from eqn (11.12) that D has a high value. To determine D at a different pressure the entire operation is repeated with the liquid in bath B maintained at a new pressure. It is worth noting how the temperature of the system is controlled. The vessel comprising the capillary and the isotope-enriched liquid is suspended inside a large dewar flask (not shown) from a long support rod. Another support rod is attached to the underneath surface of the vessel and this dips into liquid nitrogen. Since the upper end of the top support is at room temperature and since the lower end of the bottom support is at liquid nitrogen temperature, the vessel must be at some intermediate temperature.

The pressure and temperature dependence of D

Fig. 13.4(a) shows the results of a series of studies of self-diffusion in liquid argon. As predicted by eqn (13.4), graphs of $\ln D$ plotted against $1/T$ are indeed linear. Their gradients $(-E_d/k)$ remain constant, to within 5 per cent, over a tenfold change in pressure and give a mean value for E_d of $5 \cdot 0 \times 10^{-21}$ J.

Instead of plotting how $\ln D$ varies with T^{-1} along lines of constant p, the data contained in Fig. 13.4(a) may be replotted to show how $\ln D$ varies with p along lines of constant T. Fig. 13.4(b) shows such a plot at $T = 100$ K. (The plots at other temperatures have the same gradient—to within 5 per cent.) We can see a possible, if not altogether convincing, reason for this pressure dependence if we examine Fig. 13.2(a). When the liquid is subjected to an external pressure, p, extra energy will be required to move atoms p and q apart through a distance of about a so that i can escape. In fact E_d is increased by about $(pa^2) \times a$ (that is, force $\times$ distance), making the activation energy $E_d + pa^3$. Substituting this revised activation energy into eqn (13.4) predicts

$$D \propto e^{-(E_d + pa^3)/kT},$$

or a linear graph of gradient $-a^3/kT$ when $\ln D$ is plotted against p. Taking $a = 3 \cdot 6 \times 10^{-10}$ m predicts a gradient of $3 \cdot 4 \times 10^{-8}$ m^2 N^{-1}. The experimental value is $4 \cdot 1 \times 10^{-8}$ m^2 N^{-1}.

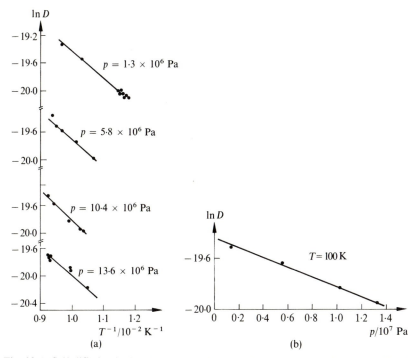

Fig. 13.4. Self-diffusion in liquid argon as a function of temperature and pressure. The data actually refers to the diffusion of ^{41}Ar into natural argon. In (a) the data is presented to show the variation of $\ln D$ with $1/T$ along lines of constant p. In (b) the variation is shown against p along a line of constant T (at $T = 100$ K). The self-diffusion coefficient is measured in units of $m^2\,s^{-1}$. (Data from Naghizadeh, J. and Rice, S. A. (1961). *J. Chem. Phys.*, **36**, 2710.)

Exercise 13.1

Make an order-of-magnitude estimate of the activation energy E_d for self-diffusion in liquid argon at a density of $1{\cdot}4\times10^3$ kg m^{-3} and a temperature of 84 K (these are the values at the triple point), comparing your answer with the measured value of $5{\cdot}0\times10^{-21}$ J. What fraction of escape attempts lead to an argon atom escaping from its cell when the liquid is at 117 K (this is midway between the triple and critical temperatures)? Assume a hard-sphere repulsive model for the interaction between two argon atoms, with the potential energy due to the (van der Waals) attractive forces given by the Lennard-Jones term

$$\mathcal{V}(r) = -4\varepsilon\left(\frac{\sigma}{r}\right)^6.$$

Take $\varepsilon = 1{\cdot}7\times10^{-21}$ J and $\sigma = 3{\cdot}4\times10^{-10}$ m.

Calculations. Working out the interatomic separation from the density in a, by now, familiar fashion gives a value of $3{\cdot}6\times10^{-10}$ m. To calculate the activation energy E_d we must consider the problems facing i as it attempts to escape from its cell—namely the fact that it must escape the backward pull of neighbours like r, s, and t (Fig. 13.2(a)) and that atoms p and q must be separated sufficiently to allow i through. The energy required to separate p and q from about $r_1 = 3{\cdot}6\times10^{-10}$ m to about $r_2 = 7 \cdot 2 \times 10^{-10}$ m is, since $\mathcal{V}(r) = -4\varepsilon(\sigma/r)^6$, given by $4\varepsilon(\sigma/r_1)^6 - 4\varepsilon(\sigma/r_2)^6 = 4 \cdot 75 \times 10^{-21}$ J. In three dimensions there will be (approximately) two more atoms to be separated; one out of the plane, and one behind the plane, of the paper in Fig. 13.2(a). The total energy required to 'open up the hole' is therefore about $9 \cdot 5 \times 10^{-21}$ J. In escaping from an atom like s, atom i moves from a distance of about $3 \cdot 6 \times 10^{-10}$ to about $7 \cdot 2 \times 10^{-10}$ m from s, calling for an energy of $4 \cdot 75 \times 10^{-21}$ J. There are in all six such atoms (the first coordination number is about ten at the triple point but we have already accounted for four of these). Hence $E_d = 9 \cdot 5 \times 10^{-21}$ J $+ 6(4 \cdot 75 \times 10^{-21}$ J$) = 3 \cdot 8 \times 10^{-20}$ J. Although this is an order of magnitude greater than the measured value of $5 \cdot 0 \times 10^{-21}$ J the discrepancy is not serious since small changes in r will lead to large changes in $\mathcal{V}(r)$.

Substituting $E_d = 5 \times 10^{-21}$ J and $T = 117$ K into $\exp(-E_d/kT)$ gives the fraction of escape attempts which are successful as $4 \cdot 5 \times 10^{-2}$.

Comment. Knowing the measured value of E_d it is tempting to recalculate it with 'improved' cell geometry. Such temptations should be resisted in making order-of-magnitude correct calculations. Knowing the answer they are expected to find, theoreticians sometimes have the knack of finding it. (Knowing what they are told to see, experimentalists sometimes have the knack of 'seeing' it too!)

13.3. Viscous flow

The dynamic viscosity

We start by picturing a liquid contained between two plates (Fig. 7.1(b)); the lower one of which is fixed, the upper one of which is moved to the right with a constant velocity. The velocity gradient present between the plates necessarily implies that one layer of atoms slips relative to its neighbouring layers. Our first task must be to find some mechanism whereby the application of a shearing force allows slip to occur. As with our discussion of viscous flow in a gas, we will direct our attention to conditions existing close to one of the plates.

The row of atoms labelled c, d, e, etc. in Fig. 13.5(a) represents the lowest layer of atoms of the moving plate. The layer of atoms i, j, k, etc. of the liquid immediately adjacent to the plate will experience a short-ranged attractive force directed towards this plate. This attractive force is so short-ranged—it often varies as the inverse eighth power of the distance from the plate—that the second layer of atoms (q, r, s, etc.) will experience no such force. Thus if a shearing force F is applied to the top plate, of area A, this will in turn exert a tangential force f on, say, atom i, given by

$$f = \frac{F}{A/(\frac{1}{4}a^2)} = \frac{Fa^2}{4A}, \tag{13.5}$$

where a is the cèll diameter. This result follows because each atom occupies an area of approximately $\frac{1}{2}a \times \frac{1}{2}a$, so that the total number of atoms like i in the layer of liquid adjacent to the plate is $A/\frac{1}{4}a^2$.

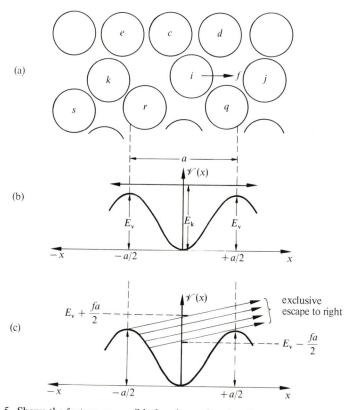

Fig. 13.5. Shows the factors responsible for viscous flow in a liquid. (a) A (shear) force F is applied to the top plate of area A (containing atoms c, d, e, etc.). (b) With no external force the activation energy is the same in the $+x$- and $-x$-directions. (c) The effect of a force f acting on atom i is to increase its kinetic energy by $\frac{1}{2}fa$ when i is at $x = \frac{1}{2}a$ and to decrease its kinetic energy by $\frac{1}{2}fa$ when it is at $x = -\frac{1}{2}a$.

When $F = 0$ the energy required to displace atom i in the x-direction is given by the potential energy curve shown in Fig. 13.5(b). We see that provided the kinetic energy E_k of i exceeds E_v, the activation energy for viscous flow, hopping is equally likely to occur in the $+x$- as in the $-x$-direction, producing no net flow in the liquid. When $E_k < E_v$ hopping occurs in neither the $+x$- nor $-x$-directions.

When F is finite atom i will experience a tangential force f given by eqn (13.5). Its effect will be to increase the kinetic energy of i as it moves from $x = 0$ to $x = +\frac{1}{2}a$ by an amount $\frac{1}{2}fa$, and to decrease its kinetic energy by $\frac{1}{2}fa$ as it moves from $x = 0$ to $x = -\frac{1}{2}a$ (Fig. 13.5(c)). Thus when the kinetic energy E_k of i (measured at $x = 0$, where $V(x) = 0$) lies in the range $E_v - \frac{1}{2}fa$ to $E_v + \frac{1}{2}fa$ it can *only* escape in the $+x$-direction (Fig. 13.5(c)). If i has $E_k < E_v - \frac{1}{2}fa$ it will remain trapped. If i has $E_k > E_v + \frac{1}{2}fa$

it can escape with equal probability to the right and to the left, producing no net flow parallel to the direction of f. The effect of f is therefore to produce a net flow to the right in the liquid when $E_v + \frac{1}{2}fa > E_k > E_v - \frac{1}{2}fa$ at $x = 0$. We have thus found the desired mechanism whereby slip can occur.

We must now calculate the escape frequency of atom i from its cell when force f is present. Since the overall flow of the liquid is in the $+x$-direction it seems reasonable to treat the problem one-dimensionally, regarding i as being constrained to move back and forth along the x-axis. On this basis, atom i will make $\bar{u}/2a = (kT/2\pi m)^{1/2}/a$ escape attempts per unit time from the right-hand side of its cell. Here we have substituted for the mean speed $\bar{u} = (2kT/\pi m)^{1/2}$ in a one-dimensional gas (see problem 4.9). Therefore the number, f_v, of escape hops per unit time of an atom like i which lead to a net flow in the direction of f is given by

$$f_v = \frac{1}{a}\left(\frac{kT}{2\pi m}\right)^{1/2} \times (\text{probability of } E_v + \tfrac{1}{2}fa > E_k > E_v - \tfrac{1}{2}fa). \quad (13.6)$$

Now according to eqn (4.59) the probability that an atom of a one-dimensional gas has an energy within a range dE of E is given by $(\pi kTE)^{-1/2}\exp(-E/kT)\,dE$. In the present context $E = E_v$ and $dE = fa$, so that eqn (13.6) becomes

$$f_v = (2m\pi^2 E_v)^{-1/2}\,e^{-E_v/kT}f.$$

Since i hops a distance a as it escapes, preferentially to the right, it follows that the hopping velocity δv of i is given by $f_v a$, that is

$$\delta v = (2m\pi^2 E_v)^{-1/2}\,e^{-E_v/kT}fa. \quad (13.7)$$

What we have actually calculated is the difference in the velocity of two adjacent layers in the liquid; the one containing i, the other q, r, s, etc. (Fig. 13.5(a)). Because these layers are separated by about $\frac{1}{2}a$ it follows that the velocity gradient in the liquid is given by $\delta v/\frac{1}{2}a = 2\,\delta v/a$. Substituting this expression for the velocity gradient, with δv given by eqn (13.7), along with the shear stress (F/A) as given by eqn (13.5), into the definition of dynamic viscosity η (eqn (7.17)) leads to

$$\eta = \frac{\text{shear stress}}{\text{velocity gradient}} = \frac{1}{a^2}(8\pi^2 mE_v)^{1/2}\,e^{E_v/kT}. \quad (13.8)$$

This may be rewritten as

$$\boxed{\eta = \eta_0\,e^{E_v/kT}}, \quad (13.9)$$

where η_0 is a constant which, in our treatment, has a value of $(8\pi^2 mE_v)^{1/2}/a^2$. This temperature dependence of η is more reminiscent of the behaviour of creep in a solid (see problem 11.23) than of viscous flow in a gas (see eqn (7.21)).

Experimental results

Until recently, most studies which have been made of the temperature dependence of η have been made along the saturated liquid line. Such experiments show that, in accord with eqn (13.9), $\ln \eta \propto T^{-1}$ for a wide variety of liquids.

In a study carried out in 1967 de Brock, Grevendonk, and Herreman measured the viscosity of liquid argon at pressures between 10^6 and 2×10^7 Pa and at temperatures between 86 and 146 K. The experimental technique consisted in comparing the amplitude of vibration of a quartz crystal in a vacuum and when immersed in the liquid. The quartz crystal (length 50 mm and diameter 5 mm) was vibrated at its (transverse) resonant frequency of $3 \cdot 84 \times 10^4$ Hz by applying an alternating voltage between silver electrodes evaporated onto the crystal. The electrical resistance of the quartz provides an (indirect) measure of its amplitude of vibration. The smaller the amplitude of vibration the greater is the viscosity of the liquid.

Fig. 13.6 shows the results obtained in this study. We see (Fig. 13.6(a)) that, as predicted from eqn (13.9), graphs of $\ln \eta$ plotted against $1/T$ are linear. The gradients (E_v/k) give $E_v = 4 \cdot 4 \times 10^{-21}$ J at $p = 0 \cdot 35 \times 10^7$ Pa and $E_v = 3 \cdot 5 \times 10^{-21}$ J at $p = 1 \cdot 9 \times 10^7$ Pa. These values compare favourably with $E_d = 5 \cdot 0 \times 10^{-21}$ J deduced from diffusion measurements (Fig. 13.4(a)). We would expect E_d and E_v to be the same since both measure the activation energy required by an atom to escape from its cell. This rough equality of E_d and E_v holds true in many liquids; in liquid carbon tetrachloride, for example, $E_d = 6 \cdot 3 \times 10^{-21}$ J and $E_v = 4 \cdot 1 \times 10^{-21}$ J (both measured along the saturated liquid line). It is also worth noting that E_d is very often about one-quarter to one-third of the enthalpy of evaporation of an atom $(H_{m,e}/N_A)$; a not-unexpected result since evaporation involves removing an atom to an infinite distance from its neighbours whereas diffusion and viscous flow only requires the atom to have sufficient energy to escape from its cell. In certain complex liquids—for example, liquids composed of long molecular chains—the molecules may align themselves along the direction of flow once flow begins. This leads to a lowering of E_v and so to a decrease in viscosity (eqn (13.9)). In their unaligned ('jumbled') state E_v, and therefore the viscosity, is high. This lowering of viscosity with increasing stress—a phenomenon known as *thixotropy*—is put to good use in certain nondrip, or thixotropic, paints and adhesives. When the stress is low (such as that resulting from gravity acting on a drop) the viscosity is high. Brushing exerts a large stress and the viscosity falls.

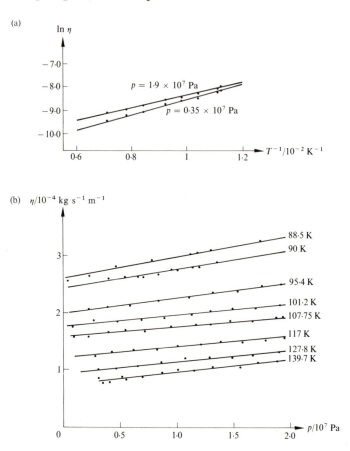

Fig. 13.6. The viscosity of liquid argon as a function of pressure and temperature. In (a) ln η is plotted against T^{-1} at two constant pressures. In (b) η is plotted against p at various constant temperatures. (Data from de Bock, A., Grevendonk, W., and Herreman, W. (1967). *Physica*, **37**, 327.)

Fig. 13.6(b) shows that η increases linearly with p over a fivefold change in pressure. In many, more complex, liquids it is ln η rather than η which increases linearly with p. This is a very useful property to have in a lubricant since it prevents it being squeezed out from between the moving surfaces as the load is increased.

Our explanation of viscous flow assumes implicitly that the force f acting on atom i in Fig. 13.5 does so for sufficient time to allow atom i to acquire momentum from it. If f acts for a time which is less than the time it takes i to traverse the cell—often called the *relaxation time*—flow cannot occur and the liquid will deform elastically. This effect is not observed in simple liquids where the relaxation times ($\approx a/\bar{u}$) are of order 10^{-10} to 10^{-12} s, but it is observed in materials of high relative molecular mass. Silicone putty ('bouncing putty') is probably familiar to you. When left on the table a ball of the putty flows to form a pool over a period of minutes. If dropped onto the table from a height it bounces. The relaxation time of a molecule of the putty is presumably greater than the time during which the ball makes contact with the table during a bounce.

Exercise 13.2

If a similar *mechanism* to that pertaining to a perfect gas applied in liquids, would the viscosity predicted for a liquid be less than, equal to, or greater than that of a perfect gas at the same density and temperature?

Discussion. In a liquid an atom jumps through a distance a—the cell size. However, unlike in a perfect gas—where the atoms are unhindered as they jump through a distance λ, and where every jump is successful—in a liquid only a fraction $\exp(-E_d/kT)$ of the attempts succeed. Close to the triple point this fraction is typically 10^{-2}. Thus the mechanism responsible for viscous flow in a perfect gas would, it it were the only mechanism, predict viscosity values for liquids several orders of magnitude *less* than the value for a perfect gas at the same temperature and density.

Comment. In retrospect we can see that our neglect of the effect of atoms hopping from one layer to another during viscous flow is justifiable.

13.4. Ionic conduction

Although the liquid inert gases normally have a near-zero electrical conductivity, they will conduct if a suitable radioactive source is placed in these liquids. The mobility of the charge carriers—that is, their drift speed per unit electric field—can be measured using time-of-flight techniques.

Experimental studies

Fig. 13.7(a) is a schematic diagram of an apparatus used by Davis, Rice, and Meyer in 1962 to measure ionic mobilities in liquid argon, krypton, and xenon. The ions were created at grid S by a layer of ^{210}Po plated on to its surface. (The emitted α-particles have such a short range that effectively all the ionization takes place in the plane of S.) When S is at voltage V_{S1} (Fig. 13.7(b)) there is a constant electric field between S and the ion collector C and a constant current will be recorded on the chart recorder. If the voltage on S is now changed to V_{S2} (Fig. 13.7(b)) no ions can pass through A. The collector current will, however, only drop to zero once the ions, created as the voltage on S changes from V_{S1} to V_{S2}, reach C. This time lapse t_d is obtained directly from the chart record (Fig. 13.7(a)). Dividing the distance l between A and C by t_d gives the drift speed v_d and hence (see eqn (7.70)) the mobility $\mu^+ = v_d/E$, where

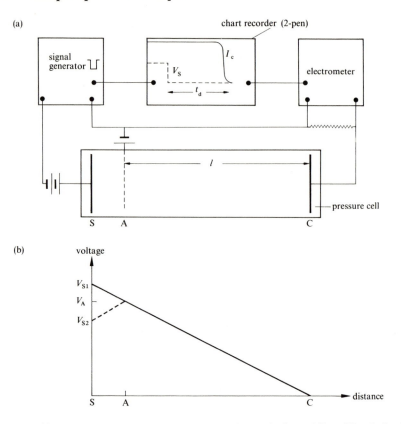

Fig. 13.7. (a) Schematic diagram of an apparatus used to study the mobility of ions in liquid inert gases. The entire grid assembly is mounted in a cell containing the liquid, whose pressure and temperature can be controlled. In addition to the grids shown here there are also intermediate grids between A and C maintained at the appropriate potential so that the field is constant between A and C. (b) The potentials on the various electrodes. By changing the potential on S from V_{S1} to V_{S2} the positive-ion current can be switched off. (Data from Davis, T. H., Rice, S. A., and Meyer, L. (1962). *J. Chem. Phys.*, **37**, 947.)

$E = (V_A - V_C)/l$. To study the mobility of negative ions the polarities of the various electrodes are reversed.

Fig. 13.8(a) shows the logarithm of the mobility μ^+ of positive ions in liquid argon as a function of pressure at three different temperatures. This data is redrawn in Fig. 13.8(b) to show how $\ln \mu^+$ varies with T^{-1} at a fixed pressure. The fact that this graph is linear shows that $\mu^+ \propto \exp(-E_i/kT)$, where E_i is a constant. The gradient $(-E_i/k)$ of the graph gives $E_i = 5 \times 10^{-21}$ J. This compares favourably with $E_d = 5{\cdot}0 \times 10^{-21}$ J and $E_v = 3{\cdot}5 \times 10^{-21}$ J as deduced from diffusion and viscosity measurements, respectively, in liquid argon.

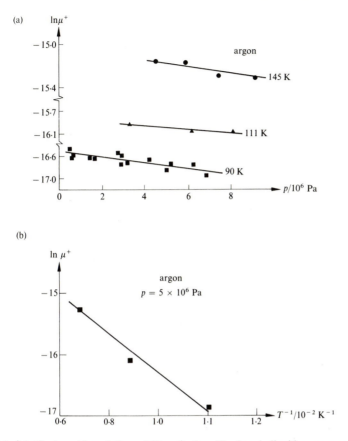

Fig. 13.8. (a) The logarithm of the mobility μ^+ of positive ions in liquid argon as a function of pressure at three different temperatures. (b) $\ln \mu^+$ plotted as a function of T^{-1} at a fixed pressure. All data in (b) was taken from the best straight lines drawn in (a). (Data from Davis, T. H., Rice, S. A., and Meyer, L. (1962). Chem. Phys. **37**, 947.)

The ionic mobility

It is not difficult to adapt our discussion of viscous flow so that it may be applied to ionic conduction. We may regard atom i in Fig. 13.5(a) as a positive ion contained within a cell formed by atoms c, d, q, and r. The number density of ions present in the liquid is so low that there is a near certainty that an ion will be entirely surrounded by neutral atoms. If the ion i has a charge q and the electric field is E, the force f acting on the ion is Eq. The rest of the analysis, summarized in Figs. 13.5(b) to (c), can be applied with $f = Eq$. In particular, eqn (13.7) gives the drift velocity v_d of the ion as

$$v_d \propto e^{-E_i/kT} Eq, \qquad (13.10)$$

where we have written E_i in place of E_v to allow for the possibility that the activation energies may differ in the two processes. The mobility $\mu = v_d/E$ is therefore given by

$$\mu = \mu_0\, e^{-E_i/kT},$$
(13.11)

where μ_0 is a constant. As we have seen, eqn (13.11) is in accordance with the experimental measurements.

There is evidence that the positive ions in liquid argon are actually Ar_2^+ and not Ar^+, but this scarcely affects the arguments. There is also evidence that the negative ions are O_2^-; dissolved oxygen may capture the electrons released when argon is ionized.

Exercise 13.3

The solid alkali halides, such as sodium chloride, have a small but measurable electrical conductivity σ. On the assumption that the positive ions have a very much greater mobility than the larger negative ions, show that $\sigma \propto \exp(\text{constant}/T)$. *Clue*: Draw on eqn (11.1) for the hopping speed of an ion in the absence of an electric field. (Diffusion in the alkali halides is mainly by hopping into adjacent vacant lattice sites.) Remember also that the probability that a vibrating atom has an energy of E, or greater, is $\exp(-E/kT)$ (see eqn (10.54)).

Derivation. The activation energy required by an ion of charge q to hop into a vacant site will be lowered from E_d to $E_d - (Eq)(a/2)$ when the ion hops in the field direction, and will be increased to $E_d + (Eq)(a/2)$ when the ion hops in the opposite direction. This follows since the force on the ion is Eq and the maximum potential energy occurs at $a/2$ (see Fig. 11.3(a)). Applying eqn (11.1) we see that the current J flowing per unit cross-sectional area of liquid is

$$J = nqv_d = \frac{nqa\omega_E}{\pi}\, e^{-E_v/kT}[e^{-(E_d - \frac{1}{2}Eqa)/kT} - e^{-(E_d + \frac{1}{2}Eqa)/kT}],$$

where n is the number density of the positive ions. The expression in brackets represents the fraction with energies between $E_d - \frac{1}{2}Eqa$ and $E_d + \frac{1}{2}Eqa$. (The first exponential gives the fraction of the ions with energies between $E_d - \frac{1}{2}Eqa$ and infinity; the second exponential gives the fraction with energies between $E_d + \frac{1}{2}Eqa$ and infinity.) Since $\frac{1}{2}Eqa \ll kT$ the exponentials can be expanded to give

$$J = \frac{nqa\omega_E}{\pi}\, e^{-E_v/kT} e^{-E_d/kT}\left(1 + \frac{1}{2}\frac{Eqa}{kT} - 1 + \frac{1}{2}\frac{Eqa}{kT}\right)$$

and thus

$$\sigma = \frac{J}{E} = \frac{nq^2 a^2 \omega_E}{kT}\frac{}{\pi}\, e^{-(E_v + E_d)/kT}.$$

Comment. It is worth noting that a graph of $\ln D$ plotted against T^{-1} (see eqn (11.3)) should have the same gradient as one of $\ln \sigma$ plotted against T^{-1}. This is so, at least at high temperatures, in solid sodium chloride.

13.5. Thermal conduction

Experimental studies

Fig. 13.9(a) shows the thermal conductivity κ of argon measured at various fixed pressures over a temperature range 90–200 K; the paths followed along the p–V_m–T surface in these experiments are indicated in Fig. 13.9(b). You will notice that the ordinate in Fig. 13.9(a) is κ and not $\ln \kappa$, and that the abscissa is T, and not T^{-1}. In other words, unlike the behaviour of D, η, and μ, the thermal conductivity of a liquid does *not* vary exponentially with the inverse temperature. This strongly suggests that a cell model cannot be used in discussing thermal conduction in a liquid.

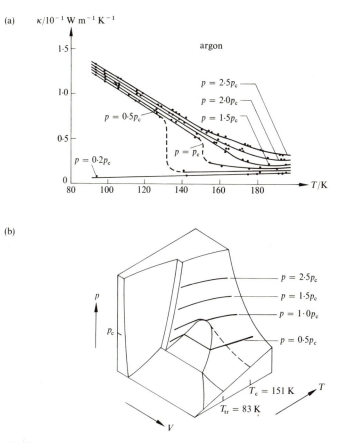

Fig. 13.9. (a) The thermal conductivity of argon as a function of temperature at various pressures (expressed as fractions of the critical pressure $p_c = 4 \cdot 9 \times 10^6$ Pa). (Data from Ziebland, H. and Burton, J. T. A. (1958). *Brit. J. Appl. Phys.* **9**, 52.) (b) The paths followed along the p–V_m–T surface in the experiments.

The thermal conductivity

In our discussions of thermal conduction in gases and in solids we saw that the speed at which energy is transmitted is roughly the speed at which sound travels in these phases. So a sensible starting point might be a consideration of the speed of sound in a liquid.

A single 'glider' on a frictionless air-track (Fig. 13.10(a)) illustrates how sound is transmitted in a (very) low-density gas; the message is transmitted with the glider's speed, u_g say, from one end of the track to the other. Now suppose that instead of a single glider we have several, all of the same mass and all initially at rest (Fig. 13.10(b)). If glider 1 is given the same speed u_g as we gave the single glider in Fig. 13.10(a), the message reaches the far end of the track much more quickly than before. The reason for this is that in a collision between any two gliders, say gliders 2 and 3, the message jumps very quickly from the right-hand side of glider 2 to the right-hand side of glider 3. In fact the speed u_g at which a message travels from a point P to a point Q, a distance l_1 apart (Fig. 13.10), is l_1 divided by the time it takes a glider to travel l_f, namely l_f/u_g. Here l_f is the 'free space' between two gliders, and l_1 is equal to the distance between the centres of two gliders (see Fig. 13.10(b)). This assumes a very much greater speed of transmission through the gliders than through the 'free space'. Therefore

$$u_1 = \left(\frac{l_1}{l_f}\right) u_g. \tag{13.12}$$

If we think of the second experiment as modelling conditions in a liquid we can explain why the speed u_1 of sound in a liquid is some ten times the speed u_g of sound in a gas; the value of l_f is of order one-tenth of l_1 in a liquid. It is not difficult to show that l_1 and l_f are related to the molar

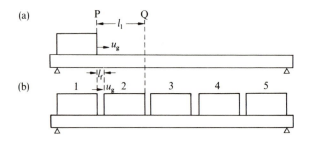

Fig. 13.10. Two possible mechanisms for transmitting energy along a linear air-track. In (a) a single glider is sent bodily from one end to the other. In (b) energy is transmitted via a sequence of inter-glider collisions.

volume $V_{m,l}$ of the liquid and to the change in molar volume ΔV_m when a solid melts to give the liquid by† $l_l/l_f = 3 V_{m,l}/\Delta V_m$. Substituting this result, along with the relation $u_g = (8kT/\pi m)^{1/2}$ (eqn (4.50)), into eqn (13.12) leads to

$$u_l = \left(\frac{3 V_{m,l}}{\Delta V_m}\right)\left(\frac{8kT}{\pi m}\right)^{1/2}. \tag{13.13}$$

The final step in the argument is to take over the gas-kinetic relation (eqn (7.42)) for thermal conductivity, namely $\kappa = \frac{1}{2}n\bar{u}\lambda k$. In the case of a gas λ measures the *overall* distance through which energy is transmitted between collisions; in a liquid this distance is $l_l = (V_{m,l}/N_A)^{1/3}$ (to a good approximation). In the case of a liquid the number of atoms per unit volume is $n = N_A/V_{m,l}$. Substituting these expressions for n, u_l, and l_l into eqn (7.42) gives

$$\kappa = \frac{3k}{2}\left(\frac{N_A}{V_{m,l}}\right)^{2/3}\left(\frac{V_{m,l}}{\Delta V_m}\right)\left(\frac{8kT}{\pi m}\right)^{1/2}. \tag{13.14}$$

Putting in the appropriate data leads to a value of the thermal conductivity for liquid argon near the triple point which agrees, to within about 30 per cent, with the experimental value. However, eqn (13.14) predicts that κ will increase with increasing temperature, whereas we know that it actually decreases (Fig. 13.9). This weakness of the model arises because it implicitly assumes that, like the gliders, the atoms are hard-sphere repulsive. Rather than attempting to rectify this aspect of the model we can always insert the measured values for the speed of sound in the liquid into $\kappa = \frac{1}{2}n\bar{u}\lambda k$ in place of u_l as given by eqn (13.13). In the case of liquid argon, the speed of sound decreases by a factor two on raising the temperature of the liquid from 84 to 140 K; as does the thermal conductivity (see Fig. 13.9(a)).

---

† Because there are N_A atoms in a volume $V_{m,l}$ of liquid the volume occupied by one atom of the liquid is $V_{m,l}/N_A \approx l_l^3$. Likewise in a solid $V_{m,s}/N_A \approx l_s^3$, where l_s is approximately the diameter of an atom. Since $l_f = l_l - l_s$ it follows that

$$\frac{l_l}{l_f} = \frac{V_{m,l}^{1/3}}{V_{m,l}^{1/3} - V_{m,s}^{1/3}}.$$

If ΔV_m is the change in the molar volume when the solid melts, that is $V_{m,s} = V_{m,l} - \Delta V_m$,

$$\frac{l_l}{l_f} = \frac{1}{1 - [(V_{m,l} - \Delta V_m)/V_{m,l}]^{1/3}} \approx \frac{3 V_{m,l}}{\Delta V_m}.$$

13.6. The liquid–gas interface

Surface tension

Fig. 13.11 shows a U-shaped frame containing a loosely-fitting slider rod S attached to a suitable force-measuring device, represented here by a Newton balance. When the frame is filled with a soap film, a force F must be applied to maintain static equilibrium. Considering a section of film between S and P, this means that the film to the right of P must exert a total force F, directed to the right, on section SP. This force might originate in two quite different ways; either through the surfaces being in a state of tension (Fig. 13.11(b)), or through a bulk tension existing throughout the film (Fig. 13.11(c)). An easy way of distinguishing between these two possible hypotheses would be to, say, halve the thickness of the liquid film. If the force F required to maintain equilibrium were to halve, this would argue for a bulk tension. If F does not change, this would argue for a surface tension. To perform the experiment one need

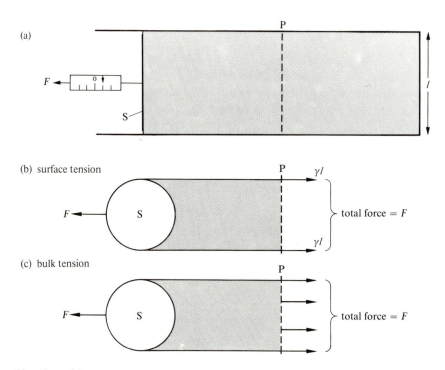

Fig. 13.11. (a) A liquid film is contained within a U-shaped frame fitted with a slider S attached to a Newton balance. (b) A section through the film, showing the existence of forces confined to the surface regions of the film. (c) Here the forces are supposed to extend across the body of the liquid.

only allow the film to evaporate, keeping a record of force F. What one discovers is that, as the film thins by evaporation, F does not change; the tension must therefore lie in the surface regions (Fig. 13.11(b)). The name *surface tension* (γ) is given to the force acting per unit length of surface. Since the line of the upper surface at P in Fig. 13.11(a) is of length l (the width of the frame), it follows that the force acting to the right of P is γl along the upper line of film and γl along the lower line (Fig. 13.11(b).) For the section of film between S and P to be in equilibrium we must have

$$F = 2\gamma l.$$

Clearly, the surface tension of liquids forming thin films can be found by measuring F. Other techniques (such as the one discussed in problem 13.14) must be applied in liquids which fail to form films. Some representative surface tensions are given in Table 13.1.

There is a further feature to note about the experiment outlined in Fig. 13.11(a). If the force F provided by the balance is increased beyond the equilibrium value, the film will expand indefinitely and eventually break. This should be contrasted with the behaviour of a solid which (unless the tensile strength is exceeded) will only extend until the bulk tension matches the new value of F. There must therefore be some mechanism operating in liquids to maintain the surface tension at a value independent of the surface area. Now when the slider of Fig. 13.11(a) moves out through a distance x an energy of $Fx = 2\gamma lx$ (recall $F = 2\gamma l$) is fed into the liquid. Since the combined areas of the top and bottom surfaces of the film increase by $2lx$ during this process an energy of $2\gamma lx / 2lx = \gamma$ has been provided per unit area of film. As just stated this is incorrect since it ignores any interchange of heat energy between the liquid and its surroundings while the slider moves out. Assuming that the heat interchanged during an isothermal expansion is negligible—this is

TABLE 13.1
Surface tensions of various liquids

Substance	Temperature /K	Surface tension[a] $\gamma/10^{-3}\,N\,m^{-1}$	Substance	Temperature /K	Surface tension[a] $\gamma/10^{-3}\,N\,m^{-1}$
Ar	85	13·12	CO_2	248	9·13
	100	9·42		293	1·16
	120	4·95	N_2	70	10·53
	130	2·99		80	8·27
	145	0·57		90	6·40
Ne	25	5·54	O_2	70	18.3
	28	4·48	CCl_4	293	27·0
He	4	0·12	Hg	293	472

[a] The liquids are in equilibria with their vapours (other gases are absent).

frequently the case—we therefore see that γ can also be defined as the surface energy per unit area of surface.

The origins of surface tension

As we shall presently demonstrate, the tension which exists at the surface of a liquid arises as a consequence of insisting—as insist we must—that as many atoms diffuse from within the bulk of a liquid to its surface per unit time as leave the surface for the bulk per unit time. To see how the surface tension arises we will begin by looking at the consequences of supposing that the interatomic separation in the surface layer of atoms is the same as in the bulk (Fig. 13.12(a)).

The problem which faces atom i as it attempts to diffuse through from the bulk to the surface region of the liquid is summarized in the portion ABC of the potential energy curve (Fig. 13.12(b)). In short, it

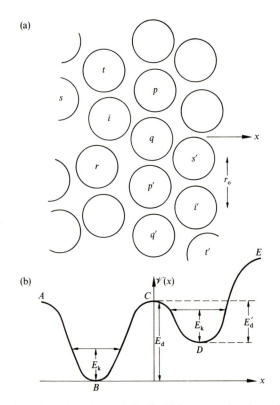

Fig. 13.12. (a) Here the surface layer of the liquid is supposed to have the same (mean) interatomic separation as in the bulk. (b) The potential energy of an atom as it is displaced from the bulk to the surface of the liquid. The potential energy is taken as zero when an atom in the bulk is in the centre of its cell. The kinetic energy of an atom is shown by E_k.

requires an activation energy E_d, the activation energy for diffusion in the bulk liquid. We have already seen (§ 13.2) that E_d is made up of the energy required to separate p and q sufficiently to allow i through, plus the energy to separate i from its neighbours r, s, and t.

Now consider the problem which faces atom i' in the surface layer as it attempts to diffuse into the bulk of the liquid. It, like i, also has to squeeze through between atoms, here p' and q'. However, it has fewer surrounding atoms pulling back on it; only t' and s' in the two-dimensional case illustrated in Fig. 13.12(a). As a consequence, the activation energy E_d' required for i' to diffuse into the bulk is lower than the activation energy E_d required for i to diffuse to the surface. (For i' to escape into the vapour it must escape entirely from the pull of all other atoms. The energy required is typically some four times E_d.) This information is summarized in portion $BCDE$ of Fig. 13.12(b).

When atoms like i and i' are given their Maxwell–Boltzmann range of energies, more atoms will leave the surface for the bulk than vice versa. To prove this point we recall eqn (5.79). This tells us that the probability of an atom (of a two-dimensional liquid) acquiring an energy of at least E is $\exp(-E/kT)$. Therefore, since E_d' is less than E_d, more atoms will succeed in escaping from the surface into the bulk than vice versa. As a result the surface layer will become depleted.

Fig. 13.13(a) shows the new condition of the liquid, where we have assumed that the depletion is confined to the surface layer. (Measurements on the reflection of polarized light from a liquid surface suggest that the decreased density is confined to a few atomic thicknesses.) On this model the value of E_d is essentially unchanged. However, the value of E_d' keeps increasing as depletion proceeds, until it equals E_d. When this happens the two escape probabilities, $\exp(-E_d/kT)$ and $\exp(-E_d'/kT)$, become identical and the same number of atoms leave the surface for the bulk per unit time as vice versa. Equality of diffusion rates means, of course, that the system has reached a state of (dynamic) equilibrium.† To see why E_d' should increase as the surface layer depletes, let us look at atom i' in Fig. 13.13(a). As in the original situation (Fig. 13.12(a)), i' has to squeeze through between p' and q'. But now in escaping from r' and s' it must do so against an increased force, the mean interatomic separation in the surface layer having increased from a value of approximately r_0 in Fig. 13.12(a) to r_s in Fig. 13.13(a). The corresponding force $F(r_s)$ which acts on

---

† A more careful analysis shows that at equilibrium E_d' and E_d will not be exactly equal. This is because escape jumps out of the surface, in the $+x$-direction, are all but excluded for a surface atom, whereas escape jumps within the bulk are equally likely to occur in $+x$- and $-x$-directions. Analysis shows that dynamic equilibrium is achieved when $E_d' = E_d + kT \ln 2$. For simplicity, we shall ignore the $kT \ln 2$ term.

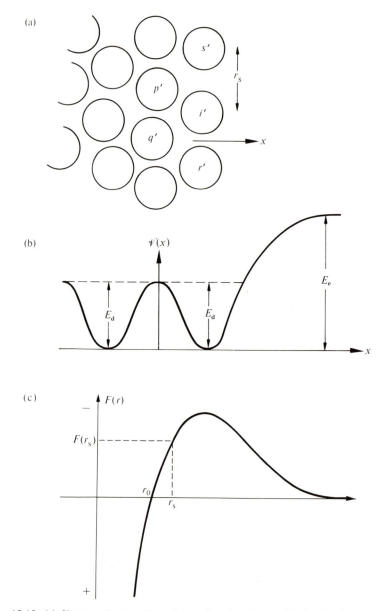

Fig. 13.13. (a) Shows a liquid with a depleted surface layer. (b) The resulting potential energy of an atom as it is displaced from the bulk to the surface of the liquid. (c) Showing how force $F(r_s)$ is obtained from the interatomic force characteristic.

i' is obtained, of course, from the form of the interatomic force characteristic (Fig. 13.13(c)).

Since atom i' is acted on by a force $F(r_s)$ from atom s' and a force $F(r_s)$ from atom r' it follows that there is a tension $F(r_s)$ along this row of surface atoms. Because the rows of surface atoms are separated by a distance r_s we conclude that the tension acting per unit length of surface—the surface tension—is given by

$$\gamma = \frac{F(r_s)}{r_s} \quad . \tag{13.15}$$

The problem of calculating γ therefore resolves into the problem of finding the value of r_s which makes $E'_d = E_d$. The calculations are not difficult but are somewhat tedious; the techniques are outlined in problem 13.15. However, when performed for neon they predict a surface tension of $0.5 \times 10^{-3}\,\mathrm{N\,m^{-1}}$, compared with an experimental value of $5.5 \times 10^{-3}\,\mathrm{N\,m^{-1}}$ close to the triple point. As another example, the model predicts a surface tension of $4\,\mathrm{N\,m^{-1}}$ for molten NaCl, compared to an experimental value of $0.11\,\mathrm{N\,m^{-1}}$.

Some refinements

Throughout this discussion we have ignored the presence of the vapour with which the liquid is in equilibrium. Instead of drawing the interface as we did in Fig. 13.12(a) we should have drawn it as shown in Fig. 13.14, where the gas atoms d, e, f, etc., pull outwards on atom i' with some resultant force f. Because of this force, atom i' will find it more difficult to move into the bulk; E'_d in Fig. 13.12(b) will increase. As a

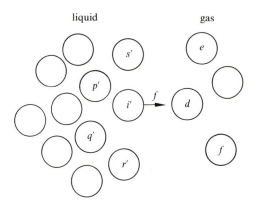

Fig. 13.14. Atoms d, e, f, etc. in the vapour phase contribute a net outward force f on atom i' in the surface layer, thereby raising its activation energy for diffusion into the bulk.

result, the surface layer need not be in such a state of tension to ensure dynamic equilibrium between the bulk and surface regions. Now when the temperature of a liquid in equilibrium with its vapour is raised, the state of the liquid moves along the saturated liquid line (line *ac* in Fig. 1.3) while the state of the vapour moves along the saturated vapour line *ec*. As the critical point is approached the two phases become indistinguishable and thus, according to our model, the surface tension should fall towards zero at T_c. This is the observed (see the data for argon in Table 13.1). By a similar argument, if a liquid is contained in an atmosphere of some foreign gas, its surface tension should fall with increasing gas pressure. It does. As a more extreme example, the surface tension of a liquid which is in contact with a solid should be lower than when the liquid is in equilibrium with a gas. It usually is. Similar arguments explain the observation that the surface tensions of two immiscible liquids are lowered when they are in contact.

The solid-gas interface

Although we have only examined the liquid-gas interface the same criteria should apply at the solid-gas interface. At *equilibrium* as many atoms must diffuse from the bulk of the solid to the surface per unit time as diffuse in the opposite direction. This should lead to the surface being in a state of tension. However, we would only expect the *surface* tension to be apparent close to the melting point, when diffusion can occur at a significant rate. In fact, a force must be applied to prevent a metal rod, maintained at a temperature between the fusion temperature T_f and about $0.9 T_f$, from contracting. This method for measuring the surface tension of a solid is, of course, the analogue of the frame method for liquids. By way of example, the surface tension of a nickel rod maintained at a temperature of 1675 K is $1.75 \, \text{N m}^{-1}$. Because of the rapid dependence of the diffusion coefficient on temperature (recall eqn (13.3)), surfaces which are formed by fracturing a solid at temperatures well below T_f may not reach equilibrium for centuries. One consequence is that the main assumption underlying the argument that the surface tension and surface energy are equal in a liquid—namely that the surface tension remains constant as the area of the liquid increases—is normally invalid in a solid. Even when a true surface tension does eventually establish itself in a solid, their high 'viscosity' at temperatures well below T_f prevents them from assuming shapes characteristic of liquid drops. Lumps of rock are not spherical! In the case of liquids any time dependence of surface tension is only evident in solutions of high molecular mass solutes, where the times required to establish the bulk and surface solute concentrations appropriate to dynamic equilibrium will be of the order of seconds, or even minutes.

Capillarity

If a small-bore glass tube is lowered into water, the water will rise up the capillary tube. This is just one example of a phenomenon known as capillary rise. Sometimes the liquid may, in fact, be depressed below the general level of the liquid; dirty mercury is a familiar example here.

The simplest explanation of capillary rise is that the surface of the glass tube is covered by a layer of adsorbed liquid which is sufficiently thick for it to exhibit the surface tension characteristic of the bulk liquid (Fig. 13.15(a)). As we have just seen, this only demands that the adsorbed layer be a few molecular diameters deep. On the basis of this model we can equate the net upward force acting on the liquid column of Fig. 13.15(a), namely $2\pi r\gamma$, where r is the radius of the capillary, to the downward force acting on the column, namely its weight $\pi r^2\rho gh$, where g is the local acceleration due to gravity, ρ is the density of the liquid, and h is the equilibrium height of the column. This leads to

$$\gamma = \tfrac{1}{2}r\rho gh \quad . \tag{13.16}$$

Capillary depression occurs when, as shown in Fig. 13.15(b), the liquid fails to wet (that is, fails to be adsorbed on) the glass. Equating the upward force acting on the liquid in the tube, namely the pressure ρgh of the head of liquid multiplied by the cross-sectional area, πr^2, of the tube to the downward force acting on the liquid, namely $2\pi r\gamma$, again leads to eqn (13.16).

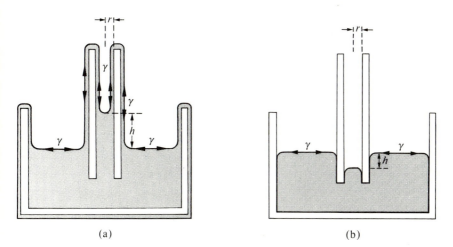

(a)   (b)

Fig. 13.15. (a) Showing how capillary rise can occur if the surface of the tube is covered by layer of liquid sufficiently thick to exhibit bulk properties. (b) Showing how capillary depression results if the liquid fails to wet the tube.

13.7. The saturated vapour pressure

Over a flat surface

We have just seen that dynamic equilibrium between the bulk and the surface regions of a liquid demands that as many atoms diffuse from the bulk to the surface per unit time as diffuse in the opposite direction. We have yet to appeal to the fact that dynamic equilibrium must also exist between the surface of the liquid and the surrounding vapour.

For atom i' in the surface of the liquid (Fig. 13.13) to escape into the vapour it must have an energy of E_e or greater, where E_e is the enthalpy of evaporation (or latent heat of evaporation) per molecule of the liquid. The number of successful escape jumps f_e per unit time into the vapour is given by eqn (13.1) with $E_e = E_d$, and with a cell diameter $a = r_s$, namely

$$f_e = \frac{1}{r_s}\left(\frac{\pi kT}{2m}\right)^{1/2} e^{-E_e/kT}.$$

Since there are $1/r_s^2$ atoms per unit area of the surface, in which r_s is the interatomic separation, it follows that the evaporation rate—the number of atoms evaporating per unit area of surface per unit time—is given by

$$\text{evaporation rate} = \frac{1}{r_s^3}\left(\frac{\pi kT}{2m}\right)^{1/2} e^{-E_e/kT}. \tag{13.17}$$

In a closed system, where the vapour is contained, evaporation will not proceed indefinitely at this rate. As it proceeds the number density n_v of vapour atoms will increase and, as a consequence, the number of atoms available to return to the liquid will increase. Now we know from gas-kinetic theory (eqn (5.4)) that the number of atoms striking unit area of surface is given by $\frac{1}{4}n_v\bar{u}$, or substituting $\bar{u} = (8kT/\pi m)^{1/2}$ eqn (4.50) by $\frac{1}{4}n_v(8kT/\pi m)^{1/2}$. However, condensation demands that the incoming atoms stick to the surface of the liquid and do not simply rebound back into the vapour. Denoting the probability that an incoming atom will stick by θ, we see that the number of vapour atoms condensing per unit area of surface per unit time is given by $\frac{1}{4}\theta n_v(8kT/\pi m)^{1/2}$. Dynamic equilibrium demands that this condensation rate equals the evaporation rate given by eqn (13.17). This leads to

$$n_v = \frac{\pi n_s}{\theta} e^{-E_e/kT}, \tag{13.18}$$

where $n_s = 1/r_s^3$ is the number of atoms per unit volume in the vicinity of the liquid's surface. In so far as we can regard the vapour as a perfect gas—this is clearly going to be very risky near the critical point—its pressure p_0 will be related to the number density n_v of the atoms and to the temperature T by $p_0 = n_v kT$ (eqn (5.19)). Eqn (13.18) therefore gives the saturated vapour pressure p_0 as

$$p_0 = \frac{\pi n_s}{\theta} kT \, e^{-E_e/kT}. \tag{13.19}$$

By comparison with the exponential term, the term kT remains nearly constant throughout the limited range of temperatures (T_{tr} to T_c) over which a substance occurs in the liquid phase. Therefore

$$\boxed{p_0 = A \, e^{-E_e/kT} = A \, e^{-H_{m,e}/RT}}, \tag{13.20}$$

where A is a constant, $H_{m,e}$ ($= N_A E_e$) is the molar enthalpy of evaporation, and $R = N_A k$ is the gas constant.

Fig. 13.16 shows $\ln p_0$ plotted against $1/T$ for liquid argon in equilibrium with its vapour at temperatures between 100 and 400 K. As predicted by eqn (13.20), this graph is linear. The measured gradient ($-H_{m,e}/R$) gives $H_{m,e} = 6.6 \times 10^3$ J mol^{-1}, which agrees closely with the experimental value of 6.5×10^3 J mol^{-1}. In view of the essential correctness of eqn (13.19) it can be used to obtain the probability θ that a vapour atom will stick on striking the surface of the liquid. Substituting for the saturated vapour pressure of liquid argon at 84 K, namely 6.9×10^4 Pa, and taking the number density n_s at the surface of the liquid to be equal to its bulk value, namely 2.15×10^{28} m^{-3}, gives $\theta = 0.09$.

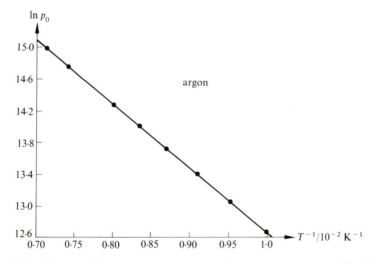

Fig. 13.16. The natural logarithm of the saturated vapour pressure p_0 of liquid argon plotted against T^{-1}. (Data from Street, W. B. and Staveley, L. A. K. (1969). *J. Chem. Phys.* **50**, 2302.)

Over a curved surface

In arriving at eqn (13.20) it was implicitly assumed that the liquid with which the vapour is in equilibrium has a flat surface. This assumption arose in considering the number of nearest neighbours to atom i' in Fig. 13.13; in particular it was assumed that these neighbours subtend a solid angle of 2π at i'. Fig. 13.17(a) shows this assumption in a more explicit form, with the range of the intermolecular force—the distance over which it exerts a significant influence—denoted by s.

If we now consider a spherical drop of radius r in equilibrium with its vapour (Fig. 13.17(b)) it is clear that atoms lying within the range s subtend a solid angle of less than 2π at i'. This necessarily implies that an atom on the surface of a small drop has fewer nearest neighbours than does an atom on the surface of a drop of larger (even infinite) radius. As a consequence, the enthalpy of evaporation per molecule will be lower in Fig. 13.17(b) than in Fig. 13.17(a). It therefore follows from eqn (13.20) that the pressure p of a vapour, in equilibrium with a drop, will be greater than the pressure p_0, say, of a vapour in equilibrium with a flat liquid. To find the relation between p and p_0 we must know how the enthalpy of evaporation is related to the radius r of the drop. This can be found most readily by considering the consequences of a single molecule, of volume v, leaving a drop of radius r. Clearly the radius of the drop will fall to $r - \delta r$, with the result that its volume changes by $v = \frac{4}{3}\pi r^3 - \frac{4}{3}\pi(r - \delta r)^3 \approx 4\pi r^2\,\delta r$, while its surface area changes by $4\pi r^2 - 4\pi(r - \delta r)^2 = 8\pi r\,\delta r$. This means the energy stored in the liquid's surface decreases by $(8\pi r\,\delta r)\gamma = 2\gamma v/r$ (recall that γ tells us the energy per unit area of surface). Assuming that this energy is given to the departing atom, the net energy $E_e(r)$ that it requires to evaporate is given by

$$E_e(r) = E_e - \frac{2\gamma v}{r}, \qquad (13.21)$$

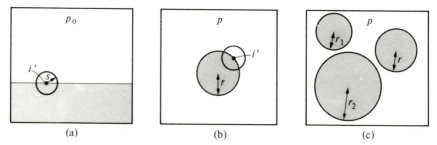

(a)   (b)   (c)

Fig. 13.17. (a) A bulk liquid in equilibrium with its vapour at pressure p_0. Here s denotes the range of the intermolecular forces. (b) A liquid drop in equilibrium with its vapour at pressure p. (c) The drop of radius r is in equilibrium with the vapour at pressure p. The drop of radius $r_1 < r$ will shrink and that of the drop of radius $r_2 > r$ will grow.

where E_e is the molar enthalpy of evaporation of a molecule from a flat surface. Substituting eqn (13.21) into eqn (13.20) in place of E_e and writing the vapour pressure as p gives

$$p = A\, e^{-E_e/kT}\, e^{2\gamma v/rkT}$$

$$\boxed{p = p_0\, e^{2\gamma v/rkT} = p_0\, e^{2\gamma V_m/rRT}}, \qquad (13.22)$$

where p_0 is the equilibrium vapour pressure over a plane surface, and $V_m = N_A v$ is the molar volume of the liquid. As an example, eqn (13.22) predicts that a water drop of radius 10^{-9} m (containing around 80 molecules) will have $p/p_0 = 2 \cdot 9$ at 300 K. This assumes $\gamma = 7 \cdot 3 \times 10^{-2}$ J m^{-2} and $V_m = 1 \cdot 8 \times 10^{-5}$ m^3 mol^{-1}.

Condensation of drops

We will now consider what may happen within a vapour at a pressure $p > p_0$; a vapour said to be supersaturated (or a gas said to be super-cooled: Fig. 1.10). In particular, let us consider the fate of three drops; one of radius r given by eqn (13.22), namely

$$\boxed{r = \frac{2\gamma V_m}{RT \ln(p/p_0)}}, \qquad (13.23)$$

one of radius $r_1 < r$, and one of radius $r_2 > r$ (Fig. 13.17(c)). Eqn (13.22) tells us that the drop of radius r is certainly in equilibrium with the vapour. However, the vapour pressure p is less than that required for equilibrium by the drop of radius $r_1 < r$ (the equilibrium vapour pressure again following from eqn (13.22)). Consequently, in the continuous interchange of molecules that goes on between a drop and the surrounding vapour this particular drop will not receive as many molecules as it loses; it will therefore shrink and disappear. Conversely, the vapour pressure p is greater than that required for equilibrium by the drop of radius $r_2 > r$. It will therefore continue to grow. We therefore conclude that if a liquid is to condense from a vapour of specified p/p_0 it must contain suitable prexisting droplets with a radius of at least that given by eqn (13.23).

Although we have considered droplets as potential growth sites, other 'condensation nuclei' are often present. C. T. R. Wilson, working at the end of the last century, found that it was necessary to make $p/p_0 = 8$ in moist dust-free air to obtain a condensation cloud, but that it was only necessary to have $p/p_0 = 4$ when ions were present in the dust-free air. It thus follows that individual ions can act as condensation nuclei. Now a droplet of radius r formed around an ion of charge q will have an electrostatic potential energy of $q^2/4\pi\varepsilon_0 r$. Since this decreases as r increases such a charged drop can lower its electrostatic potential energy by growing. However, as it grows its surface energy $4\pi r^2 \gamma$ will increase. The effect of the charge is therefore equivalent to a reduction in the surface tension of the liquid making up that droplet. This, in turn, decreases the size of the nuclei required for growth (eqn (13.23)).

PROBLEMS

13.1. With what mean speed will an argon atom move through liquid argon at a temperature of 88 K and a density of $1 \cdot 4 \times 10^3 \, \text{kg m}^{-3}$? Take $E_d = 5 \times 10^{-21} \, \text{J}$ and $A_r(\text{Ar}) = 40$.

13.2. Calculate the self-diffusion coefficient in liquid argon at a temperature of 100 K and a pressure of $1 \cdot 3 \times 10^6 \, \text{Pa}$. Comment on any discrepancy between the predicted value of D and the value given in Fig. 13.4(a). Take $a = 7 \cdot 2 \times 10^{-10} \, \text{m}$, $E_d = 5 \times 10^{-21} \, \text{J}$, and $A_r(\text{Ar}) = 40$.

13.3. (a) How does the escape probability of an argon atom from its cell change as the temperature of the liquid argon is raised from $T_{\text{tr}} = 84 \, \text{K}$ to $T_c = 151 \, \text{K}$? Assume that $E_d = 5 \times 10^{-21} \, \text{J}$ throughout this temperature range, but comment on whether this is a realistic assumption. (b) How does the number of escape *attempts* change throughout the range T_{tr} to T_c? Take $a = 7 \cdot 2 \times 10^{-10} \, \text{m}$ and $A_r(\text{Ar}) = 40$.

13.4. This easily performed experiment should provide some feel for the temperature dependence of D in a liquid. Although it may not be able to confirm that $D \propto \exp(-\text{constant}/T)$ it should enable you to rule out a perfect-gas model of a liquid, which predicts $D \propto T^{3/2}$.

Simultaneously drop one lump of sugar into a cup of cold water and another into a cup of hot water. Compare the time taken for the taste of sugar to become apparent at the surface of the cold water with the time taken for the warm water. This may (with a little thought) enable you to deduce the ratio of the diffusion coefficients at two temperatures. Are you sure you are observing diffusion?

13.5. It is implicitly assumed in eqn (13.7) that $fa \ll E_v$. Investigate the validity of this assumption by calculating fa/E_v in liquid argon when the velocity gradient is $10^4 \, \text{s}^{-1}$ and the liquid is at $p = 1 \cdot 0 \times 10^7 \, \text{Pa}$, $T = 88 \cdot 5 \, \text{K}$. Assume that the cell diameter is $7 \times 10^{-10} \, \text{m}$ and use the data given in Fig. 13.6.

13.6. Calculate the dynamic viscosity η of liquid argon at $p = 1 \times 10^7 \, \text{Pa}$ and $T = 88 \cdot 5 \, \text{K}$. This is close to the triple point (see Table 1.1) where, as we saw in exercise 13.1, $a = 7 \cdot 2 \times 10^{-10} \, \text{m}$. Use the data given in Fig. 13.6 to calculate E_v. Comment on any discrepancy between the value you calculated for η and the value given in Fig. 13.6. Take $A_r(\text{Ar}) = 40$.

13.7. It is an experimental fact that many liquids have roughly the same viscosity, as measured at their melting point. For example, the viscosity of liquid copper at its melting point (under normal atmospheric pressure) of 1356 K is $3 \cdot 4 \times 10^{-3} \, \text{kg m}^{-1} \text{s}^{-1}$; the viscosity of bromine at its melting point (266 K) is $1 \cdot 3 \times 10^{-3} \, \text{kg m}^{-1} \text{s}^{-1}$; the viscosity of lead at its melting point (600 K) is $2 \cdot 6 \times 10^{-3} \, \text{kg m}^{-1} \text{s}^{-1}$. Try to account for this experimental fact. *Clues:* Think about the exponent term.

13.8. What is the relaxation time of an atom in liquid argon at a temperature of 84 K and at a density of $1 \cdot 4 \times 10^3 \, \text{kg m}^{-3}$? Take $A_r(\text{Ar}) = 40$.

13.9. At one time it was suggested that a liquid could be considered to be a solid with a particularly large number of vacancies. Adopting such a model, derive expressions for the self-diffusion coefficient and the dynamic viscosity of the liquid. You may draw on any suitable relations already derived in the text.

13.10. In the study made by Davis, Rice, and Meyer, whose results are shown in Fig. 13.8, the ion density in the liquid argon was about 10^{11} m^{-3}. What was the mean separation of the ions in the liquid?

13.11. In the study whose results are shown in Fig. 13.4(a) the self-diffusion coefficient was measured at a series of fixed pressures over a range of temperatures. Sketch in on a $p-V_m-T$ surface the 'routes' followed in these experiments. Do the same for the study of the dynamic viscosity whose results are shown in Fig. 13.6(a) and also for the study of the ionic mobility whose results are shown in Fig. 13.8(b). Note that all of the experiments relate to argon. Make use of the data contained in Fig. 1.12; this will tell you which experiments were performed on liquid argon and which on gaseous argon. Do D, η, or μ^+ undergo any sudden change as the argon changes from a 'liquid' to a 'gas'?

13.12. Predict the thermal conductivity of liquid argon near the triple temperature (84 K), at a density of $1\cdot41\times10^3$ kg m^{-3}. Solid argon has a density of $1\cdot64\times10^3$ kg m^{-3} at the triple point. Take $A_r(Ar)=40$.

13.13. Predict the speed of sound in liquid argon at a temperature of 87 K and at a density of $1\cdot4\times10^3$ kg m^{-3}. Compare your answer with the experimental value of 836 m s^{-1}. At the triple point solid argon has a density of $1\cdot64\times10^3$ kg m^{-3}. Take $A_r(Ar)=40$.

13.14. One technique for measuring the surface tension of a liquid is to introduce a bead of the liquid into a *horizontal* capillary tube and to apply an excess pressure to one end of the tube (via a manometer) so as to drive the bead to the far end of the tube where its meniscus can be changed from concave, through plane, to convex form. Show that when the excess pressure p is such that the meniscus at the open end of the tube is flat then $\gamma=\frac{1}{2}pr$, where r is the radius of the capillary. Advantages of this technique are that it only requires a few cubic millimetres of the liquid and no determination of its density. If the excess pressure is applied via a water manometer (density 10^3 kg m^{-3}), what head of water will be required to ensure that the free meniscus of a bead of benzene ($\gamma=2\cdot89\times10^{-2}$ N m^{-1}) is flat? The capillary has a radius of 5×10^{-4} m. Take $g=9\cdot81$ m s^{-2}.

13.15. This problem outlines how to make order-of-magnitude correct predictions of the surface tension of simple liquids. The starting point is Fig. 13.13. You will recall how we argued that, at equilibrium, the activation E_d' required by a surface atom to diffuse into the bulk will equal the activation energy E_d for bulk diffusion. Now E_d' is made up of the energy required to separate p' and q', plus the energy required to escape from some *five* nearest neighbours, each a distance r_s away (remember that the surface is two-dimensional). Likewise E_d is made up of the energy to separate two atoms (like p and q in Fig. 13.12) plus the energy to escape from approximately *eight* nearest neighbours, each a distance r_0 away. As a surface atom moves into the position where it is poised midway between p' and q' its separation from each of its neighbours increases from r_s to about $2^{1/2}r_s$. In the case of a bulk atom its separation from each of its neighbours increases from r_0 to about $2^{1/2}r_0$. The corresponding changes in potential energy are obtained from the interatomic potential energy characteristic $\mathscr{V}(r)$.

Using these ideas—you may wish to refine them somewhat—calculate the surface tension of liquid argon (density 1.4×10^3 kg m^{-3}). You may find it helpful to plot out $\mathscr{V}(r)$; in so doing adopt a Lennard-Jones 6–12 potential (eqn (2.20)) with $\varepsilon = 1.7 \times 10^{-21}$ J and $\sigma = 3.4 \times 10^{-10}$ m. Once you have found the value of r_s which satisfies the requirement $E_d' = E_d$, calculate $F(r_s)$ from $F(r) = -d\mathscr{V}(r)/dr$. The surface tension then follows from eqn (13.15). Close to the triple point, liquid argon has a surface tension of 1.32×10^{-2} N m^{-1}.

13.16. The known tensile strength of a liquid and its known surface tension γ may be used to estimate the range of the intermolecular forces in that liquid. Consider a column of liquid of cross-sectional area A. In pulling it apart we create two surfaces, each of area A. This requires an energy of $2\gamma A$; recall that γ is also the energy per unit area of surface. If we now suppose that the intermolecular forces are effectively constant over a limited range s (dropping suddenly to zero at distances greater than s) the energy fed into the system by the breaking stress σ_f as the two new surfaces are created is $\sigma_f A s$. Equating this to $2\gamma A$ gives $s = 2\gamma/\sigma_f$. Given that water has a surface energy of 7.3×10^{-2} J m^{-2} and a tensile fracture stress of 3×10^7 N m^{-2}, deduce the range of its intermolecular forces.

13.17. Using the saturated vapour pressure values listed in Table 13.2, investigate whether eqn (13.20) is satisfied by water vapour. Also determine the probability θ that a vapour atom striking the surface of the water will stick when the water is (a) at 275 K, (b) 470 K, and (c) at 647 K. The density of water is 1.0×10^3 kg m^{-3} at 275 K, 0.86×10^3 kg m^{-3} at 470 K, and 0.32×10^3 kg m^{-3} at 647 K. Take $M_r(H_2O) = 18$.

TABLE 13.2
Saturated vapour pressure p_0 of water

T/K	$p_0/10^5$ Pa	T/K	$p_0/10^5$ Pa
273.16^a	0.00611	393	1.985
275.15	0.00705	413	5.614
277.15	0.00813	433	6.180
281.15	0.01027	453	10.027
288	0.01704	473	15.55
298	0.03166	513	33.48
313	0.07375	553	64.19
333	0.1992	593	112.9
353	0.4736	633	186.7
373	1.0132	647.29^b	221.2

[a] Triple point. [b] Critical point.

13.18. Bulk liquid argon in equlibrium with its vapour at 84 K has a vapour pressure of 6.88×10^4 Pa. What will be the pressure of the vapour in equilibrium with a drop of radius 10^{-9} m at 84 K? At 84 K liquid argon has a surface tension of 1.31×10^{-2} N m^{-1} and a molar volume of 2.8×10^{-5} m^3 mol^{-1}.

13.19. What diameter water droplets would have to have been present in the experiments of C. T. R. Wilson to explain his observation that moist dust-free air, devoid of ions, will condense when $p/p_0 = 8$ at 300 K? How many water molecules would be contained in such a droplet? Take $\gamma = 7 \cdot 3 \times 10^{-2} \, \text{N m}^{-1}$, $V_m = 1 \cdot 8 \times 10^{-5} \, \text{m}^3 \, \text{mol}^{-1}$, and assume that water molecules are spherical and with a diameter of $4 \cdot 7 \times 10^{-10} \, \text{m}$.

13.20. Explain, in words, the underlying physical reasons for the appearance of each of the following terms: (a) the $T^{1/2}$ and the $\exp(-E_d/kT)$ term in the final expression for D (eqn (13.3)); (b) the term $fa \exp(-E_v/kT)$ in eqn (13.7); (c) the terms $3V_{m,l}/\Delta V_m$ and $(8kT/\pi m)^{1/2}$ in eqn (13.13).

Values of selected physical constants

Quantity	Symbol	Value
Speed of light in vacuo	c	$2 \cdot 997 \times 10^{8} \, \text{m s}^{-1}$
Permittivity of free space	ε_0	$8 \cdot 854 \times 10^{-12} \, \text{N}^{-1} \, \text{m}^{-2} \, \text{C}^2$
	$1/4\pi\varepsilon_0$	$8 \cdot 987 \times 10^{9} \, \text{N} \, \text{m}^2 \, \text{C}^{-2}$
Charge of a proton	e	$1 \cdot 602 \times 10^{-19} \, \text{C}$
Planck constant	h	$6 \cdot 626 \times 10^{-34} \, \text{J s}$
	$\hbar = h/2\pi$	$1 \cdot 055 \times 10^{-34} \, \text{J s}$
Avogadro constant	N_{A}	$6 \cdot 022 \times 10^{23} \, \text{mol}^{-1}$
Rest mass of electron	m_{e}	$9 \cdot 109 \times 10^{-31} \, \text{kg}$
Rest mass of proton	m_{p}	$1 \cdot 673 \times 10^{-27} \, \text{kg}$
Gas constant	R	$8 \cdot 314 \, \text{J K}^{-1} \, \text{mol}^{-1}$
Boltzmann constant	k	$1 \cdot 381 \times 10^{-23} \, \text{J K}^{-1}$
Gravitational constant	G	$6 \cdot 673 \times 10^{-11} \, \text{N} \, \text{m}^2 \, \text{kg}^{-2}$

(These three conversion factors are now defined as exact.)

$1 \, \text{in} = 2 \cdot 54 \, \text{cm}$

$1 \, \text{lb} = 0 \cdot 453\,592\,37 \, \text{kg}$

$1 \, \text{standard atmosphere} = 1 \cdot 013\,25 \times 10^{5} \, \text{Pa}$

Answers to selected problems

Chapter 1

1.2. It may have *looked* the same as the $p-V_m$ projection before you put in the isobars, but do not be deceived by the similarity!

1.4. The mass fraction of liquid present is $(V_g - V)/(V_g - V_l)$. The mass fraction of gas present is $(V - V_l)/(V_g - V_l)$.

1.7. (a) 304.34 K. Gas phase; (b) 1.27×10^{-6} m³; (f) p_{tr} is greater than 1.0×10^5 Pa.

1.9. (a) 37.5 J mol^{-1}; (b) 1.6 J mol^{-1}; (c) 6.25 s; (d) No.

1.10. 855 J.

1.11. (a) 24.4 J K^{-1} mol^{-1}; (b) $(C_{p,m} - C_{V,m})/C_{p,m} = 3 \times 10^{-4}$ per cent.

1.12. 2.8×10^2 J.

1.14. 8.31 J K^{-1} mol^{-1}.

Chapter 2

2.1. 6.6×10^{-5} s.

2.3. The answers will clearly depend on the assumptions you have made but they will probably lie within a factor of 10 of the following: (a) 10^{28}; (b) 10^{29}; (c) 10^{11} N(!).

2.6. 3.4×10^{-10} m.

2.9. 4.7×10^{-1} m; 1.3×10^{-2} V.

2.11. 9.2×10^{-19} J.

2.12. 3.7×10^{-9} N; 5.9×10^4 N.

2.14. (a) -3.61×10^{-23} J; (b) -1.41×10^{-22} J; (c) 8.43×10^{-21} J.

2.16. $\varepsilon = 1.36 \times 10^{-20}$ J; $\sigma = 2.77 \times 10^{-10}$ m.

2.19. 6.0×10^{-52} J, which is some 10^{33} times less than the actual dissociation energy.

2.20. (a) 1.8×10^{-10} m; (b) 5.2×10^8 N C^{-1}; (c) 4.4×10^{-11} N.

Chapter 3

3.3. Because you are given data you do not necessarily need to use any—or all—of it!

3.6. In the range 450–500 m s^{-1}.

3.7. 1.41.

3.9. Assuming $\frac{3}{2}kT_f = \Delta E$ at fusion gives $T_f = 44\,500$ K, which compares badly with the experimental value of 1075 K. The disagreement is not surprising since the identity $\Delta E = \frac{3}{2}kT_f$ implies that the interatomic separation in a liquid is vastly greater than that in a solid, which we know is not the case.

3.10. 2.5×10^{-10} m.

3.13. 60 m s^{-1}.
3.15.   (a) $9{\cdot}35 \times 10^3 \text{ J}$; (b) $9{\cdot}35 \times 10^3 \text{ J}$; (c) $9{\cdot}35 \times 10^3 \text{ J}$.
3.17.   (a) $1{\cdot}99 \text{ s}$; (b) $3{\cdot}16 \text{ rad s}^{-1}$.
3.18.   (a) $1{\cdot}25 \times 10^{-2} \text{ s}$; (b) $1{\cdot}22 \times 10^3 \text{ J}$; (c) $6{\cdot}12 \times 10^2 \text{ J}$; (d) $6{\cdot}12 \times 10^2 \text{ J}$;
          (e) $4{\cdot}95 \times 10^{-2} \text{ m}$.
3.20. $1{\cdot}0 \times 10^{-6} \text{ m}$.
3.22. $1{\cdot}3 \times 10^3$; $0{\cdot}07$ per cent.

Chapter 4

4.1   (b) 14 kg; (c) 90 kg; (d) $3{\cdot}1 \times 10^4$ marks.
4.2.   The mean number of atoms at $E = 0, 1, 2, 3, 4$ units are $1{\cdot}0$, $0{\cdot}8$,
          $0{\cdot}6$, $0{\cdot}4$, and $0{\cdot}2$, respectively. These numbers must, of course,
          total to the number of atoms present (3).
4.7.   22 per cent.
4.9.   (a) $1{\cdot}25$; (b) $1{\cdot}13$.
4.11. (a) $0{\cdot}017$; (b) $0{\cdot}367$; (c) $0{\cdot}135$.
4.13. $1{\cdot}0 \times 10^{-20} \text{ J}$.

Chapter 5

5.2. $2{\cdot}46 \times 10^{-2} \text{ Pa}$.
5.5.   (a) 1200 kg m s^{-1}; (c) 1200 N s; (d) 120 N; (e) 240 N.
5.9. $2{\cdot}9 \times 10^{28}$.
5.11. $7{\cdot}5 \times 10^{-8} \text{ kg m}^{-3}$.
5.12. $10^8 \text{ m}^2 \text{ s}^{-2}$; 10^4 m s^{-1}.
5.13. $8{\cdot}24 \times 10^3 \text{ m}$.
5.14. $1{\cdot}6 \times 10^4 \text{ m}$; $3{\cdot}2 \times 10^4 \text{ m}$; $4{\cdot}8 \times 10^4 \text{ m}$.
5.17. (a) 3; (b) 5; (c) $6{\cdot}7$.
5.19. 150 m.
5.20. (a) 400 K; 32 m^3; $6{\cdot}6 \times 10^6 \text{ J}$; $6{\cdot}6 \times 10^6 \text{ J}$; 0.
          (b) 175 K; 14 m^3; $2{\cdot}7 \times 10^6 \text{ J}$; 0; $-2{\cdot}7 \times 10^6 \text{ J}$.
5.23. $0{\cdot}09 \text{ m}$; $3{\cdot}6 \times 10^{-2}$.
5.24. $5{\cdot}8 \times 10^{-3} \text{ m}$.
5.26. 482.

Chapter 6

6.1.   (a) $3{\cdot}74 \times 10^7 \text{ Pa}$; (b) $6{\cdot}35 \times 10^7 \text{ Pa}$.
6.3.   132 K.
6.5. $1{\cdot}38 \times 10^{-5} \text{ m}^3 \text{ mol}^{-1}$.
6.8. $p_c = 7{\cdot}3 \times 10^6 \text{ Pa}$; $T_c = 300 \text{ K}$; $V_{c,m} = 1{\cdot}28 \times 10^{-4} \text{ m}^3 \text{ mol}^{-1}$.
6.9. $2{\cdot}4d$; $2d$.
6.12. $2/e^2$.
6.14. $0{\cdot}913$; 138 K, 277 K, 30 K.
6.15. $1{\cdot}3R$.

6.16. 260 J; 32 J.
6.21. 3·8 K.

Chapter 7

7.3. $3·1 \times 10^{-10}$ m.
7.6. $1·67 \times 10^3$ N; $1·67 \times 10^3$ N.
7.7. $2·5 \times 10^{-5}$ Pa s.
7.9. $3·63 \times 10^{-10}$ m; $3·50 \times 10^{-10}$ m.
7.12. (a) $4·27 \times 10^{-8}$ Pa m^3 s^{-1}; (b) $9·64$ Pa m^3 s^{-1}.
7.13. $2·02 \times 10^{-10}$ m.
7.17. $6·24 \times 10^{11}$ s^{-1}.

Chapter 8

8.3. 11°; 35°.
8.6. 1·633.
8.8. (a) 0·52; (b) 0·74; (c) 0·68.
8.10. $U_m = -N_A \alpha e^2 / 4\pi\varepsilon_0 r$.
8.12. $-6·1 \times 10^5$ J mol^{-1}.
8.14. $1·9 \times 10^{-6}$ per cent.
8.15. 3 J m^{-2}.

Chapter 9

9.1. 10·5.
9.3. $1·6 \times 10^9$ Pa.
9.5. $2·2 \times 10^{-13}$ s.
9.6. $5·1 \times 10^3$ Hz.
9.8. $1·93 \times 10^{-4}$ m.
9.9. $9·0 \times 10^7$ Pa; 0·11.
9.12. 0·11.
9.13. $6·4 \times 10^{-4}$ N m.
9.17. 3×10^{-8} m.
9.18. $2·1 \times 10^{-2}$ m.

Chapter 10

10.1. 5 per cent.
10.2. 118 N m^{-1}; $1·13 \times 10^{-13}$ s.
10.4. $2·3 \times 10^{13}$ rad s^{-1}; 176 K.
10.5. 2·76 R.
10.6. 2·5 per cent.
10.7. $V(r - r_0) = k(r - r_0)^2$, where $k = 18·9$ N m^{-1}.
10.9. $q = 1·74, 3·49, 5·24, 6·98, 8·73$ rad m^{-1}; $\omega_q = 0·28, 0·53, 0·72, 0·85, 0·89$ rad s^{-1}.
10.14. $6·33 \times 10^{13}$ rad s^{-1}.

10.16. $1 \cdot 1 \times 10^{-18}$ J; $1 \cdot 6 \times 10^{6}$ m s^{-1}, $5 \cdot 3 \times 10^{4}$ K.
10.17. $5 \cdot 3 \times 10^{-10}$ m.
10.19. $8 \cdot 0 \times 10^{-30}$ J.
10.21. 340 K; $0 \cdot 21$ J K^{-1} mol^{-1}.

Chapter 11

11.2.   13 hours.
11.5. $2 \cdot 5$ s; $2 \cdot 5 \times 10^{4}$ s.
11.6. 1×10^{-8}; $1 \cdot 5 \times 10^{-3}$.
11.7. $4 \cdot 3 \times 10^{4}$ s (12 hours).
11.8. $5 \cdot 3 \times 10^{23}$.
11.11. $1 \cdot 6 \times 10^{-8}$ m.
11.12. 2.
11.16. $3 \cdot 2$ m s^{-1}; $3 \cdot 2 \times 10^{-2}$ m.
11.20. 170.
11.23. 25 K.

Chapter 12

12.3. $N - 1$.
12.4.   (a) 7; (b) $5 \cdot 5$.
12.7. $x = 2 \, \Delta C_{V,m}/3R$.
12.9. $1 \cdot 351 \times 10^{8}$ Pa.
12.13. $-2 \cdot 8 \times 10^{-20}$ J.
12.15. $x = 4 \cdot 56 \times 10^{-12}$ m; $y = 9 \cdot 03 \times 10^{-13}$ m; $v_x = 306 \cdot 5$ m s^{-1}; $v_y = 90 \cdot 4$ m s^{-1}.
12.16. $2 \cdot 18 \times 10^{-12}$ m.

Chapter 13

13.1. $2 \cdot 76$ m s^{-1}.
13.2. $1 \cdot 7 \times 10^{-9}$ m^{2} s^{-1}.
13.3.   (a) $1 \cdot 33 \times 10^{-2}$; $9 \cdot 08 \times 10^{-2}$; (b) $2 \cdot 30 \times 10^{11}$ s^{-1}; $3 \cdot 08 \times 10^{11}$ s^{-1}.
13.5. 6×10^{-8}.
13.6. $7 \cdot 4 \times 10^{-3}$ kg s^{-1} m^{-1}.
13.8. $1 \cdot 7 \times 10^{-12}$ s.
13.10. $2 \cdot 2 \times 10^{-4}$ m.
13.12. $0 \cdot 24$ W m^{-1} K^{-1}.
13.14. $1 \cdot 18 \times 10^{-2}$ m.
13.18. $1 \cdot 97 \times 10^{5}$ Pa.

Index